Canals in a Changing Britain

Environment and Society

Series Editor

Douglas Vakoch

As scholars examine the environmental challenges facing humanity, they increasingly recognize that solutions require a focus on the human causes and consequences of these threats and not merely a focus on the scientific and technical issues. To meet this need, the Environment and Society series explores a broad range of topics in environmental studies from the perspectives of social sciences and humanities. Books in this series help the reader understand contemporary environmental concerns, while offering concrete steps to address these problems.

Books in this series include both monographs and edited volumes that are grounded in the realities of ecological issues identified by the natural sciences. Our authors and contributors come from disciplines including but not limited to anthropology, architecture, area studies, communication studies, economics, ethics, gender studies, geography, history, law, pedagogy, philosophy, political science, psychology, religious studies, sociology, and theology. To foster a constructive dialogue between these researchers and environmental scientists, the Environment and Society series publishes work that is relevant to those engaged in environmental studies, while also being of interest to scholars from the author's primary discipline.

Recent Titles in the Series

Canals in a Changing Britain: Construction, Culture, and Environment, 1760–1968, by Jules P. Gehrke

Urban Narratives About Nature: Socio-Ecological Imaginaries Between Science and Entertainment, edited by Carlos Tabernero

The Age of Capitalism, Consumer Culture, and the Collapse of Nature in the Anthropocene, by Jack Thornburg

Sustainable Energy Development: Technology and Investment, edited by Kimarie Engerman, Elizabeth A. Keenan, Dmitry Kurochkin, Elena V. Shabliy

Making Nature Social: Towards a Relationship with Nature, by Rembrandt Zegers

Earth Polyphony: Law, Ecocriticism, and Eco-Activism in Postcolonial India, by Suhasini Vincent

Staying Together: Natureculture in a Changing World, Edited by Kaushani Mondal

Learning in the Anthropocene: Reimagining Education in the Twenty-First Century, by Carl A. Maida

The Social Life of Unsustainable Mass Consumption, by Magnus Boström

Everyday Life Ecologies: Sustainability, Crisis, Resistance, by Alice Dal Gobbo

Environmental Legacies of the Copernican Universe, by Jean-Marie Kauth

Anticipatory Environmental (Hi)Stories from Antiquity to the Anthropocene, edited by Christopher Schliephake and Evi Zemanek

Mapping the Environmental Humanities: The Emerging Role of GIS in Ecocriticism, edited by Mark Terry and Michael G. Hewson

Canals in a Changing Britain

Construction, Culture, and Environment, 1760–1968

Jules P. Gehrke

LEXINGTON BOOKS
Lanham • Boulder • New York • London

Published by Lexington Books
An imprint of The Rowman & Littlefield Publishing Group, Inc.
4501 Forbes Boulevard, Suite 200, Lanham, Maryland 20706
www.rowman.com

86-90 Paul Street, London EC2A 4NE

British Library Cataloguing in Publication Information Available

Library of Congress Cataloging-in-Publication Data

Names: Gehrke, Jules P., author.
Title: Canals in a changing Britain : construction, culture, and environment, 1760-1968 / Jules P. Gehrke.
Description: Lanham : Rowman & Littlefield, [2024] | Series: Environment and society | Includes bibliographical references and index.
Identifiers: LCCN 2024031956 (print) | LCCN 2024031957 (ebook) | ISBN 9781793652331 (cloth) | ISBN 9781793652348 (ebook)
Subjects: LCSH: Canals--Great Britain--History. | Technology--Social aspects--Great Britain--History.
Classification: LCC TC657 .G44 2024 (print) | LCC TC657 (ebook) | DDC 627/.130941--dc23/eng/20241009
LC record available at https://lccn.loc.gov/2024031956
LC ebook record available at https://lccn.loc.gov/2024031957

Contents

Acknowledgments vii

Introduction 1

1 Re-Making the Landscape: Canals in the Late Eighteenth and Early Nineteenth Centuries 13

2 An Independent Spirit: Life Afloat in the Nineteenth Century 49

3 An Era of Decline: The Debate Over Canals in the Early Twentieth Century, 1906–1945 109

4 An Independent Spirit Reclaimed?: The *Cressy* and the Re-Imagining of Canals Before 1951 145

5 Revival and Regeneration: Canals and a Transforming Britain, 1951–1968 199

Conclusion 249

Bibliography 257

Index 277

About the Author 283

Acknowledgments

I would like to offer my thanks to the staffs of the British Library and the National Archives of the United Kingdom for making their resources available, and also to Linda Barley and Carl Kenneally of the National Waterways Archive of the Canal & River Trust for offering assistance both in Ellesmere Port and remotely. In addition, I would like to thank the Office of Academic Affairs, the College of Arts and Behavioral Sciences, and Zahnow Library at Saginaw Valley State University for directing their respective institutional resources to this project. Near its end, Joseph Boughey brought the critical eye of a canal researcher and my wife Anne Huebel offered significant editorial insight. Remaining errors are, of course, my own. It is to Anne, who along with my daughters Margaret and Elizabeth allowed me to sequester myself at repeated intervals, that this book is dedicated.

Introduction

Britons' two and one-half century relationship with their nation's canals and inland waterways is a complex narrative marked by evolving understandings of canals' construction and cultural impact as well as relationship with the environment. This study focuses largely upon canals, artificial navigations which were constructed primarily between the 1760s and 1820s to meet local and regional needs, but which later came to be identified with a nationally significant system of water routes. Canals' basic narrative is familiar to many who have traveled on or near them or watched contemporary adventures on them unfold in the popular media.[1] Built to transport the raw materials and goods of the industrial era, they were eclipsed by railways and then by lorries, and then left largely forgotten until activists' efforts at restoration in the late twentieth century. Revived, they came to be closely aligned with Britons' relationship with the natural environment, associated with leisure and heritage rather than commerce and industry, and retreat from contemporary urban life in favor of immersion in the countryside. This book explores crucial chapters in this narrative, as well as its varied interpretations, until the late twentieth century. It addresses early efforts to build canals and the significance of natural barriers that affected understandings of their place in the landscape. It evaluates the creation of cultural space on canals and the societies which took root in that space. It delves into questions of canals' declining economic competitiveness and juxtaposes this with arguments for their preservation amidst countervailing steps toward reduced maintenance and official closure. And it examines transitions in attitudes toward canals that by 1968 reflected the legacies of two centuries of evolving use and understanding. At the nexus of relationships between technology, culture, and the environment in modern Britain, canals' significance continues to unfold in the twenty-first century.

Canals continue to be one of the most substantive applications of engineering knowledge to Britain's landscape—a landscape notable, even by the time of the industrial era, as one altered by thousands of years of human intervention.[2] Built to re-direct naturally occurring sources of water through newly navigable channels, canals represented a significant intensification of human intervention in the landscape in the service of commerce and industry. Practices of water drainage and transport had been well-established in Britain and the rest of Europe by the late eighteenth century and would continue to evolve in succeeding eras.[3] Yet, Britain's mostly private endeavors quickly added hundreds of miles of navigable waterways by tapping new supplies of capital and engineering knowledge and operated in a synergistic fashion to create a system of waterborne trade eventually stretching across England and affecting portions of Wales and Scotland. In conjunction with locks, bridges, and reservoirs, these artificial waterways fostered an amalgamation of technology with the natural and built environments through which they passed, navigating around, through and sometimes over obstacles as they negotiated the landscape in innovative ways. Some projects were enhancements of existing river systems. Many others were artificial "cuts," supplied with redirected surface waters.[4] All sought the advantages of smooth transport on water and most were reliant on human and animal muscle to manage locks and pull boats. Canal projects were generally constructed in accord with parliamentary legislation passed in response to entrepreneurial initiatives that sought to collect tolls from private carriers in return for responsibility in managing and maintaining the infrastructure.

The ambitious plans of canal companies could be confounded. Engineers could find their grandest designs falter when confronted with obstacles requiring additional financial outlays and reconstruction. Over time canals, like their naturally occurring partners, rivers, could change, often much more rapidly. Weeds could grow, and water supplies dwindle without continued human attention. When both traffic and canal company income became light, an ensuing lack of maintenance exacerbated degradation caused by external forces. In canals, the impact of technology was (as many have written of other systems) mediated by the environment.[5] Beginning in the 1830s, steam railways offered a greater independence from local geography and hydrology, reaching more widely than canals across commercial networks for supplies of iron, steel, and coal that would collectively mark an end to reliance on human and animal muscle in inland trade. Investment money soon followed the rails. By the mid-nineteenth century, many canals, even if identified as remunerative routes, were associated with a bygone era, assumed to have potential little different from that of riverine transport. Where maintenance ceased as a result of sharp declines in traffic, nature returned as banks eroded, weeds increased, channels narrowed, and towpaths disappeared into the brush.

There was little willingness to invest in the massive reconstruction needed to help canals compete more effectively as a linked network. In the short term, they became symbols of technological obsolescence and a changing industrial context. In the long term, however, straddling the divide between the natural and built environments, the physical decay of many prompted a greater association with the natural world. Not easily decommissioned or removed, they became intimately linked with economic, social, and cultural changes taking place around them.

In both the waxing and waning of their engineering might, the cultural impact of canals was determined significantly by those inhabiting spaces on or near the water. Early on, thousands of "navvies" (navigators) recruited to build canals sowed suspicion among local populations. Later, some navvies may have joined the ranks of the male labor force working the boats, who became associated with drunkenness, lawlessness, and violence. Boat workers, otherwise known as "boatmen," were independent and mobile but became focal points for missionary efforts and campaigns against drink and for policing. After the middle of the nineteenth century, fears of restiveness and violence did cool in many areas as canal traffic lightened. Yet, boatmen, sometimes traveling and living in cramped cabins with their families (who often took on the roles of crew) ran afoul of new concerns regarding labor, health, and education. Two parliamentary canal boats acts of the late nineteenth century attempted to erect mechanisms of control, but enforcement could be uneven. Along the waterways, brightly painted boats adorned with images of roses and castles suggested a life free of the confines of urban tenements. Canal boatmen and their families could at once be identified as alien, nomadic, and prone to criminality, but also free to exercise the traditions of bygone craftsmen. A few Victorian travelers, embarking on lesser-used canals in voyages of domestic discovery, found boatmen a source of fascination. Changing understandings and interpretations of the canal boat population were later to be a significant factor in shaping Britons' understanding of canals not only as a source of escape but as an interaction with a pre-modern world.

In addition, despite a once robust presence as symbols of commercial and industrial advance, the declining use of many canals in the late nineteenth century transformed them into symbols of the nation's apparently faltering struggle against its European rivals. Rail had been critical in continued economic growth for several decades, but the nation had been eclipsed in some economic and industrial spheres by rivals who had invested more heavily in waterways. Between 1906 and 1911, a Royal Commission on Canals and Inland Waterways came to support a plan of restoration and national systematization. Yet, little action followed before World War I. In the interwar period, some interest in canals was revived, but declines in traffic continued through World War II. In the minds of most Britons, many canals became forlorn holdovers of bygone industrial days, noted for accumulations of rubbish and childhood drownings. Such perceptions were later difficult to dislodge and became a component in arguments over the waterways' future.

It was from this much-reduced position that this book emphasizes canals began to claim a new cultural and environmental position. In the late nineteenth century, some travelers had found peace and independence on underutilized waters that blended into the landscape and offered an alternative to an increasingly urbanized and congested Britain. By the twentieth century, some associated life on the waterways with ethnic Romani ("Gypsies"/"gypsies"). Later, as canals struggled in the face of further deterioration, remaining canal boatmen became, in the minds of some, cherished reminders of workers who had escaped the deadening effects of mass industrialization and urban life. By the interwar era, a new age of writing and travel pressed for the re-discovery and appreciation of rural spaces in the midst of accelerated technological change and the growth of suburbs. In the simplified slower pace of village life, some writers argued, were to be found markers of the true character of the nation now being overwhelmed by forces of governmental centralization and global commercial networks. Canals were largely absent from these interwar "ruralist" writings—still associated for some with the detritus of the industrial age. Nonetheless, the contrast they provided with a world focused on automobiles and suburbanization suggested to a few observers that in their quiet waters, populated by a workforce maintaining independent rhythms of work, living, and dress, could be found an essential counterbalance to contemporary Britain. When the writer and engineer-craftsman Lionel Thomas Caswall (L.T.C.) Rolt published *Narrow Boat* in 1944 as a mournful reflection on the disappearing traditions of the waterways, he associated canals and boatmen with Britain as it had been before rampant technology had distanced workers from the traditions of their manual crafts.[6]

Rolt's work helped to forge the first links in a transformed relationship between Britons and their waterways. In the late 1940s, a majority of existing canal mileage was placed under the authority of the government's British Transport Commission (BTC). Officials sought to reduce expanding deficits on the waterways account by facilitating increases in trade where possible and moving for the closure of some routes. Yet, a small group of outside observers saw in canals a national resource whose remaining industrial and recreational possibilities were being squandered in the face of a narrow-minded bureaucracy. These voices came through loudest in the activities of the Inland Waterways Association (IWA), for whom the waterways comprised a system which offered a comfortable amalgam of landscape and technology and, in their continuing leisure and industrial uses, could prove a foil to a desultory reliance on roads and suburban expansion. In the Transport

Act of 1968, government officials, increasingly affected by these new perspectives, as well as the costs associated with closing and eliminating unused waterways, embraced a future for canals as a largely recreational resource. With their infrastructure once a prime example of engineering innovation and power over the landscape, canals had begun to be woven into understandings of the natural world. This repositioning would continue after 1968 in a more specifically environmental vein, with new official considerations given to ecosystems in and alongside canal waters as well as canals' links with engineering and cultural heritage. Canals would thus be at the center of a reawakening environmental engagement in the late twentieth century.

In the midst of these rich narratives, academic histories of Britain's canal system have not yet reached their full potential. Many late eighteenth and early nineteenth-century assessments saw them as tools of the new industrial age, and thus focused on their engineering, construction, and commercial use. Significant among these was the *Historical Account of the Navigable Rivers, Canals, and Railways Throughout Great Britain*, written by Joseph Priestley, the head clerk of the Aire and Calder Navigation.[7] Later in the nineteenth century, Samuel Smiles chronicled significant components of canals' construction in his multi-volume *Lives of the Engineers*, bringing to the fore the artificial waterways as an achievement of individual initiative.[8] The decline of canals as symbols of the nation's strength brought little direct assessment beyond the continuing publication of *Bradshaw's Canals and Navigable Rivers of England and Wales,* which chronicled their mechanisms and capacities as a largely commercial reference. Some cultural testaments, including missionary literature and fictional and non-fictional travel narratives, did emerge, such as Temple Thurston's *The Flower of Gloster.*[9] Both Bradshaw's guides and *The Flower of Gloster* offered reference works for the first true generation of canal historians, led by Rolt and Charles Hadfield in the immediate post-World War II era.[10] Both Rolt and Hadfield were heavily involved in the movement to save the canals and continued to exercise significant influence on later interpretations of them. In recent decades, economic historians have continued to examine the interplay between canals and railways as dueling transportation mechanisms of the nineteenth century, while non-economic assessments have offered both technical accounts of canal history and appreciations of their significance from the perspective of contemporary users.[11] Yet, many of these non-economic histories reflect an era prior to the 1980s, a point at which more intensive consideration began to be given to how interrelationships between technologies, cultures, and environments have affected historical understandings in all three categories.[12]

Modern studies of the overlap between technology, culture, and the environment have set the stage for further study of Britain's canals. Studies of technology have examined the application of large engineering projects to the landscape and the impact of derelict and disused sites.[13] They have assessed shifting cultural and environmental impacts over time in systems where autonomous functions have operated alongside those of human intervention. In notable instances, the creation of cultures surrounding or dependent upon such works has also been a focus.[14] In the broad plain of environmental history, the porous boundaries between technology and the environment have been well-studied on land and upon the water in the form of natural river systems, dams, reservoirs and the oceans.[15] In some cases, these studies have given primacy to governing structures and their control over provision and access to water, and at other times to the development of cultures and societies using and interacting with the waterways.[16] Historians of water have been enormously broad in their coverage of technology and water, giving attention to issues of health, pollution, and sanitation, in addition to those of water control and management.[17] In many cases, the physical linkages of rivers and canals have meant an examination of the techniques of water management within hybrid systems, often canalized rivers.[18] Yet canals and canal systems, as sites in which boundaries of technology and the environment have been re-negotiated over time, and where such negotiations have been significantly affected by canals' industrial origins, are still due further analysis. In Britain, canals' rapid construction by commercial and industrial interests, significant diversity in size, early challenge by railways, and unique positioning at the intersection of the natural and industrial worlds, all suggest the importance of continued study. Social and cultural examinations have not lain dormant, with studies undertaken in recent decades by Harry Hanson, Wendy Freer, Joseph Boughey, and, most recently, Jodie Matthews.[19] This study seeks a new direction in emphasizing themes of construction and technology, culture, and environment through two critical centuries, bringing understanding of these waterways up to the cusp of the emergence of modern environmentalism at the close of the 1960s.

Bringing together several stages in the history of canals, including a focused examination of their post-World War II transformation, this study posits that many of the meanings of canals today may be best interpreted as part of a 200-year arc of development. The study focuses on evolving industrial definitions and interpretations of canals in chapters 1 through 3 and turns to the cultural transformation of canals by the mid-twentieth century in chapters 4 and 5. In the first set of chapters, it argues that an understanding of the forces which led to the revitalization of canals in the late twentieth century must be seen in the context of the social and commercial forces which effected their construction, and the cultural understandings associated with those who first worked them. Chapter 1 accesses pamphlets and newspapers to review the motivations and strategies of those who first invested in the waterways, and the methods of engineers attempting to complete their tasks within the allowances of the physical environment and to the satisfaction of their employers. Chapter 2 highlights canals' intimate links with the social and cultural worlds of Victorian Britain. It examines early nineteenth-century newspaper accounts associating boatmen with alcohol, vice, and violence, and the later writings of the middle-class reformer George Smith ("of Coalville") who heightened anxieties regarding moral and sanitary conditions on the waterways.[20] It then moves to the records of inspectors under the canal boats acts, as well as the records of education officials, missionaries, and early recreational travelers and cultural observers. Chapter 3 emphasizes that environmental and cultural understandings of canals today are inseparable from real and perceived fears of industrial decline as they existed in the early twentieth century.[21] It assesses the investigations and conclusions of the Royal Commission on Canals and Inland Waterways as well as the contributions of two post-World War I committees on canals headed by Neville Chamberlain.[22] In the midst of the latter investigations, boatmen were shorn of the reputation of vice and violence that had once dominated public discussion and reintegrated into the cultural space of canals as honest and beleaguered workers struggling amidst a rapidly shifting economic and industrial world.[23] Finally, chapter 3 reviews a host of published writings and government documents to assess the significant shifts in understandings of the countryside occurring as a result of the increased mechanization and militarization of rural areas brought about by World War II.

In the second set of chapters, the study outlines how a re-building and re-envisioning of canals emerged against the backdrop of a nadir of their economic and industrial fortunes in the mid-twentieth century. Chapter 4 argues that the place of canals, as well as boatmen and their families, within the cultural landscape of Britain was affected by ruralist and organicist visions of the countryside, in particular the unease felt regarding the impact of modern technology between the end of World War I and 1951.[24] Drawing upon accounts of travel on the waterways since the late nineteenth century, analyzing the place of canals (or, initially, their lack of a place) within much ruralist and organicist literature, it describes their first substantive entrance into social and cultural discourse with the writings of Rolt. It then assesses the earliest forays of the Inland Waterways Association into issues of policy regarding canals through the organization's own records, as well as those of the government, between 1945 and 1951. Critically, chapter 5 argues that it is within the period between 1951 and 1968 that the place of canals in modern British society was reshaped and renewed. In Rolt's writings as well as those of the IWA in its *Bulletin*, emerged a more complex understanding of the countryside in which traditions of the canals were understood to be woven into rural society.[25] On canals, many supporters believed, was a culture of life and labor that ran counter to a new age of industrialization and centralization. At the same time, Ministry of Transport records reveal transport officials were more functional, pointing to studies of the waterways with their struggling accounts and calling for reduced commitments to maintenance. Yet, in a host of publications and personal accounts, waterways enthusiasts saw forays onto the canals as a valued alternative to a nation they believed dominated by motorways and an increasingly homogenized national culture. Evolving official responses were tentative and resistant to the most determined IWA campaigns. Yet, as revealed in government communications, by the late 1960s, Britain's Labour government chose to make a commitment to working for the preservation of the nation's canals as a system. Its decisions both reflected and reinforced the significance of canals and the experiences they engendered in an era that was to be more widely environmentally aware than any before.

Historical understanding of the technological and cultural impact of canals, as well as their integration into the evolving relationship of Britons with their environment, may thus be best witnessed as part of a 200-year arc. Canals' physical infrastructure had been defined by the location of natural resources including coal and water, the location of towns, and the contours of the landscape at the close of the eighteenth century. In the course of the nineteenth century, the meaning of the waterways had developed an extra dimension by virtue of the human communities working on their waters and along their banks. By the twentieth century, these factors began to transform canals, in some interpretations, from engineered constructs of industrial society into components of the countryside, shifting their place at the boundary of technology and the environment. By 1968, canals offered not only a platform for multi-faceted use as a linear resource allowing access to the natural environment and leisure but also, for many travelers, a means of engaging with more complex histories of industrialization.

Canals' renewed, contemporary meaning is diverse and dynamic. Artificial cuts in the landscape dependent upon engineering to maintain supplies of water, they are often capable of supporting complex ecologies integrating vegetation and wildlife. Once boldly seeking to provide efficient pathways through the landscape, they have often been limited by variances in topography and water supply. Built to facilitate the demands of industrialization, some continued to use horsepower well into the modern era, and many rely upon human muscle for the operation of locks in the twenty-first century. In the eighteenth century, observers feared that their intrusion sounded the death knell of rural Britain. Now, canals and the built structures alongside them serve those seeking escape from the intensity of urban life for the peace of the waterways. In an era when significant shifts in sites of work and mobility are re-shaping the contours of work and leisure, canals will continue to occupy a critical place in the evolving relationship of Britons to their built and natural environments. This study may better inform that discussion by assessing the impact of canals from the early stages of industrialization to the era of environmentalism.[26]

NOTES

1. Perhaps most significantly, Prunella Scales and Timothy West hosted *Great Canal Journeys* over several years between 2014 and 2021, building an international reputation with their journeys along canals in Britain as well as internationally. Portions of the program are available on DVD. *Great Canal Journeys*, Series 1–5 (London: Spirit Entertainment Ltd., 2018).
2. Virtually no portion of the contemporary British landscape has remained untouched by human technology. I. G. Simmons has identified the creation of a "humanized" landscape in the expansion of agriculture over 5000 years ago in *An Environmental History of Great Britain: From 10,0000 Years Ago to the Present* (Edinburgh: Edinburgh University Press, 2001).
3. See Jan Nijman, ed., *Amsterdam's Canal District: Origins, Evolution, and Future Prospects* (Toronto: University of Toronto Press, 2020).
4. The first use of the term "cut" with reference to canals is uncertain, though as an engineering term, it has been used to refer to the artificial channels of canals, as well as spaces for transport more generally where rock or soil has been removed to allow for a smooth gradient.
5. See James C. Williams, "The Technology Junction: Exploring Technology and the Environment," *ICON: Journal of the International Committee for the History of Technology* 6 (2000): 7–20.
6. Lionel Thomas Caswall (L.T.C.) Rolt, *Narrow Boat* (London: Eyre Methuen, 1944).
7. Joseph Priestley, *Historical Account of the Navigable Rivers, Canals, and Railways, of Great Britain* (London: Longman, Rees, Orme, Brown, and Green, 1831). This Joseph Priestley should be distinguished from the more well-known scientist and philosopher Joseph Priestley. See E. Kilburn Scott, "Three Joseph Priestleys," *Transactions of the Newcomen Society* 20, no. 1 (1939): 147–149.
8. See, in particular Samuel Smiles, *Lives of the Engineers, with an Account of Their Principal Works; Comprising Also a History of Inland Communication in Britain*, Vol. 1 (London: John Murray, 1862).
9. See, for instance, Henry Rodolph de Salis, *Bradshaw's Canals and Navigable Rivers of England and Wales. A Handbook of Inland Navigation* (London: Henry Blacklock & Co, Ltd., 1928). *The Bradshaw's* handbooks were published in several updated editions in the first half of the twentieth century. See also, Temple Thurston, *The "Flower of Gloster"* (New York: Dodd, Mead and Co., 1912).
10. Here, L.T.C. Rolt's *The Inland Waterways of England* (London: George Allen & Unwin, 1950) remains a highly readable one-volume work, while the multi-*volume The Canals of the British Isles* series, edited and substantively written by Hadfield from the 1960s to the 1970s, remains a highly detailed (if less casually read)

reference. The series closed in 1977 with John Boyes and Ronald Russell, *The Canals of Eastern England* (London: David & Charles, 1977).

11. In economic literature, see in particular, Peter Maw, *Transport and the Industrial City: Manchester and the Canal Age, 1750–1850* (Manchester: Manchester University Press, 2013). In non-economic contributions, Anthony Burton and Derek Pratt have offered an updated survey which includes an examination of canals' multi-faceted impact in the late twentieth century, though they do not reflectively focus on themes of technology, culture, and the environment. See Anthony Burton and Derek Pratt, *The Anatomy of Canals: The Early Years* (Stroud: Tempus Publishing 2001), *The Anatomy of Canals: The Mania Years* (Stroud: Tempus Publishing, 2002), and *The Anatomy of Canals: Decline and Renewal* (Stroud: Tempus Publishing, 2004). More recently, Liz McIvor's *Canals: The Making of a Nation* (London: Penguin Random House, 2015) was published to accompany a BBC television series; though valuable, it is not comprehensive. Julian Dutton's *Water Gypsies: A History of Life on Britain's Rivers and Canals* (Cheltenham: The History Press, 2021) is a thoughtful and highly readable overview of the lives of those who have worked and lived on Britain's waterways, but is not focused on an analysis of primary source and archival material.

12. This process of change reached full stride in the 1990s when Jeffrey Stine and Joel Tarr were able to delineate a steadily accelerating roster of new studies related to cities, industries, and natural resources. See Jeffrey K. Stine and Joel A. Tarr, "At the Intersection of Histories: Technology and the Environment," *Technology and Culture* 39, no. 4 (October 1998): 601–640.

13. Bill Luckin, *Questions of Power: Electricity and Environment in Inter-War Britain* (Manchester: Manchester University Press, 1990).

14. The importance of understanding the interpretive creation, through an evolving human presence, of the spaces in which technology is constructed and applied is suggested by Christopher Tilley in *A Phenomenology of Landscape* (Oxford: Berg, 1994).

15. See Christof Mauch and Thomas Zeller, *Rivers in History, Perspectives on Waterways in Europe and North America* (Pittsburgh, PA: University Pittsburgh Press, 2008); and David Blackbourn, *The Conquest of Nature: Water, Landscape, and the Making of Modern Germany* (New York: W. W. Norton, 2006). In Britain, cultural and environmental stories of the Thames River became a new subject of interest just after the turn of the millennium. See Peter Ackroyd, *Thames: The Biography* (New York: Anchor Books, 2007).

16. Significant studies of early modern state-building and the introduction of major water management schemes are provided by Chandra Mukerji in *Impossible Engineering: Technology and Territoriality on the Canal du Midi* (Princeton, NJ: Princeton University Press, 2009) and Eric A. Ash in *The Draining of the Fens: Projectors, Popular Politics, and State Building in Early Modern England* (Baltimore, MD: Johns Hopkins University Press, 2017). Ash is able to highlight the significance of evolving cultures of use after the seventeenth century, 298–299.

17. See, in total, the three series and nine volumes of *A History of Water*, published between 2006 and 2014, in particular, volume 2 of series 2 giving particular attention

to river management. (Terje Tvedt, Terje Oestigaard, et al., eds. *A History of Water*, Series 1, vols. 1–3; Series 2, vols. 1–3 (London: I. B. Tauris, 2006–2014)).

18. Here, see Mark Cioc, *The Rhine: An Eco-Biography, 1815–2000* (Seattle, WA: University of Washington Press, 2002); and Sara B. Pritchard, *Confluence: The Nature of Technology and the Remaking of the Rhône* (Cambridge, MA: Harvard University Press, 2011).

19. See, in particular, Jodie Matthews, *The British Industrial Canal: Reading the Waterways from the Eighteenth Century to the Anthropocene* (Cardiff: University of Wales Press, 2023). Matthews has offered significant insight into links between canals as agents of domestic industrial and economic change and the development of the colonial world. The world of canal boatmen has been unevenly studied, with significant room for additional work. Among existing contributions are Harry Hanson, *The Canal Boatmen, 1760–1914* (Manchester: Manchester University Press, 1975); Wendy Freer, *Women & Children of the Cut* (Gwernymynydd: Railway & Canal Historical Society, 1995); and Wendy Freer and Gill Foster, *Canal Boatmen's Missions* (Oxford: Railway and Canal Historical Society, 2004). For an examination of Rolt's approach to history and the canals, see Joseph Boughey, "From Transport's Golden Ages to an Age of Tourism: L.T.C. Rolt, Waterway Revival and Railway Preservation in Britain, 1944–54," *The Journal of Transport History* 34, no. 1 (June 2013): 35.

20. See, for instance, George Smith (of Coalville), *Our Canal Population: The Sad Condition of the Women and Children,—with Remedy* (London: Haughton and Co., 1875), 8.

21. The study of canals can offer an important contribution to discussions of British economic stagnation and decline, and its cultural effects, though the overall argument is now an old one. See Martin J. Wiener, *English Culture and the Decline of the Industrial Sprit, 1850–1980* (Cambridge: Cambridge University Press, 1981, 2004) and W. D. Rubinstein, *Capitalism, Culture, and Decline in Britain 1750–1990* (London: Routledge, 1993).

22. The Royal Commission eventually produced several volumes of material through its reports and supporting material. See, in particular, United Kingdom, Royal Commission on Canals and Inland Waterways, *Fourth and Final Report the Royal Commission Appointed to Enquire into and to Report On The Canals and Inland Navigations of the United Kingdom*, Vol. VII (London: Printed for H. M. Stationary Office by Wyman and Sons, Ltd., 1909).

23. See Committee on Living-in on Canal Boats, Ministry of Health, *Report of the Departmental Committee* (London: HMSO, 1921).

24. In particular, see Philip Conford, *The Origins of the Organic Movement* (Edinburgh: Floris Books, 2001), and David Matless, *Landscape and Englishness*, 2nd ed. (London: Reaktion Books, 2016).

25. The early history of the IWA has continued to attract attention. Surveys of the organization's efforts include Roger W. Squires, *Canals Revived: The Story of the Waterway Restoration Movement* (Bradford-on-Avon: Moonraker Press, 1979) and David Bolton, *Race Against Time: How Britain's Waterways were Saved* (London: Methuen, 1990). These organizational narratives are complemented by the present author's examination of links between interwar ruralism and organicism, and the

early IWA. See Jules Gehrke, "Countryside, Recreation, and the Transformation of Canals in Britain in the Mid-Twentieth Century," *Journal of Tourism History* 11, no. 2 (2019): 167–186. Evolving understandings of "country" and "countryside," and the influence of perceived "Golden Ages," should be placed at the heart of discussions of Britain's canals in the mid-twentieth century. See Raymond Williams, *The Country and the City* (New York: Oxford University Press, 1973).

26. Andrew Wallace and Katy Wright have emphasized the ambivalence with which canals have been approached in the twentieth century, as well as canals' significance for understanding the post-industrial landscape. As they have written in the results of their ethnographic study conducted in London, "The canal sections we knew were compelling contradictions; they were murky and rubbish-strewn relics but also ecologically and materially rich. They were mute and sunken liminalities one minute, polished and carnivalesque the next." See Andrew Wallace and Katy Wright, "The Post-Industrial English Canalscape: From Enclosure to Enlacement," *Cultural Geographies* 29, no. 2 (2021): 188.

Chapter 1

Re-Making the Landscape

Canals in the Late Eighteenth and Early Nineteenth Centuries

Between the mid-eighteenth century and the late twentieth century, Britain's canals and navigable rivers were transformed from sites of audacious industrial and capitalist drive to sites celebrating both the environment and the nation's pre-industrial heritage. In the 1760s, a quickly evolving engineering profession accelerated the amelioration, re-direction, and removal of land and water-based obstacles to establish artificial channels rationalizing water for transport. This process stood in the shadow of Britain's much wider prowess on the open seas but was critical for the health and development of the nation's domestic economy, the core of a widening empire.[1] Parliamentary power authorized the formation of companies, the selling of shares, the taking on of debt, the collection of tolls, and the framing of rights in access to, and control of, water and land. Administrative and business acumen were honed through processes of soliciting surveys, overseeing construction, organizing staff to collect and supervise tolls and maintenance, and the distribution of dividends. The engineering profession also gained its first heroes as hills, mountains, and other watercourses were transcended or altered—with cuttings, locks, tunnels, bridges, aqueducts, and reservoirs—to ensure steady access to, and supplies of, water within artificial workings. Projects involving canals and other inland navigations were never static. In their early stages, investors quickly realized their risks as legislative and legal battles took time, inexact knowledge of the natural world could bring expensive and time-consuming design changes, profits could be uncertain (particularly as the number of independent projects increased), and inter-modal competition continued to evolve, particularly in rail transport. Yet, in re-imagining the potential of water as a natural resource and embedding artificial channels within the landscape of urban and rural areas, canal engineers laid the bedrock for an evolving technological, cultural, and environmental system.

Britain's era of canal-building was shaped by precedents both foreign and domestic. In Europe, France and the Netherlands were significant.[2] In 1666, work began on the first portion of what in its entirety is now known as the *Canal du Midi*, linking the French Mediterranean and the Atlantic. Rooted in the financial and administrative abilities of Jean-Baptiste Colbert and the organizational capabilities of Pierre-Paul Riquet, the first stage of the canal stretched from Toulouse to Sète on the Mediterranean by the early 1690s.[3] It represented one of the most significant engineering achievements in early modern Europe and included one of the first known artificial reservoirs for the supply of its water.[4] Later, in the nineteenth century, the Canal de Garonne provided the link from Toulouse to the Atlantic.[5] In the Netherlands, advances in the construction of a system of inland transportation helped the Dutch economy to thrive in the 1600s on the basis of both coastal trade and inland communication linking important cities. The system of inland navigation was marked by the building of *trekvaarten*, inland canals (with towpaths) which could operate regardless of the season. Their construction came largely at the behest of individual cities. Well into the era of the railways, the *trekvaarten* represented an important and economically successful undertaking generating profits for Dutch municipal owners.[6] As the Irish architect and engineer Richard Castle wrote in his *Essay on Artificial Navigation* of 1730,

> The Dutch have exceeded all nations we know of, in works of this kind, and the benefits they have and do receive thereby are evident to the most careless passenger that travels the country. The Province of Holland in effect is but one continued cluster of cities by means of their canals alone, which afford the most inland towns the same advantages with the most commodious seaports.[7]

In Britain, transport was important on both tidal and non-tidal rivers. In the early modern era, major tidal rivers for transport included the Thames, the Severn, the Mersey, and the Dee. Important inland, non-tidal rivers, and the major towns located on them included the River Ouse (York), the River Trent (Nottingham), and the River Nene (Peterborough). Navigable inland rivers stretched for about 650 miles before the age of engineered navigation improvements.[8] Travel was subject to a number of environmental uncertainties. Tidal rivers could be affected by coastal storms and deposits of silt, while non-tidal rivers could be affected by frost and ice; all could be affected by low water flows at certain times of the year.[9] Rivers were the focus of the most significant of Britain's early navigation works, and efforts at "improvement" would bring the total length of navigable inland waterways in England to just over 1,100 miles in the 1720s.[10] Among the most significant projects was that on the River Exe, where a canalized channel begun in the 1560s was cut for approximately 1.75 miles from Exeter to a point rejoining the Exe

River.[11] Although small and making use of an existing river channel, the Exe improvement, a century later, incorporated pound locks in which two sets of gates created a level body of water (known as the pound), which was raised or lowered through the opening or closing of either one set of doors or the other, thus raising or lowering the boat inside.[12] Pound locks allowed boats to move from higher to lower elevations (or vice versa). Their maintenance, as well as the need to ensure a steady supply of water for them, marked a critical dividing line between engineered and natural waterways.

The Fens, a region of low-lying and periodically inundated land in England, had remained an intensive space for water engineering in Britain. For centuries, human settlement in the region waxed and waned with rises and declines in sea levels. In the mid-seventeenth century, a major effort at drainage was undertaken by the Earl of Bedford, who employed the Dutch engineer Cornelius Vermuyden for the effort.[13] Here, the completion of the New Bedford River allowed for a straighter waterway inland from the Wash, England's major inlet on the North Sea.[14] Though dynamic, the drainage efforts were of limited utility as the drying of the lands led to a shrinking of peat, a subsequent lowering of the land, and renewed flooding. In the early nineteenth century, the application of steam engines helped to ensure continued drainage but did not solve the difficulties of shrinking soil. Efforts to manage the Fens, including initiatives seeking to return the lands to their state prior to the advent of agriculture, continue in the twenty-first century. In the southeast of England, efforts to manage existing river flows in the early modern period continued as acts of Parliament authorized improvements on the Lea River in 1571 and 1588.[15] In 1635, the Thames was made navigable inland as far west as Oxford, and in 1641, navigation was extended westward once again from Eynsham to Lechlade and Cricklade.[16] Later, in the west of England, improvements were also made to the tributaries of the River Severn.[17] As would be the case with canals, many river improvements were engineered and managed with local business and industrial interests in mind rather than a plan of national transport. In 1755, the Corporation of Liverpool, seeking to provide better access to the St. Helens coal fields, departed from an initial plan of simply making the Sankey Brook a navigable waterway and used the powers it had been granted by Parliament to create "side cuts" to fashion an entirely new channel that would provide protection from the impact of flooding and provide a series of locks accommodating a 90-foot drop (over the course of 12 miles) to the River Mersey.[18] Though its roots lay in the processes of river improvement, the creation of the Sankey Brook Navigation expanded commercial enterprise into regions considered previously inaccessible and represented a significant intensification of water engineering.

River improvements, even if backed by local interests, required the legal sanction of Parliament and prefigured the complex politics that were to be connected with water engineering in the era of canals. Riparian law in England had generally recognized lines of property ownership as extending from a shoreline through a portion of the riverbed abutting a neighboring property.[19] Although there was substantive precedent for widened use of some rivers, unless sanctioned by Parliament, navigation rights within many riverbeds were not guaranteed.[20] Rights to undertake and maintain changes on rivers (as well as to profit by them) have generally been assigned through legislative authority, sometimes in the interest of a local governing authority and sometimes in the interest of a private business. In 1699, a group of "undertakers" was allowed to make changes to the rivers Aire and Calder, allowing for the integration of Leeds into a transport network linked with the east coast.[21] The improvements to the Aire and Calder would eventually total over 30 miles and include over a dozen locks by the early nineteenth century.[22] In the upper reaches of the Midlands in the early eighteenth century, parliamentary legislation allowed for the Mersey and Irwell Navigation Company to use those two rivers to connect Manchester to the Irish Sea.[23] In the eastern Midlands, efforts to improve navigation on the River Don had begun early in the 1700s, but had not been successful until early in the eighteenth century. The Company of Cutlers of Sheffield, a manufacturers' organization, prevailed against the arguments of the Duke of Norfolk to create cuts that would circumvent obstacles near the city, and in 1729 created a new company of undertakers that would continue to move forward with improvements down the river as far as Doncaster.[24] By mid-century, these undertakers had delegated operation of the river to a group of lessees and, later, to two individuals who managed it on behalf of the supervisory committee in return for salaries.[25] In return for undertakers' efforts in re-engineering rivers, they were rewarded with the right to collect tolls.[26]

Though inland water transport had been seen as an alternative to the inefficiency of an ill-maintained road network, continuing demand for transport on land prompted the first legislation for turnpike roads in the 1660s. In general, roads were at a disadvantage in the transport of some bulk goods, but they could deliver goods door-to-door, avoiding the costs of transshipment (i.e., loading goods from one form of transport to another) where rivers failed to pass by industrial works. The building and maintenance of roads had often been assigned to parishes, which supported the work through property taxes and labor that could be claimed from local residents.[27] In the mid-eighteenth century, in an effort to grapple with the pitfalls of long-distance roads (including poor construction, maintenance, and over-use), Parliament began to assign their maintenance to turnpike trusts. These trusts exercised powers similar to those that organized river improvements, though their commissioners were generally representatives of local governing bodies.[28] The most significant period of trust creation was in the 1750s and 1760s when 300 trusts were created and over 10,000 miles of road addressed.[29] By the 1830s, the turnpike system was about 20,000 miles, or about 17 percent of Britain's road network.[30]

Yet, as an endeavor with a wide potential impact, the unwieldy management and localized interests of groups of turnpike administrators were a source of criticism for those reflecting upon the trusts' utility at the end of the eighteenth century. The writer of an article on roads for the Board of Agriculture remarked in 1797 upon the violent quarreling that could emerge at meetings of commissioners, noting

> a proposed new line of road or, perhaps, the repair of an old one, will sometimes be contested with as great keenness and vehemence as if the parties were contending whether Great Britain shall be a monarchy or a republic.[31]

The transport of heavy loads could make routes both treacherous and muddy from the damage caused by increased weight, leading to a good deal of inconsistency in roads. Amidst his wide travels in northern England, the agriculturalist Arthur Young offered brief assessments of roads ranging from excellent to intolerable, with candid comments on those responsible. The road to the hamlet of Kiplin he found "much superior to most of the turnpikes in this country; and owing to the spirited conduct of Christopher Crowe, esq; who, in the capacity of Justice of the Peace, has given much attention to roads."[32] The road to Wigan he found much worse, writing,

> Let me most seriously caution all travellers, who may accidentally purpose to travel this terrible country, to avoid it as they would the devil; for a thousand to one but they break their necks or their limbs by overthrows or breakings down.[33]

Though sometimes following a similar business pattern, promoters of river and, later, canal improvements could argue that the smooth path across the countryside provided by water prevented both travelers and their products from being smashed by jarring road transport.

In contrast to the re-engineering of existing rivers, canal-building was generally defined by the construction of new water channels for either water conveyance or navigation. In the first case (inclusive of drainage and irrigation), canals were built with a declining bed to facilitate the flow of water from a source, while in the second, they were built with level beds.[34] The advantages of canals over river improvements included linking disconnected bodies of water, efficient transportation between otherwise separated regions and towns, and the creation of water channels less affected by droughts, flooding, and underwater obstacles.[35] Moreover, canals generally lacked currents, could maintain standard depths closer to their banks, and made use of aqueducts and tunnels to overcome geographic barriers. Canals fell into two categories: those crossing boundaries separating river valleys, sometimes called arterial canals, and those generally following established waterways within the same river valleys, often referred to as lateral canals.[36] Roads (and later, railways) could be built to follow gradual undulations in the landscape, while transport canals, with the need to maintain a level bed, were often forced to tack around landscape contours to avoid the costs of either cuttings or a complex series of locks. Canal costs generally included clay lining (often referred to as "puddling") to prevent water leakage and the building and maintaining of large earthen banks. Moreover, bridges, aqueducts, tunnels, locks, and reservoirs added to construction costs.

As with river improvements, canals often began with local rewards in mind. Francis Egerton, third Duke of Bridgewater, opened Britain's first truly modern canal (rather than river improvement) to link his coal mines to Manchester in the early 1760s. As a young man, Bridgewater had visited continental Europe, observed the canals of Italy and the *Canal du Midi* in France, and decided to take the first steps toward familiarizing himself with the engineering challenges of artificial waterways. He worked with his agent John Gilbert and engineer James Brindley to develop a plan for a canal that, initially receiving parliamentary approval in 1759 and then later amended, provided drainage for his mines at Worsley and transported coal just over 10 miles to Castlefield Basin, Manchester.[37] Brindley spent several weeks surveying the canal's route which, boldly, crossed the River Irwell by means of an aqueduct. At Castlefield Basin, the Duke invested in a series of warehouses that stored goods as part of a multi-faceted industrial enterprise. At Worsley, a tunnel excavated into the hillside allowed access to the coal face and was part of a system of some 40 miles of underground waterways. The tunnels were ventilated by a "water bellows," a current of air produced by the rush of water through a large "trunk" pipe.[38] Soon after the completion of the initial canal, legislative approval was given for an extension that took the canal to the River Mersey at Runcorn, by which there was access to Liverpool.[39] The extension to the River Mersey took several years to complete; however, after overcoming a difference in elevation of 90 feet between the canal and the terminus on the Mersey, the extension, with operational locks, was ready in 1776.[40]

For contemporaries, the Duke's engineering feat was a triumph over the impediments of nature and the creation of perhaps one of the most "extraordinary" things in Europe.[41] Moreover, in August of 1762, an auction advertisement pointed to the regional economic implications of the Duke's efforts. It indicated the state of a particular home for sale, the agricultural value of the 50 acres that accompanied it, and noted that it was, "within one Mile of Altringham, seven from Manchester, four from Knutsford, and seven from

Warrington, all good Market Towns, and within one Mile of the Duke of Bridgewater's intended Canal."[42]

Such undertakings, given that they were bestowed by human ingenuity rather than by nature, opened the door to international comparison. A 1763 letter to the *St. James's Chronicle* placed canals at the intersection of the natural and engineered worlds. The author wrote,

> Tis not long since I viewed the artificial curiosities of London, and now have seen the natural wonders of the Peak; but none of them have given me as much pleasure as I now receive in surveying the duke of Bridgwater's [*sic*] navigation in this country.[43]

Furthermore, the author considered the adaptation of the waters of the canal to serve the Duke's mines and their engineered path across the River Irwell sources of awe.

> I almost trembled to behold the large river Irwell underneath me, across which this navigation is carried by a bridge, which contains upon it the canal of water, with the barges in it, drawn by horses, which walk upon the battlements of this extraordinary bridge. This navigation begins at the foot of some hills, in which the duke's coals are dug, from whence a canal is cut through rocks, which day-light never enters.[44]

As a nod to the rarified heights to which the Duke had ascended, the letter writer added, "Nay, it is even said, that some Dutch engineers are coming over hither to perfect themselves in the art of inland navigation."[45] Though the visit of the Dutch engineers may remain unsubstantiated, King Christian VII of Denmark integrated a visit to the Bridgewater Canal into a longer tour near Manchester in 1768.[46] In October 1767, the Venetian resident and the brother of the Spanish ambassador were both reported to have visited.[47]

The Duke's and Brindley's achievement ushered in a new era of resource exploitation with fully artificial water channels accelerating the exit of raw materials and importation of finished goods from, to, and through non-coastal areas. It also represented the interlocking capacities of ambition, personal capital, and commitment to technical advancement that forged critical links in the chain between water transport and professional engineering in modern Britain.[48] Brindley, who at an early age had received little formal education but demonstrated significant skill as a millwright, became key to defining the range of the modern engineer's responsibilities. His career accelerated in the late 1760s, and he became known for his vision of linking Britain's major rivers—the Thames, the Severn, the Mersey, and the Trent—through the creation of a series of navigable cross-country links known as the "great cross" (or, "silver cross"), a vision of coordinated waterway development that was to be recognized [if not respected] well into the twentieth century.[49] In addition to his work on the Bridgewater, Trent and Mersey, Coventry, and Oxford canals, Brindley soon led projects including the Birmingham and Droitwich canals in the West Midlands and the Chesterfield Canal in the East Midlands.[50] His characteristic lateral movement around major changes in elevation helped to keep construction costs to a minimum and offered what was understood to be wider economic gain through servicing a wider swath of the countryside between major ports.[51] The eighteenth-century antiquarian Reverend Stebbing Shaw wrote,

> An all-contriving power was given us in the great Mr. Brindley, sufficient to encounter all difficulties and to remove the most perplexing obstacles. To his perforating hand the immense hills and stubborn rocks were no insurmountable difficulty, and he could with the greatest ease carry water over waters.[52]

Between 1760 and 1769, a total of twenty-nine parliamentary acts in Britain supported inland navigation, of which sixteen were new promotions.[53] Despite Brindley's aspirations of a great "cross," most canal initiatives, both in his time and after, were conceived and initiated in response to local economic and geographic concerns rather than national ones; design and construction then proceeded with the narrow interests (often financial) of an individual group of backers. In keeping with Brindley's decision to build around major obstacles, most canals also remained constrained by local geographic and water resources, their varying engineering solutions often meaning their infrastructure (particularly the size of the vessel used) was just partially compatible with other waterways.[54] Though wider links did emerge in accord with the demands of an emerging national market, the local and distinctive environmental, engineering, commercial, and cultural histories of many British waterways had emerged by the turn of the nineteenth century.

Brindley died in 1772 and thus saw only the opening stages of an acceleration in British canal-building. The work he had begun on an effort to link the port cities of Kingston Upon Hull and Liverpool in the Midlands through a waterway between the Trent and Mersey rivers, however, was to continue for several years after his death. In 1765, an essay writer in the *Gentleman's Magazine* emphasized the significance of raw materials for transport such as rock salt, coal, limestone, marl, iron, and timber, as well as the commercial advantages for merchants at Birmingham, Liverpool, and Hull, in promoting the Trent and Mersey route.[55] Parliamentary acts for the Trent and Mersey (soon to be known as the "Grand Trunk") and Staffordshire and Worcestershire Canal (which would provide a link from the Trent and Mersey to the Severn) were passed in 1766.[56] Both waterways were significant for their engineering challenges and inter-regional scope, as well as for increasing the significance of Birmingham to canal transport. Engineers pushed the Trent and Mersey through notable obstacles. For instance, the construction of a flight of thirty-five locks was needed to carry the Trent and Mersey up to Harecastle Hill, through which a tunnel of a mile and two-thirds carried it through the hill. Construction then included a series of forty locks that would take the canal down to the River Trent. Along the route would be four further tunnels and a series of crossings of the Dove and Trent rivers.[57] In the west, the canal established links with the cities of Liverpool and Manchester, as well as the Bridgewater Canal. Josiah Wedgwood, whose potteries stood to gain much from the smooth path offered by the Trent and Mersey, became one of its prime supporters and emphasized the public benefit he believed could be obtained through its construction. He was joined in his support by Erasmus Darwin and others within the intellectual, scientific, and business circles of Birmingham.[58] In September of 1767, just months after construction began on the Trent and Mersey, a letter writer from Burslem, just north of Stoke on Trent, highlighted Brindley's multi-faceted accomplishments in almost messianic terms, signaling ways in which canals would continue to be recognized for their integration of local materials, even amidst an evolving national transport web:

> [Brindley] has cut a mile through bogs, which he binds up, embanking them with stones which he gets out of other parts of the navigation, besides about a quarter of a mile into the hill Yelden; on the side of which he has a pump, which is worked by water, and a stove, the fire of which sucks through a pipe the damps that would annoy the men, who are cutting towards the centre of the hill. The clay he cuts out serves for brick, to arch the subterraneous part, which we heartily wish to see finished to Wilden Ferry, when we shall be able to send Coals and Pots to London, and to different parts of the Globe.[59]

While promoters often suggested the materials of the English landscape were entirely malleable in the hands of canal engineers, the waterways' construction often meant a careful legal renegotiation of the claims of others upon the land. Acts of Parliament delineated canal routes and addressed the management and/or re-direction of local streams, as well as the tolls upon the transport of certain products.[60] In the first act empowering the building of the Duke of Bridgewater's Canal, the Duke's company had been given permission

> to enter lands, dig and remove obstructions, and then to make towing-paths on the side of his canal, gutters and water-courses, locks, sluices, &c. &c. weighing-beams, bridges, arches, &c. for setting up posts, rails, &c. first making satisfaction to the owners of, or persons interested in, such lands or grounds &c. &c.[61]

Enabling legislation for the Trent and Mersey Canal provided for the building of (non-locomotive) railways up to 1,000 yards in length to carry building materials to the construction site, the accessing of local streams no more than 1,000 yards from the canal, the creation of towpaths for horses working the canal, and the building of accessory structures or works such as bridges necessary to support the canal. Provision was made for the protection of existing structures which might include a "Dwelling-house or Outhouse . . . or a Garden Yard, Park, Paddock, planted Walk, or Avenue to Any House, Lawn inclosed or adjoining to a Mansion-house."[62] Companies were generally held as indemnified as long as reasonable efforts were made to compensate for damage done. Some prices demanded by landowners could be high, but there is little evidence of general opposition.[63] At the same time, evidence also indicates that landed, commercial, and industrial interests often joined in the investment in canals.[64]

Not surprisingly, canal promoters argued that disruptive landscape changes would bring revitalization. For instance, eighteenth-century promoters suggested the Trent and Mersey offered an expansion not only of trade in England's Midlands but also an opportunity to revive and transform the people and their environs. These themes were a part of *The History of Inland Navigations*, a unique compendium of letters, newspaper excerpts, and observations gathered to support and encourage the building of canals, and potentially edited by Brindley.[65] It was likely compiled, amended, and recirculated as canal promoters attempted to gain the support of Parliament. Thus, it professed interest on behalf of the nation, but also assuredly sought to stimulate the interest of investors.[66] The editor noted in the Introduction that there were a number of inland areas that had no natural means of water communication,

> and many still, whose inhabitants have never thought of any such conveyance; by which large tracts of country are locked up in impenetrable gloom, no verdure chears [*sic*] the eye, and the industrious peasant beholds the fruits of his daily toil confined to his own, or sent to a distant market, which such a load for carriage as eats up his little gains, and cramps him in obtaining the necessary articles of subsistence.[67]

Canals would thus benefit inland manufacturing, as well as support the establishment of new concerns and benefit merchants living in port cities. Landed gentlemen, the editor argued, would benefit much like the Duke of Bridgewater, adding,

> In many instances their lands have been improved to tenfold value. Without this artificial communication, corn, timber, iron, coals, stone, &c. would be of little use to the landowner . . . They occasion the discovery of mines and minerals, which would be else useless to the owners; they aid the cultivation of poor,

> barren, and waste lands; . . . [they] must of course be the cause of employing vast numbers of our poor.[68]

Though built by private concerns, they were part of an ennobling public endeavor:

> It would be happy for this country, if private interest, prejudice, ignorance, or obstinacy, were not employed to discredit such patriotic undertakings, as must redound so greatly to the honour and welfare of the kingdom . . . But it is time, in the present critical circumstances of the nation, when rivals in trade and manufactures are taking every advantage over us, when enormous taxes, and the advanced prices of the necessaries of life oppress our manufacturers and our poor, that we unite as one man in promoting those designs, which will contribute to raise our drooping commerce, to find employment for our labourers, and enable us to bear the burden of our numerous taxes and with some degree of chearfulness [*sic*] and patience.[69]

The connection of canals to Britain's public endeavors and the nation's ability to compete with its European rivals in the eighteenth century prefigured the system's symbolism for national economic and competitive struggles in the late nineteenth and early twentieth centuries (as will be seen in chapter 3). The significance of canals in opening up new domestic territories also made its way to North America, where just a few decades later, political leaders in the United States would see canals as an opening to a newly acquired frontier.[70] A contributor to *The History of Inland Navigations*, suggested that the arguments in favor of the Trent and Mersey Canal might be marshaled under: "1. Natural productions of the countries that lie near the canal. 2. Cultivated commodities and manufactures. 3. Imported raw materials, and general commerce."[71] From rock salt to coal to limestone, and from coal to marl to iron ore, the commodities along the proposed route of the canal were enumerated. Links with the changing agricultural world were significant, the writer arguing, "Agriculture is an inexhaustible source of plenty and riches, which can never be so much enlarged, or its streams so widely diffused, as by the means of inland navigations."[72] The benefits to be brought to the population by the importation of grain in times of scarcity were expected to trump the concerns of farmers who might lose out on higher prices during those times, and manufacturers were expected to benefit from the ease of transport of their finished goods. Various objections to canals—such as disruptions to estates, petty theft by workmen, and reductions in the volume of Britain's coastal transport—were also addressed. In the nineteenth century, after canals had established themselves as corridors for the circulation of products and wealth, they were noted for the dangers they offered in the circulation of unwanted boatmen and disease and danger. Yet, in the eighteenth century, they held overwhelming value for the nation and its commerce in this writer's eyes,

> How far these favourable circumstances must contribute to enhance the value of our lands; to promote the wealth, strength, and splendour of this nation; and to confirm, and perpetuate, the peculiar blessings and privileges of its inhabitants; is referred to the imagination of every intelligent reader.[73]

The pace of canal legislation continued to accelerate in the early 1770s. Between 1770 and 1774, a further twenty-three acts were passed, of which twelve were new proposals for waterways.[74] Among the most significant was the 1770 act for the Leeds and Liverpool Canal, which provided a route from the western end of the Aire and Calder Navigation at Leeds to the western coast of Britain.[75] The project was notable for the momentous crossing of the Pennines and an over four-decade gap between the beginning of construction and the final link with the rest of the canal system at Manchester. The length of the route was 127 miles and, from its highpoint, it dropped 525 feet into Lancashire and 446 feet into Yorkshire.[76] In speculating on the potential value of the Leeds and Liverpool in 1769, a letter writer argued optimistically that the undertaking would have benefits both for the local regions through which it passed, and the kingdom as a whole. The canal had the potential to, "connect the whole [of the nation], and make every part partake, at an easy expence, of the natural and acquired advantages and conveniencies of the rest."[77]

> The whole intermediate country between Leeds and Liverpool, and for a considerable extent on each side of the canal, will be supplied with wool, woolen-yarn, corn and provisions, hides, tallow, &c. from Ireland, the produce of America, and whatever else is imported at Liverpool.[78]

In addressing potential objections, the letter writer noted protections that would be offered to "houses, gardens, yards, parks, paddocks, planted walks or avenues, or lawns inclosed, and adjoining to any house."[79] The company would be organized on a sound financial basis.

> May we not therefore, presume to hope, that if public virtue, and the love of our country, or even private interest, has not lost its influence, we shall unite upon this interesting occasion; and then this grand and beneficial undertaking, (which on a private view, may appear insuperable,) will be effected with ease.[80]

Canals now began to connect inland England with both the coasts and the wider commercial world. Canals' financial strength, and thus their impact upon the growth and development of inland areas, came to be linked with wider international concerns. The years of the American War of Independence represented a decline in parliamentary legislation with thirteen acts between 1775 and 1779, eleven between 1780 and 1784, and eleven immediately following between 1785 and 1789.[81] However, with eighty-two pieces of inland navigation legislation enacted between 1790 and 1794, Britain entered what has come to be known as its "canal mania."[82] In 1792, a set of Standing Orders indicated that all canal bills were to be advertised in both the *London Gazette* and local papers, that maps should be on deposit with the local Clerk of the Peace, that the costs of the projects should be worked out, and that lists of property owners along the routes should be delineated.[83] An increase in the number of canal bills was accompanied by both increases in the paperwork accompanying them as well as the testimony provided. The *Times* of London reported that nearly 200 witnesses were brought forward for the Birmingham Canal Bill of 1792.[84] The Grand Junction Canal, constructed between 1793 and 1805 linking the Midlands to London, was one of the most significant undertakings of the 1790s, initiating a north–south link that would remain important well into the twentieth century.[85]

The creation of canal companies was only the first step in the construction of the wider commercial and social world that remained integral to them. Workers, many likely part of a wider arrival of rural Irish, joined English laborers displaced from rural estates in moving the earth that was the first step in the construction of canals.[86] As they arrived in small communities, a reputation for drunkenness and violence soon set in, likely intensified by uncertainty over whether canal companies would make good on the pay they owed. For example, in April 1811, a group of approximately 300 men employed on the Grand Western Canal, who were seeking a prompt payment of wages, began drinking at a fair at Sampford Peverell, resulting in a confrontation with an official which led to the loss of one life and the severe beating of two other men.[87] A local newspaper expounded,

> It is a most extraordinary circumstance that the whole village and neighbourhood should have been kept in a state of the greatest terror and commotion for more than twenty-four hours, and no efforts of the police or Military made to quell the tumult. In the name of Justice, where are the Magistrates?[88]

Once the canals were in water, some of those who had been navvies likely turned to working the boats. Like river improvements, canal schemes allowed the public to use the new waterways as long as tolls were paid. Though some canal companies actively entered the world of transport, owning perhaps 20 percent of boats in one set of surveyed counties, most waterways were plied by a variety of other transport companies headed by merchants and industrialists who had a stake in a region's mining and industrial works.[89] However, canal carriers could range from large firms employing multiple boats and crews— with some boats operating as "flyboats" which traveled twenty-four hours a day and included shifts of crewmen—to individual proprietors working their own boats.[90] Although in the twentieth century a belief took root that in their heyday canal boats had been largely manned by a sturdy independent class of owner-boatmen, this was not the case. Approximately, 60 percent of boats may have been controlled by entrepreneurs or firms with more than five boats.[91] The divisions between canal companies and their carriers could complicate efforts to coordinate and manage the waterways.

Moreover, boats came in a variety of different shapes and sizes, all scaled to the locks and waterways through which they traveled. Most noticeable was the "narrowboat," which often measured just under seven feet in width and about 70 feet in length, and carried 20–30 tons.[92] Narrowboats originated with Brindley's earliest canal efforts and were associated with the Oxford, Trent and Mersey, and Coventry canals, among others, the dimensions of which might allow just enough room for two boats to pass.[93] Elsewhere, a wide variety of craft sizes emerged encompassing both "wide boats" and barges. Wide boats, generally from 70–72 feet long and about 10–11 feet wide, were used primarily in the south, notably on the Grand Junction.[94] Barges could range up to just over one hundred feet in length with a 17-foot width, though the largest were generally used on river navigations rather than canals.[95] A host of other permutations existed on other canals; rivers with histories of navigation might have distinctive craft of their own. Most canal craft were hauled by individual horses traveling along the banks on towpaths, setting the maximum speed at just a few miles per hour. The complexities of canal and lock sizes ensured business managers had to plan through routes with great care. In the late eighteenth and early nineteenth centuries, disparate ownership of canals led to a series of ongoing and innovative engineering efforts to link mines, factories, markets, and ports at a regional level, but without the consistency that would have made it easier to introduce new technology and infrastructure at a national level.

In contrast to the early excitement generated by Brindley's success on the Bridgewater Canal, failures in canal works almost always resulted in accusations leveled at their engineers. Many canal undertakings involved multiple supervisory engineers who were employed over several years of construction, meaning that multiple strategies could affect a single canal before its completion. Brindley was only the first among several engineers who made canal-building central to their careers. John Smeaton, William Jessop, John Rennie, and Thomas Telford were among those whose careers either continued or began, after his. In the late 1760s, the Birmingham Canal Company began work on a line which was to link Birmingham with the Staffordshire and Worcestershire Canal, then still under construction, near Wolverhampton. Brindley, who had already been placed in charge of the Staffordshire and Worcestershire Canal, was selected as the engineer; although initial plans had called for a tunnel through high ground at Smethwick, a series of locks carrying the canal over this portion of the route was substituted.[96] When water supply later became a problem, John Smeaton was called in and supported the elimination of three locks on each side of the summit and the creation of a cutting approximately 46 feet deep.[97] Later, increasing use of the canal prompted the company to seek the advice of Thomas Telford who, upon investigating the route, found the towpath a shambles and groups of boatmen fighting to secure their passage at the locks.[98] Telford's subsequent plan called for a straightening and widening of the canal with revamped towpaths and new bridges, leaving Brindley's old "tortuous navigation" to survive only as "a series of loops and branches winding away to right and left of the new straight cut."[99] Water supply problems were to be addressed by the building of a new reservoir at Rotton Park, later to be known as the Edgbaston Reservoir.[100] Farther to the north, the Huddersfield Canal contained one of the most significant achievements within the canal system, the three-mile-long Standedge Tunnel, which came with its own dilemmas and was not opened until 1811, after Benjamin Outram had resigned as a consulting engineer on the project and Telford had devised a new plan for the tunnel.[101]

Thus, in addition to being affected by the decisions and priorities of various promoters, the diversity of Britain's canals was shaped by multiple and sometimes conflicting engineering responses to promoters' ideas and the challenges of local terrain. Engineers' efforts at the opening of the nineteenth century were central to the widening impact of canals upon the engineering profession, which had taken significant steps forward from the less defined skills in mathematics and physics of those who had supervised river improvements. The canal engineer's job in planning was diverse, encompassing concerns of water supply, terrain, and property ownership. The 1819 entry on canals in Abraham Rees's *Cyclopaedia* included the following on a waterway's engineer:

> He should also be conversant with the general principles of trade and commerce; with the various operations and improvements in agriculture; with the interests and connection of the different owners and occupiers of land, houses, mills, &c.; and with all the general laws and decisions of courts, pertaining to the objects connected with his profession. By an extensive acquaintance with the disposition, inclination, and thickness of the various strata of matter, which compose the soil or land of the British islands, he will be able to avoid many errors incident to those who are destitute of this knowledge, and of the courses and causes of springs, to which it leads.[102]

Canals were to lack many of the consistent elements of design, use of materials, and building that characterized the later steam railway industry. Instead, their construction represented an amalgam of processes relying upon both existing local resources and new resources becoming available through an emerging national economy. They both initiated and reflected changes brought by industrialization. Anthony Burton has written that North Wales had a vibrant trade in slate and that shipments through the Montgomeryshire Canal [today, the Montgomery Canal] to Ellesmere and then to the Midlands canal network helped to facilitate the wider replacement of thatch by slate.[103] He also argues that canal builders more directly challenged environmental conditions as greater emphasis came to be placed on direct, efficient lines of transport between manufacturing cities. Ultimately, canal builders challenged Brindley's conservative approach that sought to navigate around major landscape barriers. For example, there was increasing pressure to improve the northern part of the Oxford Canal given its significance in the Grand Junction route between London and Birmingham, he writes. A new and straighter line of the canal was laid out in the 1830s that included embanking low-lying areas, digging through rising ground, and cutting through what had been the previously winding path of the canal. "The new bridges, that carry the towpath across the old line, no longer use old materials but the new building material of the Industrial Revolution – cast iron," he observes.[104] These cast iron bridges began to be produced according to a standard plan. The process of standardization (only begun in the canal era) would also be significant in the era of railway-building, including the basic process of the "cut and fill," or the making of cuts and using the excess to build needed embankments, Burton writes.[105]

Unbeknownst to those trumpeting the dramatic power over the landscape represented by canal-building was what that landscape was revealing about the brief chapter of human beings in the unfolding of geologic time. William Smith, primarily self-educated as a surveyor, had begun to consider the significance of the layering he saw in his work in Somersetshire when he was employed by Rennie for work upon the Somerset Coal Canal.[106] Subsequent canal work, including a wide-ranging geographical tour, allowed him to continue collecting evidence informing both an early thesis regarding evolution and the first major geological map of Britain.[107] He was dismissed from the Somerset project due to "financial irregularity" and appears to have continued to run into other problems, being briefly confined to a debtors' prison in 1819.[108] Nonetheless, he was finally awarded the Wollaston Medal by the Geological Society of London in 1831 and became known as "Strata" Smith, the "Father of English Geology."[109]

Although it would take some time for Smith to find his recognition, the geographic challenges of canals and the resources deployed to address them could more immediately be seen as points of pride. Two years after the completion of all phases of the Leeds and Liverpool Canal, a process that took nearly 50 years, the *Hull Packet* remarked that there had been "great and apparently insuperable obstacles" that stood opposed to its progress.

> It was necessary to pierce for it a passage through mountains, to erect vast mounds, to enable it to traverse deep vallies, and to carry it over unsubstantial and deceitful bogs. These difficulties have, however, been overcome by the resources of ingenuity, and the exercise of unwearied perseverance.[110]

In the southeast of England, initially sought for the aid in military preparation it offered, work on the Thames and Medway Canal had begun in 1800 and stopped in 1804, and begun once again only to be completed in 1824. At its opening, a newspaper article emphasized the width of the 4,090 yard (3.5 km) long Higham and Strood Tunnel which, with its towpath, the writer believed compared quite favorably with the Harecastle Tunnel on the Trent and Mersey (which was dependent upon legged labor to push boats through).[111] The article downplayed both the canal's time to completion and its estimated £250,000 cost:

> Upwards of 1,700 yards [of the tunnel] are arched up with bricks, the remainder of the arch is formed in solid chalk. The greater part of the excavations was effected by blasting, and the materials removed by shafts, sunk for the purpose . . . The number of persons employed on it for the last four years has been, on an average, 250; . . . We can say, the execution, both as to neatness and solidity, is quite worthy of the vastness.[112]

Despite their impact upon the landscape, engineering, and role in transforming how building materials were transported and used, difficulties associated with the integration of canals into a national network and the checkered ability of companies to address unforeseen costs emerged as early as the late eighteenth century.[113] Initial cost estimates for some canals could fall short of the amounts needed, and the number of times canal committees returned to their investors for additional funds were prominent criticisms. In Scotland, where canal construction was focused upon two major routes that bisected it, construction of the Forth and Clyde Canal began in 1768 and eventually connected Edinburgh with Glasgow. Smeaton offered the first survey of the route, and over the course of the next decade the canal opened in stages.[114] In the late 1770s, as the project ran short of funds, a loan of £50,000 came from the Forfeited Estate Fund.[115] Concerns associated with requests for further funding soon arose, however. Farther south, in the West Midlands, the Coventry Canal Company began work in late 1768 but ran out of money after the construction of the first 17 miles of the waterway. Later, work was again begun but halted until 1782, when it was completed by the proprietors of the Trent and Mersey Canal and the Birmingham & Fazeley Canal (then itself under construction).[116] The Oxford Canal, begun in 1769 and plagued by financial difficulties, had to obtain powers to raise more money and did not successfully link with the Thames until 1790.[117] In the eastern Midlands, the shareholders of the Cromford Canal each contributed an additional £35 a share, and a further £20,000 was raised through a loan before the project was finished.[118] Nearby, in August of 1776, the committee of the Chesterfield Canal asked its shareholders for an additional £12,000, and when such was not forthcoming, borrowed a total of £53,000, including sums from banks in York and Nottingham.[119] In some cases, canal committees appear to have responded by reaching further down the economic ladder to cover costs, in particular by issuing shares well below the £100 standard that had once been common.[120]

Between 1795 and 1825, overall canal construction slowed significantly. Inflation during the wars against France raised costs substantially. Construction on a significant proportion of the canals originally envisioned in the 1790s had, however, begun and would eventually establish the connections Brindley had imagined in his "great cross." Involving the assistance of both Telford and Jessop, the Grand Junction was finally completed in 1805 with the opening of the Blisworth Tunnel, which was approximately 1.74 miles in length. The [Old] Grand Union Canal, linking Leicester with the Grand Junction and providing a more direct route between London and the eastern Midlands, took several years to complete.[121] In 1808, company leaders had to admit that they could no longer afford to complete the project, and a group of new investors (some of whom came from the Grand Junction Company) took over.[122] The route was finally opened in 1814. The [Old] Grand Union filled in a gap in the eastern connections of England's waterways as it helped to forge a thus far unrealized link between London and Nottingham. The Regent's Canal, begun in 1812, amended this just a little. It began at a link with the Grand Junction Canal at Paddington Basin in London and threaded its way across north London, eventually turning south to meet the River Thames at Limehouse. In the west of England, the most significant goal was to link east to west, and here the Thames and Severn Canal first provided a link between the Stroudwater Navigation near Stroud and the River Thames near Lechlade. It was marked by the Sapperton Tunnel, which, at just over two miles long, was for a time the longest tunnel in England. The Thames and Severn was followed by the Kennet and Avon Canal, which eventually provided a link between the River Thames and Bath and Bristol in the west.

At the same time that these uncertainties in canal construction intensified, so did confidence in Britain's rise as a formidable commercial and naval power. The nation could point to the taxing effects of world conflict with

France as the century dawned, but after 1814, global expansion without a major European foe seemed within reach. In 1804, one contemporary perspective offered encouragement to investors in the then uncompleted Grand Junction Canal, which would one day link London with the canals of the Midlands, arguing that they would be rewarded for their "long patience and spirit of perseverance in assisting a work of this national consequence."[123] Moreover,

> What a splendid contrast do these proved facts furnish to the condition of our enemy. France is driven to acts of unqualified robbery to collect the momentary means for carrying on the war . . . while in England, the confidence and the wealth of the people are such, that after every sacrifice which the safety of the empire demands, we find millions embarked in commercial speculations.[124]

Although the construction of canals in Britain had significantly slowed by the early nineteenth century, at least one voice noted that they still might be critical in the nation's struggle against Napoleonic France. In 1810, Joshua Jepson Oddy, a London merchant, wrote,

> Spite of the darkened cloud which hovers over our horizon, let us . . . give weight to our commercial importance. . . . The streams, rivers, *inland-navigations*, and the several fisheries on its coast, all which, with proper management and industry, can be made contributary to our opulence and prosperity.[125]

Nonetheless, tensions generated by the era of "canal mania" in the 1790s—including disputes regarding canals' size, financial viability, and ability to overcome geographic obstacles —had taken their toll. In the early part of the decade, as efforts began to secure parliamentary approval for what would become the Rochdale Canal, there were significant concerns regarding the ability of mill owners to retain substantial enough supplies of water. Parliamentary approval was secured by the mid-1790s, but there was still concern regarding the availability of water at its highest elevations and the use of reservoirs. In 1804, after the canal's completion, one of its critics, John Sutcliffe, a mill owner, penned a critique of canals which was not published until 1816, one year after his death.[126] Sutcliffe offered a sobering view of canals and their relevance to the nation, writing, "The great advantage which the public were supposed to derive from the making of them, has been painted in the most glowing colours, while the real state of the subscribers has been passed by in silence"[127] He concluded,

> No man can be more desirous than I am of promoting canals, where there is a fair prospect of the subscriber having 5 per cent interest; but should the country be so unfriendly to a canal, as not to justify the expense of making one, a

> [non-locomotive] railway may be made instead of it; and the public I think have as yet but contracted ideas of what may be carried upon a railway made upon the best principle, or of the easy expense of carrying a ton of goods per mile upon them;[128]

Estimating that approximately £30 million had been expended on canals over the last 25 years, he argued that when they were most needed by farmers, they turned out to be little more than "dry ditches."[129] He said the Rochdale Canal was undersupplied by the 318 acres of reservoir water that had been set aside for it and that some canals could never be beneficial to their subscribers, "having been designed in defiance of nature, and under difficulties and obstacles which rendered them unworthy of being executed."[130] Others, he said, passed through productive areas that might provide tonnage, "but the principal cause of their failure, has been the want of skill in the execution of them."[131] Central to Sutcliffe's argument were the limitations created by nature; here, one of the canals he highlighted was the Huddersfield Narrow Canal. In particular, its Standedge Tunnel, built through the Pennines, had encountered significant problems and was not fully opened until 1811. Severe floods in 1799 had done significant damage to aqueducts, and in 1810 a reservoir failure led to flooding and the loss of six lives.[132]

Attitudes toward canals could thus be ambivalent as the nineteenth century entered its second decade. In the 1790s, the Romantic writers and siblings John Aikin and Anna Laetitia [Aiken] Barbauld had ruminated on the meeting of two spirit figures, the Genius of the Canal and the Spirit of the Brook, and given voice to the Genius of the Canal, which pronounced it was the "noblest child of art" who was able to "pierce the solid rock for my passage and connect unknown lands with distant seas." The Canal had concluded,

> Behold my channel thronged with capacious vessels for the conveyance of merchandise, and splendid barges for the use and pleasure of travelers; my banks crowned with airy bridges and huge warehouses, and echoing with the sounds of industry. Pay then the homage due from sloth and obscurity to grandeur and utility.[133]

Yet, even before the major era of steam railways and the competition they engendered, the dynamic vision of canals as efficient and assured generators of income and economic prosperity had been blunted. Their construction had moved tons of earth, redirected waterflows and established communities of boatmen who continued to operate horse-drawn boats for the next several decades. Yet, despite their significance in carrying bulk loads, their gains for individual canal companies were uncertain. They had added to previous technologies of water transport, but not necessarily transformed them. The movement of watercraft was based upon naturally occurring supplies of water and labor provided by horses and people. Locking mechanisms simply adjusted the level on which canal boats traveled, and they did not provide motive power. Boats could be stopped by a turn in the weather. Inland water transport had been severely affected by frost in 1813–1814, 1819–1820, and 1830–1831 and had prompted some businesses to stock coal as a backup to that provided by the waterways.[134] The time taken for locks was of concern; alternative ways of raising and lowering boats had been introduced but rarely became widespread, such as the inclined plane near Coalbrookdale which could raise two small tub boats in four minutes in the link between the Shropshire Canal and the River Severn.[135] In addition to heavy rain which could cause flooding, no or little rain could threaten industrial transport on the canals. In September of 1854, over 200 miles of canals faced closure as a result of a lack of rain.[136]

Locomotive railways soon triumphed over some of the barriers against which canals had struggled, and in the process, re-wrote the relationship of workers, passengers, and goods to the landscapes through which they traveled. Railways were freed from accepting natural forces on their own terms and turned to the extraction of energy from the transformation of carbon (coal). Wolfgang Schivelbusch has argued that such control over the nature of energy began a process of alienation in which travelers no longer understood precisely the power moving them forward but began to see it as something wondrous.[137] Moreover, he has argued, the railway functioned as a machine in which the man-made products of rails, cars, and engine operated together as a unified whole, transcending the world of canals in which boats traveled largely as individual units once tolls were paid. Railway construction also focused on travel on level terrain, and in an effort to reduce friction, as well as travel time, pressed through obstacles that early canals had navigated around.[138] Although the bending of the landscape to suit travel and transport had certainly begun with the age of canals, railways in a bolder fashion made cuttings through hills and placed bridges both over waterways and depressions in the land, making more obvious their ability to transcend rather than negotiate obstacles.[139] The railways also saw the end of a freedom and independence that had been transferred from roads to canals. The inability to allow traffic to travel freely over a single railway line led first to company control over both tracks and transport services and then the centralization of control over usage and the consolidation of rail companies.[140] Between Manchester and London, there were seven and eight canal companies on two competing routes (with eleven on a third), but only one railway company in 1846.[141] If a business sought transport to a particular location by rail, the railway could immediately quote a price; the process on canals was more extended. Soon, the railways also counted on the proceeds drawn from passenger services, which canals provided only on a limited basis.

Many canal companies offered reductions in tolls in response to the locomotive railways and, drawing in some further traffic, were able to maintain dividend payments.[142] Yet, achieving the necessary coordinated responses among carrying companies to provide an integrated response to the railways proved difficult. In the Midlands, the Erewash, Cromford, and Derbyshire canals, as well as the [Old] Grand Union and Grand Junction canals, all reduced their rates. While indicating that the canals would no longer be the source of plenty that they had once seemed to be, documented results after the introduction of steam railways could include outliers, as in the case of the Erewash Canal where tonnage increased from 166,506 tons in 1830 to 390,524 tons in 1840; a dividend as high as 61 percent was paid in 1837.[143] In his study of canals in Manchester, Peter Maw has emphasized that canals continued to be the primary conveyance for coal in Manchester in the first half of the nineteenth century, as opposed to the burgeoning railways.[144] Generally, though, the advent of railways provided the key component of speed. Many boats on inland waterways pulled by horse and "road wagons" could carry freight at just two miles an hour, though "sprung wagons" and "fly" boats on canals could improve efficiency. For their part, railways could attain speeds between 25 and 30 miles an hour. In 1830, the standard travel time between the closely linked cities of Manchester and Liverpool was 17–18 hours; by rail, the trip was reduced to under two hours.[145] The annihilation of space by time, which was to form such an important part of the cultural history of rails, had begun.[146] Even if time was of little concern to a client, a significant lowering of costs for water transport between general destinations did not include the costs that would be associated with the transshipment of some products to road transport for distribution away from a canal.[147] For some merchants, the risks of pilferage on canals also argued in railways' favor.[148]

By the 1840s, canal traffic had begun to decline on a number of routes. The Cromford Canal saw tonnage reduced from 322,724 to 284,693, and income on the Nottingham Canal declined by about one-third.[149] Investment money rushed to steam-powered railways creating investment frenzies which resembled those of canals, including the organization of sets of proprietors, the employment of engineers (some of whom had cut their teeth in canal construction), and the planning and building of tunnels and bridges. To a greater degree than canals, however, steam railways included the building of stations, warehouses, and engine works that transformed rural and urban spaces. Individual stages of railway "mania" in the 1830s and 1840s followed some of the steps familiar in the canal-building process, such as application to Parliament for appropriate legislation.[150] In their turn, some of the same challenges emerged, such as the need to manage the opposition of property owners, the problems created by intense speculation which ignored realities of geography, the complexities of multiple companies each crafting their own portion of a wider network, and the anxieties created by "navvies" whose presence was often distrusted.

Attention had turned to steam railways as a powerful new force that would provide efficiencies in transport, primarily through speed and a consistent infrastructure, across the nation. Railways were dependent upon the iron and coal produced by the new industrial era and in their turn magnified the power of industrial Britain at mid-century. In 1830, a proposal to build a railway from Sheffield to Goole along the River Don valley in southern Yorkshire was met with opposition from the Don Navigation company, which marshaled opposition and argued that "the projected railway is uncalled for and is unnecessary to the Interests of the Country and will materially injure the interest of the Don Proprietors."[151] Yet, railways had deep pockets and could use the proceeds of passenger services to weather price wars with the canals and eventually to offer lucrative terms for the buyout or leasing of canal routes.[152] By the mid-1840s, the Don Navigation found it impossible to stand against multiple rail interests, and its committee agreed to be absorbed into the South Yorkshire Coal Railway with a generous payout.[153] Many similar agreements would be reached in the 1840s and 1850s.[154] In 1845, five navigations comprising 78 miles came under railway control, with seventeen navigations and 774 miles in 1846, and six navigations and a total of 96 miles in 1847.[155] After a further fifteen canals and 300 miles of waterway were acquired by railways in subsequent years, the total amount of railway-controlled canals was over 1200 miles. In financial amalgamation with railways, canal shareholders saw an opportunity to ameliorate what they believed would be inevitable losses as a result of competition. In 1830, there were approximately 100 miles of steam railway lines open in Britain, but by 1852 about 6600.[156] Though some canal lines remained successful well into the nineteenth century, the railway appeared to have replaced canals as the most significant mode of industrial transport.

Eighteenth and early nineteenth-century canal-building was part of a wider application of capitalist interests to the environment, along with enclosure, scientific farming, and mining. The creation of the Duke of Bridgewater's Canal marked the beginning of a two-hundred-year period in which the place of canals within the nation's fabric of inland waterways evolved from one of industrial transit to one of countryside, heritage, and leisure. Much of the infrastructure of the Bridgewater Canal was built upon the deeper experiences of water management both in Europe and in Britain. On continental Europe, the Dutch example suggested the correlation of inland transit with commercial and global power, and the French example indicated what might be achieved with central government direction. In comparison, in Britain, river management suggested the advantages to be gained by removing blocks to navigation between important cities and sites of production. Drainage efforts in the Fens suggested the potential gains from creating dry land but also limits to transcending nature, particularly notable with the continued sinking of land in the Fens. From their earliest stages, canals were thus linked with Britain's ability to deploy techniques of "improvement" to meet the challenges of the environment and the examples of international competitors. The mid to late nineteenth century would be an era in which the power of British rails signaled the nation's industrial and commercial dominance at a global level. For many, at first, the eclipse of canal transit seemed of little importance. However, the collapse in the canal system's traffic as the twentieth century dawned and international competition was reinvigorated, ensured Britain's canals and their potential (or lack of it) remained an object of reflection as the nation considered its more constrained leadership in the world.

By the early nineteenth century, canals occupied a significant place in bridging the natural and man-made worlds. They crossed river valleys and bridged existing regional divides in water drainage. Their distinctiveness was not only in the localities and regions served but in their design, construction, and application of new schemes for locks, tunnels, reservoirs, and aqueducts. They proved to be a seedbed from which the modern profession of engineering would emerge within Britain. Surveys of routes and available water resources, the delineation of geographic challenges to be overcome and, perhaps most importantly, the ability to produce accurate cost estimates and respond to the demands of committees of proprietors, helped establish the responsibilities of the modern engineer and the engineering profession. An important marker in the professionalization of engineering was the founding of the Society of Civil Engineers in 1771. Upon their completion, most canals saw relatively few overarching technological changes associated with water management for over two centuries. In their length, breadth, and depth, they had been designed to meet the needs of regional business interests in the most efficient way possible. These parameters governed the size of the craft that could be used and limited the ability to ship large loads across the country in a single-size boat. Travel time was predicated on the pace of a walking horse. The much later application of motorized power to canal boats was ultimately limited by the need to maintain canal banks through limiting wave actions. . Canals thus became associated with small crews and antiquated-looking craft pulled by horses, as the scale and speed of mines, factories, cities, and competing forms of transport increased. For some, this would eventually be a reason to reject new investment and even maintenance of the canal system. But for others, it suggested a level of technology that, in its rudimentary qualities, offered stability in the face of disruptive innovation.

The relatively static nature of canal technology, as well as the trials that had been experienced in the building of many routes, quickly doused enthusiasm for canals with the arrival of locomotive railways. Few rail companies invested significantly in their canals after their takeovers. This transition meant some canals faced a slow decline in the nineteenth century as traffic diminished, maintenance lapsed, and administration remained less than innovative. Many observers concluded that canals represented an anachronistic industrial infrastructure whose closure was inevitable. At the opening of the twentieth century, new doubts emerged about Britain's ability to compete at a global level and manage the taxing effects of two world wars. For those who raised the flag for canals, the belief that a once valuable technology had been suppressed under the dominance of railways proved a provocative one. Arguments that the nation was ignoring the potentially significant advantages of returning heavy freight to inland waters harkened back to the vibrant era of investment and construction at the close of the eighteenth century.

Canals' evolving place in British society, from pathways of industrial and commercial expansion to neglected and often maligned back-channels of industry and society, and then to renewed spaces of cultural and environmental experience in the late twentieth century, offers a critical mechanism for viewing Britons' relationship with the environment. Some planned canals did not make it past the drawing board, blocked by an inability to stave off the opposition of landowners and other canal owners, or blocked by a significant geographical barrier. Limitations could quickly become apparent as misjudgments in the composition of the landscape led to cost overruns or the abandonment of a canal project altogether. Yet, new routes were generally lauded for their ability to propel the nation's economic advance and improve the efficiency of transport across the landscape. Significant for canals, however, was that new technology was applied to the landscape while old forms of human and animal labor prevailed. The stage was thus set for later environmental, social, and cultural interpretations of canals. Waterborne access to inland regions, with its economical and smooth transport of heavy goods and distinctive infrastructure and labor practices, all entrenched canals within the modern technological, cultural, and environmental histories of Britain.

NOTES

1. In the early 2020s, new attention was focused on links between the building of Britain's industrial empire and the profits from the trans-Atlantic slave trade. Early investigation of these links was undertaken in Jodie Matthews, "Canals and Trans-atlantic Slavery: A Preliminary Literature Review," Canal & River Trust, accessed December 10, 2023, https://canalrivertrust.org.uk/media/original/42453-canals-and-transatlantic-slavery.pdf?v=445116.

2. The opening of the early modern era, in which peoples in western Europe and Britain in particular were able to bring greater resources to the exercise of control over the natural world, is emphasized in Keith Thomas, *Man and the Natural World* (Oxford: Oxford University Press, 1983).

3. Chandra Mukerji has distanced Riquet from the status of heroic engineer of the *Canal du Midi*, emphasizing instead the multiple sources of knowledge gathered to build the canal. Chandra Mukerji, *Impossible Engineering* (Princeton, NJ: Princeton University Press, 2009), 8.

4. Charles Hadfield, *World Canals: Inland Navigation Past and Present* (New York: Facts on File, 1986), 43.

5. Alexandre Doin, *Mémoire sur le Canal Latéral a la Garonne, Établissant la Jonction Définitive des deux mers* (Paris: Éverat, 1832).

6. Hadfield, *World Canals*, 50–51. See also, Jacobus de Vries, "Water Control in the Netherlands with a Focus on the Period 1750–1900," in *A History of Water*, Series 2, Vol. 2, eds. Terje Tvedt and Richard Coopey (London: I. B. Tauris, 2010).

7. Quoted in Hadfield, *World Canals*, 53. Richard Castle was born David Riccardo (or, David Ricardo), the son of an English-born man who had been Director of Munitions and Mines to the Elector of Saxony and King of Poland. Little is known about the time or place of his birth, although in the mid-1720s, he appears to have been active in London. His writing became one of the first significant essays in English on the modern potential of canals. He later established a substantive engineering and architectural practice in Ireland. (Dictionary of Irish Architects, 1720–1940, s.v. "Castle, Richard," accessed December 10, 2023, https://www.dia.ie/architects/view/347/CASTLE%2C+RICHARD#tab_biography. See also, "Richard Castles," *Anthologia Hibernica* 2 (October 1793): 242.) "Castles" was another last name by which he was known. The *Anthologia Hibernica* indicates he was born in Kassel, Germany.

8. Mark Baldwin and Anthony Burton, eds., *Canals: A New Look* (Chichester: Phillimore & Co., 1984), 23.

9. Both the number of navigable rivers in Wales and England and their ability to handle heavy loads may have been in decline by the seventeenth century. Recent research indicates that the city of Boston, situated on the Witham River at the Wash and providing a route to inland Lincoln, was facing significant difficulties with silt. See Max Satchell, "Navigable Waterways and the Economy of England and Wales: 1600–1835," 2017, 14–15, accessed July 5, 2022, https://www.campop.geog.cam.ac.uk/research/projects/transport/onlineatlas/waterways.pdf.

10. T. S. Willan, *River Navigation in England, 1600–1750* (Oxford: Oxford University Press, 1936), 133.

11. Baldwin and Burton, *Canals: A New Look*, 24.

12. Henry Rodolph De Salis, *A Handbook of Inland Navigation* (London: Henry Blacklock and Co., Ltd., 1928), 6–9.

13. William Dugdale, *The History of Imbanking and Drayning of Divers Fenns and Marshes, Both in Forein Parts, and in this Kingdom; and of the Improvements thereby* (London: Alice Warren, 1662), 413.

14. Eric H. Ash, *The Draining of the Fens: Projectors, Popular Politics, and State Building in Early Modern England* (Baltimore, MD: Johns Hopkins, 2017), 257.

15. 13 Elizabeth, c. XVIII, "An Act for the Bringying of the Vier of Lee, to the North Side of the Citie of London," *At the Parliament begunne and Holden at Westminster the Second of April, in the xiii yere of the raigne of Our Most Gratious Soveraigne Lady Elizabeth, . . .* (1571). See Max Satchell, "Navigable Waterways and the Economy of England and Wales: 1600–1835," 2017, 18, accessed July 5,

2022, https://www.campop.geog.cam.ac.uk/research/projects/transport/onlineatlas/waterways.pdf.

16. R. B. Peberdy, "Navigation on the River Thames between London and Oxford in the Late Middle Ages: A Reconsideration," *Oxoniensia* 61, no. LXI (1996): 312–313, and Satchell, "Navigable Waterways and the Economy of England and Wales," 18.

17. T. S. Willan, "The River Navigation and Trade of the Severn Valley, 1600–1750," *The Economic History Review* 8, no. 1 (November 1937): 68–79.

18. Edwin A. Pratt, *A History of Inland Transport and Communication in England* (London: Kegan Paul, Trench, Trübner & Co., Ltd., 1912), 166.

19. Anthony Scott and Georgina Coustalin, "The Evolution of Water Rights," *Natural Resources Journal* 35, no. 4 (Fall 1995): 849.

20. There is still significant dispute about whether the public maintains rights of access and navigation on many inland waterways. See, for instance, the controversies generated in public forums in the early twenty-first century regarding Douglas Caffyn's argument that the public has maintained rights on navigable rivers in England for centuries. Douglas Caffyn, *The Right of Navigation on Non-Tidal Rivers and the Common Law* (Self-Published, Thesis Master of Laws, 2004). Caffyn's position has been embraced by some organizations advocating for public rights but has been rejected by a variety of legal and governing authorities.

21. Joseph Priestley, *Historical Account of the Navigable Rivers, Canals, and Railways, Throughout Great Britain* (London: Longman, Rees, Orme, Brown, and Green, 1831), 9.

22. Baldwin and Burton, *Canals: A New Look*, 30.

23. Willan, *River Navigation in England, 1600–1750*, 59–61.

24. G. G. Hopkinson, "The Development of Inland Navigation in South Yorkshire and North Derbyshire, 1697–1850," *Transactions of the Hunter Archaeological Society* 7 (1951): 229–251. See also Pratt, *A History of Inland Transport and Communication in England*, 148–149.

25. Hopkinson, "The Development of Inland Navigation in South Yorkshire and North Derbyshire," 234.

26. See, for instance, "William III, 1698: An Act for the Makeing and Keeping Navigable the Rivers of Aire and Calder in the County of Yorke. [Chapter XXV. Rot. Parl. 10 Gul. III. p. 4. n. 10]," in *Statutes of the Realm: Volume 7, 1695–1701*, ed. John Raithby (s.l: Great Britain Record Commission, 1820), 534–538. *British History Online*, accessed September 10, 2023, http://www.british-history.ac.uk/statutes-realm/vol7/pp534-538.

27. Pratt, *A History of Inland Transport and Communication in England*, 64.

28. Remarking upon the clay soils which proved extraordinarily difficult, once one got past Dunstable just to the north of London, Daniel Defoe in his *A Tour thro' the Whole Island of Great Britain* (written in the 1720s) argued in favor of turnpikes, remarking,

> It is perfectly frightful to Travellers, and it has been the Wonder of Foreigners, how, considering the great Numbers of Carriages which are continually passing with heavy Loads, those Ways have been made practicable; indeed the great Number of Horses every Year kill'd by the Excess of Labour in those heavy Ways, has been such a Charge to the

> Country, that new Building of Causeways, as the Romans did of old, seems to me to be a much easier Expence.

Daniel Defoe, *A Tour Through England and Wales* (London: J. M. Dent & Sons, Ltd., 1928), 2:118.

29. Dan Bogart, "Did Turnpike Trusts Increase Transportation Investment in Eighteenth-Century England?" *The Journal of Economic History* 65, no. 2 (June 2002): 440.

30. Bogart, "Did Turnpike Trusts Increase Transportation Investment in Eighteenth-Century England?," 440.

31. Great Britain. Board of Agriculture, *Communication to the Board of Agriculture; on Subjects Relative to the Husbandry, and Internal Improvement of the Country*. Vol. 1, Parts 3 and 4 (London: W. Bulmer and Co., 1797), 121.

32. Arthur Young, *A Six Months Tour Through the North of England*, Vol. 4 (London: W. Strahan, 1770), 577.

33. Young, *A Six Months Tour Through the North of England*, 580.

34. Leveson Francis Vernon-Harcourt, *Rivers and Canals* (Oxford: Clarendon Press, 1896), 2: 343–345.

35. De Salis, *A Handbook of Inland Navigation*, 1–2.

36. Vernon-Harcourt, *Rivers and Canals*, 2: 347–350. For a more recent reference to "arterial canals," see Alice Harvey-Fishenden and Neil Macdonald, "The Development of Early Reservoirs to Supply Water to Arterial Canals in England and Wales," *Landscape History* 42, no. 2 (2021): 79–98.

37. L.T.C. Rolt, *The Inland Waterways of England* (London: George Unwin and Allen, 1950), 40.

38. Rolt, *The Inland Waterways of England*, 40.

39. An Act Enabling Francis Duke of Bridgewater to make a canal from Longford Bridge in Stretford (Lancashire) to the River Mersey at a place called The Hemp Stones in Halton (Cheshire), 2 Geo III, *c*. 11.

40. A. L. Thomas, "The Duke of Bridgewater and the Canal Era," *Geography* 21, no. 4 (December 1936): 299.

41. [Letter], *The Gentleman's Magazine, and Historical Chronicle* [1760], 30 (London: D. Henry and R. Cave, 1760), 623.

42. "To be Sold by Auction," *Manchester Mercury*, August 3, 1762, 3. British Newspaper Archive (BNA).

43. [James Brindley], *The History of Inland Navigations. Particularly Those of the Duke of Bridgwater* [*sic*] *in Lancashire and Cheshire . . .* 2nd ed. (London: T. Lowndes, 1769), 36–37. This particular letter also appears to have appeared in *The Annual Register*. See "Projects," *The Annual Register or a View of the History, Politics, and Literature for the Year 1763* (London: Dodsley, 1764), 99–100.

44. [Brindley], *The History of Inland Navigations, Particularly Those of the Duke . . .*, 37.

45. [Brindley], *The History of Inland Navigations, Particularly Those of the Duke . . .*, 37.

46. "Country News," *Salisbury Journal*, September 19, 1768, 1, BNA.

47. "Newcastle, Oct. 3," *Newcastle Current*, October 3, 1767, 2, BNA.

48. Here, see Joel Mokyr's discussion of the critical connections between technology and the incentive for profit that developed during the "Industrial Enlightenment" in *The Enlightened Economy* (New Haven, CT: Yale University Press, 2009). The influential environmental historian Carolyn Merchant has argued that the eighteenth century represented a profound shift, as the balance between human beings and the natural world that had existed in the pre-modern era was undone by a new male-centered emphasis upon dominance of the land. Carolyn Merchant, *The Death of Nature: Women, Ecology, and the Scientific Revolution* (New York: Harper and Row, 1980).

49. Rolt, *Inland Waterways of England*, 42. See also, Anthony Burton and Derek Pratt, *The Anatomy of Canals: The Early Years* (Stroud: Tempus Publishing, 2001), 49.

50. *A Biographical Dictionary of Civil Engineers in Great Britain and Ireland*, s.v. "Brindley, James" (by John H. Boyes) (London: Thomas Telford, 2002), 78.

51. *A Biographical Dictionary of Civil Engineers in Great Britain and Ireland*, s.v. "Brindley, James," 78, and Rolt, *Inland Waterways of England*, 45. See also, Burton and Pratt, *The Anatomy of Canals: The Early Years*, 31. The engineer Thomas Telford took a skeptical view of Brindley's work, and the business of canals. In describing a project he undertook following upon Brindley's early engineering, he wrote,

> The Birmingham Canal was originally planned by the celebrated Mr. Brindley, and its numerous and inconvenient curvatures can only be accounted for by the consideration, that having rendered carriage by canals cheaper than by the bad roads then in use, it was not judged advisable to incur expense on deep cuttings and embankments, so as to shorten the distance, especially as the toll-dues were levied by the mile; and this effect of monopoly was not discontinued until rival navigations were opened, turnpike-roads improved, and railroads in immediate contemplation.

Thomas Telford, *Life of Thomas Telford*, ed. John Rickman (London: James and Luke G. Hansard and Sons, 1838), 82.

52. Rev. S. Shaw, *A Tour to the West of England in 1788* (London: Robson and Clarke, 1789), 539–540.

53. Baron F. Duckham, "Canals and River Navigations," in *Transport in the Industrial Revolution*, eds. Derek H. Aldcroft and Michael J. Freeman (Manchester: Manchester University Press, 1983), 101.

54. Anthony Burton and Derek Pratt, *The Anatomy of Canals: The Mania Years* (Stroud: Tempus Publishing, 2002), Introduction.

55. Becket, "A View of the Advantages of Inland Navigations; with a Plan of a Navigable Canal, Intended for a Communication between Liverpoole [*sic*] and Hull," *The Gentleman's Magazine and Historical Chronicle* [1765], 35 (London: D. Henry and R. Cave, 1765), 532–533.

56. An act for making a navigable cut, or canal, from the River Trent, at or near Wilden Ferry in the County of Derby, to the River Mersey at or near Runcorn Gap, 6 Geo. 3, *c*. 96 [1766]; and, Anact for making and maintaining a navigable Cut or Canal from the River Severn between Bewdley and Titton Brook to cross the River Trent

near Haywood Mill and to communicate with a canal intended to be made between the River Trent and the River Mersey, 6 Geo. III, *c*. 97 [1766].

57. Pratt, *A History of Inland Transport and Communication*, 174.

58. See Desmond King-Hele, "The 1997 Wilkins Lecture: Erasmus Darwin, the Lunaticks and Evolution," *Notes and Records of the Royal Society of London* 52, no. 1 (January 1998): 161–162. Though Wedgwood and Darwin were among the canal's early supporters, they were not among the original shareholders. See Charles Hadfield, *The Canals of the West Midlands* (Newton Abbott: David & Charles, 1969), 29.

59. [Brindley], *The History of Inland Navigations, Particularly Those of the Duke of Bridgwater* [*sic*] *in Lancashire and Cheshire* . . . , 88–89.

60. [Brindley], *The History of Inland Navigations, Particularly Those of the Duke* . . . , 15.

61. [Brindley], *The History of Inland Navigations, Particularly Those of the Duke* . . . , 15. Legislation providing for improvements such as these generally empowered the undertakers to negotiate with property owners for damages incurred in the process of building the waterways. If no accommodations could be made, legislation sometimes provided for jurors who were called upon to render an impartial decision regarding the issue.

62. An act to enable the Company of Proprietors of the Navigation from the Trent to the Mersey, to make a navigable canal from the said navigation, on the South Side of Harecastle, in the County of Stafford, to Froghall, and a railway from thence to or near Caldon, in the said county; and to make other railways. [1776]

63. J. R. Ward, *The Finance of Canal Building in Eighteenth-Century England* (Oxford: Oxford University Press, 1974), 149.

64. See Charles Hadfield, *The Canal Age* (Newton Abbott: David & Charles, 1969), 38–41. Ward notes the degree to which landed and commercial wealth in Britain could overlap, and the difficulties of this for delineating investment in canals. Initially, having been able to trace roughly £5 million out of an estimated £17 million invested in canals, he finds that landed interests comprised just over 24 percent of the capital he could follow. The presence of landed wealth behind a number of other categories of investment, however, is a distinct possibility. See Ward, *The Finance of Canal Building in Eighteenth-Century England*, 23.

65. [Brindley], *The History of Inland Navigations. Particularly Those of the Duke of Bridgwater* [*sic*] *in Lancashire and Cheshire* . . . Brindley is commonly listed as the author of the work, though no clear connection to his editorship is apparent. The compendium, first published in 1766, moved through a number of editions and accumulated additional material promoting canals as it did so. It is a significant representation of the promotional literature associated with canals, literature which was geared toward unlocking public and private support of individual undertakings.

66. Of material directed to Members of Parliament and potential investors in this period, Ward writes, "There is, for example, an extensive pamphlet literature ostensibly devoted to the statement by promoters of their aims, but because this was directed as much to the members of parliament whose legislative sanction was sought as to potential investors, the eloquent protestations of disinterested concern for the public welfare of which it largely consists cannot be usefully brought to bear upon

the problem under consideration here [assessing the aspirations of canal investors]." Ward, *The Finance of Canal Building in Eighteenth-Century England*, 126.

67. [Brindley], *The History of Inland Navigations. Particularly Those of the Duke* . . . , 2.

68. [Brindley], *The History of Inland Navigations, Particularly Those of the Duke* . . . , 10–11.

69. [Brindley], *The History of Inland Navigations, Particularly Those of the Duke* . . . , 11–12.

70. As Peter Coates has written, "The roads and bridges that Americans were so keen to show off to visiting Europeans more interested in seeing what was scarce back home—the works of nature gloriously unimproved—in the vanguard of 'opening up' the country." Peter Coates, "Can Nature Improve Technology?" in *The Illusory Boundary*, eds. Martin Reuss and Stephen H. Cutcliffe (Charlottesville, VA: University of Virginia Press, 2010), 43.

71. [Brindley], *The History of Inland Navigations, Particularly Those of the Duke* . . . , 60.

72. [Brindley], *The History of Inland Navigations, Particularly Those of the Duke* . . . , 65.

73. [Brindley], *The History of Inland Navigations, Particularly Those of the Duke* . . . , 75.

74. Duckham, "Canals and River Navigations," 101.

75. An Act for making and maintaining a navigable Cut or Canal from Leeds Bridge to North Lady's Walk Liverpool and from thence to the River Mersey, 10 Geo. III, *c*. 114 [1770].

76. Pratt, *A History of Inland Transport and Communication*, 180.

77. [Brindley], *The History of Inland Navigations, Particularly Those of the Duke* . . . , 105. A portion of this letter also appears to have been repeated later in John Phillips, *The General History of Inland Navigation; Containing a Complete Account of All the Canals of the United Kingdom, with Their Variations and Extensions, According to the Amendments of Acts of Parliament to June 1803, and a Brief History of the Canals of Foreign Countries*, 4th ed. (London: C. and R. Baldwin, 1803), 241.

78. [Brindley], *The History of Inland Navigations, Particularly Those of the Duke* . . . , 108. See also, Phillips, *The General History of Inland Navigation* . . . , 245.

79. [Brindley], *The History of Inland Navigations, Particularly Those of the Duke* . . . , 109.

80. [Brindley], *The History of Inland Navigations, Particularly Those of the Duke* . . . , 111.

81. Duckham, "Canals and River Navigations," 106.

82. Duckham, "Canals and River Navigations," 106.

83. These Standing Orders seem to have been issued in 1792 and then revised again by 1794. See Anthony Burton, *The Canal Builders* (Stroud: Tempus, 2005 (1972)), 50; and *A Collection of Rules and Standing Orders of the House of Commons, Relative to the Applying for and Passing Bills for Inclosing and Draining of Lands; Making Turnpike Roads, Navigations, and Aqueducts; . . .* A New Edition (London: Luke Hansard, 1799), 13. The standing orders of 1792 are also recorded in

J. Debrett, *The Parliamentary Register; or History of the Proceedings and Debates of the House of Commons . . . Second Session of the Seventeenth Parliament . . .* , (London: J. Debrett, 1792), 33:231–236; and the orders of 1794 in *The Parliamentary Register; or History of the Proceedings and Debates of the House of the Commons . . . During the Fourth Session of the Seventeenth Parliament . . .* (London: J. Debrett, 1794), 37:757–759. Some stipulations regarding information on canal bills had already been resolved in 1774. See J. Debrett, *The Parliamentary Register; or History of the Proceedings and Debates of the House of Lords During the First Session; and of the House of Commons During the Second Session of the Fourteenth Parliament of Great Britain* (London: John Stockdale, et al., 1802), 2:347–349.

84. Burton, *The Canal Builders*, 51.

85. Concessions to large landowners were substantive. The Earl of Clarendon received £5,000 "for the accommodation of carrying the canal through his Lordship's Park and Grounds" while a payment of £15,000 was made to the Earl of Essex for many of the same reasons. (Ward, *The Finance of Canal Building*, 144.)

86. For a discussion of Irish labor, see Great Britain. Royal Commission into the Condition of the Poorer Classes in Ireland. Appendix G: *Report on the State of the Irish Poor in Great Britain* (London: W. Clowes and Sons, 1835), 142–143, 163.

87. "Alarming Riot at Sampford Peverell," *The Taunton Courier*, April 25, 1811, 1088.

88. "Alarming Riot at Sampford Peverell."

89. Hanson, *The Canal Boatmen*, 16.

90. Hadfield, *The Canal Age*, 73–77.

91. Hanson, *The Canal Boatmen*, 16–17.

92. De Salis, *A Handbook of Inland Navigation*, 16.

93. De Salis, *A Handbook of Inland Navigation*, 16, and Hadfield, *The Canal Age*, 121–122.

94. De Salis, *A Handbook of Inland Navigation*, 17.

95. De Salis, *A Handbook of Inland Navigation*, 17.

96. Hadfield, *The Canals of the West Midlands*, 65.

97. Hadfield, *The Canals of the West Midlands*, 72.

98. Telford, *Life of Thomas Telford*, 82–83.

99. L.T.C. Rolt, *Thomas Telford* (Harmondsworth: Penguin Books, 1985 (1958)), 181.

100. Telford, *Life of Thomas Telford*, 85.

101. Despite Outram's later dismissal, he had argued for the creation of "capacious" reservoirs on the canal and that every effort be made to avoid infringing on the interests of mill owners. See "The Report of B. Outram," *Manchester Mercury*, November 19, 1793, 3. BNA. For the opening of the canal, see "Commercial Herald," *Carlisle Journal*, April 20, 1811, 3. BNA.

102. "Canals," in *The Cyclopaedia: or, Universal Dictionary of Arts, Sciences, and Literature*, Vol. 6, ed. Abraham Rees (London: Longman, Hurst, Rees, Orme, & Brown, 1819), 120. Richard Coopey has used the term "pre-emptive engineering" in reference to engineers who could address scientific and technical issues, as well as "social, cultural or political obstacles and considerations." (Richard Coopey, "The

Severn, 1750–1850," in *A History of Water*, Series 2, Vol. 2, eds. Terje Tvedt and Richard Coopey (London: I. B. Tauris, 2010), 404).

103. Anthony Burton, "Canals in the Landscape," in *Canals: A New Look*, eds. Mark Baldwin and Anthony Burton (Chichester: Phillimore, 1984), 13–14. Other evidence of materials transported on canals affecting the overall built environment of Britain, Burton finds more difficult to trace—particularly the increasing use of "red" bricks which came in the late nineteenth century. For a primary discussion of the change from thatch to slate, see Thomas Pennant, *The Journey from Chester to London* (London: Printed for Wilkie and Robinson, et al., 1811), 75.

104. Burton, "Canals in the Landscape," 12. See also W. G. Hoskins, *The Making of the English Landscape* (London: Hodder and Stoughton, 1955), 253.

105. Burton, "Canals in the Landscape," 12.

106. Simon Winchester, *The Map That Changed the World: William Smith and the Birth of Modern Geology* (New York: HarperCollins, 2001), 84–85.

107. Winchester, *The Map That Changed the World*, 90–91.

108. Winchester, *The Map That Changed the World*, 257.

109. Winchester, *The Map That Changed the World*, 282–283.

110. "Completion of the Leeds and Liverpool Canal," *Hull Packet*, November 5, 1816, 4. BNA.

111. "The Thames and Medway Canal," *Morning Post*, October 16, 1824, 2. BNA.

112. "The Thames and Medway Canal."

113. For a good discussion of the complex and ongoing efforts to establish canals amidst the competing interests of various landowning and commercial groups, see Hopkinson, "The Development of Inland Navigation in South Yorkshire and North Derbyshire 1697–1850," 229–251.

114. Priestley, *Historical Account of the Navigable Rivers, Canals, and Railways, Throughout Great Britain*, 268.

115. Priestley, *Historical Account of the Navigable Rivers, Canals, and Railways, Throughout Great Britain*, 270. The Forfeited Estate Fund had been established to re-direct funds from the estates of those identified as participating in, or supporting, the Jacobite rising of 1745.

116. Rolt, *Inland Waterways of England*, 43.

117. Rolt, *Inland Waterways of England*, 43.

118. G. G. Hopkinson, "The Inland Navigations of the Derbyshire and Nottinghamshire Coalfield, 1777–1856," *Derbyshire Archaeological Journal* 79 (1959): 31.

119. This sum also included monies for wharfs and warehouses. See Hopkinson, "The Development of Inland Navigationin South Yorkshire and North Derbyshire," 239.

120. Duckham, "Canals and River Navigations," 115.

121. The designation [Old] Grand Union allows a distinction to be made between the original Grand Union Canal, completed in the early nineteenth century and extending from near Braunston to Leicester, and the later Grand Union Canal, established as an amalgamated canal incorporating the Grand Junction Canal and the original Grand Union, in the early twentieth century. Today the [Old] Grand Union is often referred to as the Leicester arm of the twentieth century's Grand Union Canal.

122. A. Temple Patterson, "The Making of the Leicestershire Canals," *Transactions of the Leicestershire Archaeological Society* 27 (1951): 93.

123. "London, Jan. 8," *Saunders's News-Letter, and Daily Advertiser*, January 12, 1804, 1. BNA.

124. "London, Jan. 8," 1.

125. Joshua Jepson Oddy, *A Sketch for the Improvement of the Political, Commercial, and Local, Interests of Britain, as Exemplified by the Inland Navigations of Europe in General, and of England in Particular; Including Details Relative to the Intended Stamford Junction Navigation, to Unite the Eastern with the Midland and Western Counties of the Kingdom* (London: J. J. Stockdale, 1810), 23.

126. See Mike Clarke, "John Rennie and the Rochdale Canal, UK," *Engineering History and Heritage* 164, Issue EH3 (August 2011): 150. Clarke suggests that the delayed publication of Sutcliffe's tract indicates such publication may have been one of his last wishes.

127. John Sutcliffe, *A Treatise on Canals and Reservoirs, and The Best Mode of Designing and Executing Them . . .*, (Rochdale: John Sutcliffe, 1816), ii.

128. Sutcliffe, *A Treatise on Canals and Reservoirs, and The Best Mode of Designing and Executing Them . . .*, vi.

129. Sutcliffe, *A Treatise on Canals and Reservoirs, and The Best Mode of Designing and Executing Them . . .*, 73. See also Phillips, *A General History of Inland Navigation . . .*, xiv.

130. Sutcliffe, *A Treatise on Canals and Reservoirs, and The Best Mode of Designing and Executing Them . . .*, 76.

131. Sutcliffe, *A Treatise on Canals and Reservoirs, and The Best Mode of Designing and Executing Them . . .*, 76.

132. "Inundation," *Saint James's Chronicle*, December 11, 1810, 4. BNA.

133. John Aikin and Anna Laetitia Barbauld, "The Canal and the Brook: An Apologue," in *Miscellaneous Pieces of Prose,* ed. John Aikin and Anna Laetitia Barbauld, 82–83 (London: J. Johnson, 1792). Jodie Matthews has recently assessed the complexity of the poem's response to canals as a powerful new force in the landscape whose ultimate effects might harbor violence and be restricted to a few. (See Matthews, *The British Industrial Canal: Reading the Waterways from the Eighteenth Century to the Anthropocene* (Cardiff: University of Wales Press, 2023), 45).

134. Hopkinson, "The Inland Navigations of the Derbyshire and Nottinghamshire Coalfield," 37–38. Beyond canals, ice could significantly affect river transportation. See, for instance, "Sunday, Feb. 26," *Gentleman's Magazine: And Historical Chronicle* (London: Nichols, Son, and Bentley, 1814), 84 (Part 1): 193.

135. Duckham, "Canals and River Navigations," 127.

136. "Interruption of Canal Navigation," *The Daily News (London)*, September 21, 1854, 3. BNA.

137. Wolfgang Schievelbusch, *The Railway Journey: The Industrialization of Time and Space in the 19th Century* (Berkeley, CA: University of California Press, 1986 [1977]), 12.

138. For a wider discussion, see James Winter, *Secure from Rash Assault: Securing the Victorian Environment* (Berkeley, CA: University of California Press, 1999), 107.

139. Michael Freeman, *Railways and the Victorian Imagination* (New Haven, CT: Yale University Press, 1999), 15.

140. Schivelbusch, *The Railway Journey: The Industrialization of Time and Space in the 19th Century*, 27–29.

141. Hadfield, *The Canal Age*, 157–158.

142. A. J. Arnold and S. McCartney, "Veritable Gold Mines Before the Arrival of Railway Competition: But Did Dividends Signal Rates of Return in the English Canal Industry?" *Economic History Review* 64, no. 1 (2011): 229.

143. Hopkinson, "The Inland Navigations of the Derbyshire and Nottinghamshire Coalfield," 37. In 1825, the returns of eighty companies which had a capital investment of approximately £13.2 million were assessed. The ten most successful companies with a capital investment of £1.1 million offered a dividend of 27.6 percent. The remaining companies, with an investment of just over £12 million, paid an average of less than four percent (with some paying none). The average return on investment thus stood at 5.75 percent. (Duckham, "Canals and River Navigations," 123.)

144. Peter Maw, *Transport and the Industrial City: Manchester and the Canal Age, 1750–1850* (Manchester: Manchester University Press, 2013), 53–54.

145. Maw, *Transport and the Industrial City*, 84.

146. Schievelbusch, *The Railway Journey*, 33.

147. Duckham, "Canals and River Navigations," 132.

148. Maw, *Transport and the Industrial City*, 85.

149. Hopkinson, "The Inland Navigations of the Derbyshire and Nottinghamshire Coalfield," 39–40.

150. Freeman, *Railways and the Victorian Imagination*, 100.

151. Hopkinson, "The Development of Inland Navigation in South Yorkshire and North Derbyshire 1697–1850," 249.

152. Maw, *Transport and the Industrial City*, 98–99.

153. Hopkinson, "The Development of Inland Navigation in South Yorkshire and North Derbyshire 1697–1850," 250.

154. In the course of the nineteenth century, greater state-directed efforts to build, expand, and maintain canals would emerge in France in contrast to Britain. See Gerald Crompton, "The Tortoise and the Economy," *Journal of Transport History* 25, no. 2 (2004): 6–7.

155. Roger Squires, *Canals Revived* (Bradford-on-Avon: Moonraker Press, 1979), 13.

156. Freeman, *Railways and the Victorian Imagination,* 1.

Chapter 2

An Independent Spirit

Life Afloat in the Nineteenth Century

Canal boat workers, often referred to as "boatmen," or "boat people" when traveling with their families, became central to cultural understandings of the waterways in the nineteenth and twentieth centuries. In the nineteenth century, their transience and association with the vices and dangers of the canals ensured their social marginalization. Later, in the twentieth century, both their labor, integral to the waterways' industrial use, and their cultural traditions, were seen as countercurrents to Britain's twentieth-century modernization. As such, they prompted efforts to restore dignities in both labor and independent living that canal supporters felt had been stripped away. This chapter argues that emerging links between canal boat workers and immorality, violence, and vice in the early nineteenth century set in motion a series of reinterpretations of canal life that affected the development of the waterways until at least the late twentieth century. It suggests that as cultural and social outsiders in the nineteenth century, canal boat workers became symbols for those who saw them as maintaining a domestic and occupational independence that had withstood the onset of capitalism, technical modernization, and governmental growth. Evolving interpretations were central to a twentieth-century revival of canals, which emphasized canals' place as—most often rural—paths of exploration situated at the boundary of the natural and built environments, occupying a nuanced position between the pre-industrial and industrial worlds of modern Britain.[1]

The origins of canal boat workers continue to be poorly understood. Many likely came from among tenant farmers who at first offered their labor seasonally, while others came from among the "navvies" who had built the canals.[2] The application of the term "boatman" and its concurrent use as a descriptor for those working in the coastal trade and established river routes suggests some traced their origins to work on the water. In January 1795, William Pitt the Younger indicated that the Commons must take up consideration of how to secure men for the navy. He argued that owners of merchant and trading ships should be subject to a quota of men and that a second class of seamen could be taken from vessels working upon rivers and canals. Although canal and river boatmen might not have skills equal to seamen, they knew much about the management of vessels, he said.[3]

On canals, male workers were split between those who worked "fly-boats," moving nearly round-the-clock in the care of four or more men working shifts, and "slow-boats," moving under the care of one or two men who rested at night.[4] At first, most canal boatmen likely stayed close to home and continued to be rooted in their local communities. In the late eighteenth century, among counties where it has been possible to discern ownership, Harry Hanson has determined that perhaps 60 percent of the boats were under the control of individual owners, while companies and partnerships controlled more than five boats. As more canal links were established in the early nineteenth century, boatmen ventured farther from home. Those who may have had charge of boats they owned were perhaps 6 percent of the captains in the noted counties.[5] Women and families might join boatmen on the slow boats, offering both their labor and the ability to reduce or avoid housing costs ashore, despite having to contend with cramped cabins perhaps no more than 10 feet (3 m) of the boat's length. Standards of work were only partially regulated. In 1771, the proprietors of the Droitwich Canal linking Droitwich Spa with the River Severn published a notice of intent to prosecute all offenses mentioned in the legislation permitting the construction of the canal.[6] Among those prosecuted would be those who would "maliciously Damage or Destroy any of the Banks or other Works" belonging to the company.[7] Captains were to ensure that their names were prominently printed on the sides of their vessels and owners were to be responsible for any damage or mischief created by the vessels. A host of other regulations addressed the throwing of rubbish into the canal, the use of lock mechanisms, and the right of the company to have boats removed.[8] Nationally, the total population of canal boat workers remains uncertain, though boatmen working either canals or natural waterways may have totaled just over 32,000 in England and Wales in the 1851 census.[9] This number would have included watermen and lightermen of London.[10] Numbers of women are uncertain due to the fact that they were likely not included in occupational numbers unless captains of their own boats (a limited occurrence).[11] Overall, women and girls over 16 years of age may have added 20 percent of the number of male workers.[12]

Few records of boat people's observations and understandings of their own world have survived, leading to a reliance on state, missionary, and newspaper-based sources to provide insight into that world. Among these outside accounts, some of the most widely available are those linked to crime, violence, and death, printed in pamphlets and provincial newspapers. Determining overall levels of crime remains difficult due to the limitations of available national data before the early part of the nineteenth century and the disjunct between what might have been the actual incidence of crime and its reporting and prosecution.[13] Despite the lack of hard statistical evidence regarding the degree to which boatmen were involved in crime in the course of the nineteenth century, at least three overriding factors affected their place before the law. First, crime was often associated with marginalized ethnic and racial categories, as well as the understanding that dangerous criminal classes were generated by industrial and urban life.[14] For instance, migrant Irish laborers had already raised alarms along the waterways as navvies; Romani ("Gypsies"/"gypsies") continued to raise concerns of "vagrancy" on land.[15] In the latter case, Romani were often included within a wider "traveler" population and understood by some to have inherited predispositions to movement or "wanderlust."[16] In the course of the nineteenth century, the activities of the "traveler" population were increasingly found to run counter to settled industrial society. As boatmen, similarly to Romani, became associated with a particular "race" inhabiting the waterways, so too would they have been subjected to concerns linking the socially marginalized with crime.[17] Second, crime was associated with a particular set of moral failings in the early nineteenth century, an outlook which transitioned to a more complex and interrelated set of factors, including social and economic status and familial inheritance, by the late nineteenth century.[18] In this respect, the uncertain links many male boatmen were believed to have with the stabilizing moral institutions of church and family would have been significant early on, and the cramped living conditions (with concurrent concerns for sanitation and sexual immorality) would have been significant later. Third, crime could be associated with the generation of a particular criminal class within society in which habits of theft and violence were embedded within social groups and locations, and passed from one generation to the next.[19] The boating community appeared to draw in more family units by the mid-nineteenth century, and families often married amongst themselves and maintained a distinct set of social traditions. By mid-century, there were assumptions that theft, violence, and drinking were passed down from one generation to the next within the canal boat community, not unlike within the Romani community.

The use of newspapers to document the incidence and perception of crime remains problematic due to inconsistent reporting and bias as well as the difficulties of accessing and evaluating broad sweeps of newspaper data.[20] Nonetheless, property crime, accidental deaths, violent assaults, infanticide, and suicide represent five major categories of violence depicted in provincial newspaper stories associated with boatmen between approximately 1800 and 1840.[21] Overall, as canals became associated with violence, crime, and death from the early nineteenth century onward, a path was paved for an understanding of canals as marginalized social and cultural spaces well into the twentieth century.

Reports of property crime appear to have become more prominent with the acceleration of canal completion.[22] For instance, in 1800, William Jacob and Co. offered a reward of fifty guineas for anyone offering information on the stealing of salt from barges on the River Thames or the Thames and Severn Canal.[23] The next year, two men were convicted of stealing a bag of cotton from a boat belonging to the carriers Thomas and James Rainsford on the Bridgewater Canal, while a boatman was convicted for having received the goods.[24] In January 1807, a boatman charged with robbing a boat was allowed to enter the army rather than being imprisoned, and the next year a boatman was convicted of stealing hay which was cut out of a meadow by the side of a canal. In passing sentence in the latter case, the judge indicated farmers frequently suffered depredations on their property located near canals.[25] In 1808, John Staines, a boatman on the Grand Junction Canal, after having been caught cutting down ash trees, was forced to both make financial amends for the trees and ask forgiveness in a column of a newspaper.[26] Most reports in which a boatman was a perpetrator involved the theft of property from boats, such as one in 1809 where a husband and wife were reported to have stolen from two boats.[27] In 1811, at the Chester Assizes, a boatman named Abraham Wood, employed on the Staffordshire Canal, was found guilty of stealing earthenware under his care and sentenced to transportation for seven years. In pronouncing his sentence, the judge noted:

> The frequency of such robberies are notorious, and the annual loss is immense. The Pottery Association aided by the Boat proprietors, were the prosecutors of Wood, and deserve much praise. . . . Were the perpetrators of similar depredations to be prosecuted and severely punished, as in the present instance, we should certainly not have to record so many systematic acts of villany [*sic*].[28]

In one General Quarter Session of Peace for Warwick in 1814, three different sentences ranging from seven days to three months were handed down to boatmen for thefts.[29] In some cases, the mobility offered by canals made it difficult to track perpetrators or those who might give evidence against them. In 1813, an ad was placed in a Staffordshire newspaper offering fifty guineas' reward for information that might show that several articles found at the home of Thomas Dixon, which were suspected as having been taken from boats passing on the Grand Trunk Canal [Trent and Mersey Canal], were indeed stolen.[30] Boatmen could also be victims of crime, as well. In 1801, Joseph Malin was charged with having stolen from a boatman Thomas Andrews on the Grand Junction Canal.[31] In 1810, Benjamin Booth, a drum-boy, was convicted of stealing a silver watch from Robert Baske, a boatman on the Oxford Canal after having taken passage with him from Coventry to Paddington Basin in London.[32] Though it is unclear how rates of crime along canals compared with those of the local communities the canals crossed, boatmen could lack permanent homes, adding complexity to efforts to prosecute and defend them.

Intentional physical violence in and alongside canals often marked a variety of crimes. Though in only some cases were boatmen recognized as perpetrators, the presence of dangerous individuals alongside canal banks certainly emphasized what seemed to be the waterways' inherent danger. In 1803, in a case tried before the Old Bailey, George Foster was convicted of drowning his wife and infant child in Paddington Basin; in responding to this case, which had been built largely upon circumstantial evidence, the defense argued that the deaths had been accidental.[33] In 1809, "a fellow of hideous appearance and above the common stature" began assaulting women near Paddington Basin, at the intersection of the Regent's and Grand Junction canals. In one instance, after a struggle of about 10 minutes involving two women in a field near the canal, boatmen responded to calls of distress, at which the attacker fled but was not apprehended.[34] Further to the east within London in 1821, Joseph Wilshaw, a boatman, was brought before the Bow Street Magistrates Court for having been part of a group of "ruffians" who had attacked several people near City Road.[35] More disturbing, in 1820, Abel Hill was executed for the murder of Mary Martin (with whom he had lived for six years) and her 16-month-old son Thomas Martin (of whom he was the father) by drowning them in a canal near Bilston.[36] Violent sexual assault committed by boatmen was a common theme. In 1817, a boatman working the Grand Junction Canal, apparently in the company of two younger males, seized and raped a 16-year-old girl named Sarah Somers, who had been walking along the towpath in the company of another girl.[37] In 1826, witnesses at an inquest reported that six drunk boatmen had attacked, brutalized, and killed a young woman, Mary Whitaker near Manchester.[38]

Accidental deaths could touch the lives of boatmen in several ways. In March 1802, the death of a boatman came at Bradford as he was attempting to shut a lock and fell in, and a death was recorded for a young child who was trapped in a boat cabin when it caught fire.[39] While fire seems to have been rare, drowning was not. In 1803, William Bird, a fifteen-year-old boatman, fell into the Nutbrook Canal and was drowned.[40] That same year, William Hedge drowned as he went to wash in a canal, and a boatman nearby (who could not swim) was unable to get him out in time.[41] In 1806, Sarah Kendrick, seventy years of age, was reported as "much intoxicated" and, while walking along the towpath, fell into the Birmingham Canal.[42] In 1810, a boatman named John Barnes, in letting down a drawbridge after having passed it, had a pole break and part of the bridge fall on one of his legs, lacerating it severely. He died about eight hours later after a significant loss of blood.[43] In 1812, John Hayes, a boatman working on the Leeds and Liverpool, fell overboard and drowned in the process of investigating a potential leak on his boat.[44] The intensity of danger seems to have been exemplified by a single night in 1818 near Leeds, when four people fell into the Leeds and Liverpool Canal, with one drowning.[45] Among accidental deaths involving boatmen, alcohol was sometimes cited, as in the 1822 case of a boatman named Samuel Wild who drowned in the basin of the Leeds and Liverpool Canal near Bingley after returning from getting some ale.[46] Two years later, Stephen Harrison, a boatman on the Leeds and Liverpool, died in much the same way, perhaps near the same site.[47] The recorded deaths of women canal boat workers appear to have been less numerous than those of men, as would accord with the smaller percentage of boat workers they represented. In the Fall of 1812, however, J. Radford, the captain of a Chesterfield canal boat then on the River Trent, found his craft overwhelmed by a tide and, filling with water, lost his wife of only five weeks as well as a servant.[48] Children appear to have been at substantive risk onboard boats. In 1825, an eight-year old girl drowned after falling from a canal boat of which her father had charge.[49] In 1828, two children died by drowning after falling into the Trent and Mersey Canal near Stone.[50] And, later that year, an infant was found drowned in the Grantham Canal as his father worked nearby.[51]

Infanticide and suicide were particularly unsettling reminders of the dangers associated with canals. Though stories connecting boat families directly with infanticide are not apparent in the local press, the act could be a part of the narrative of violence and death along the waterways. In infanticide, Victorian ideals of wife and mother could be significantly compromised. In 1801, Charlotte Turner was convicted of abandoning her "bastard" child along the banks of a canal and imprisoned for 12 months. The child was found by a boatman while the mother apparently stayed at a nearby inn, making no mention of the child.[52] In 1816, in a prominent case covered by a number of papers, a young woman was charged with murder for drowning her child in the River Trent near Burton.[53] In 1819, a young woman threw herself into the Kennet and Avon Canal near Devizes in an apparent suicide but was rescued shortly after by a young man working nearby.[54] Less fortunately, in 1827, a fourteen-year-old girl was found dead in a canal after giving indications of being depressed and mentally unstable. An inquest returned a verdict of "Insanity."[55] In 1823, a newborn was found along the towpath of the Oxford Canal, having likely been in the water for some weeks.[56] In 1827, Nancy Jones was arraigned after a coroner's inquest for the murder of her young daughter, after the daughter had been found dead in a canal and Jones and her eldest daughter were represented as behaving suspiciously in the time just beforehand. After testimony which both indicated the steepness of the descent into the canal and the affection the mother had previously shown, no charges were pursued.[57]

Generally, sudden deaths would be ruled accidental until evidence indicated otherwise. Designations of accidental drownings could cover suicide in Victorian Britain, particularly when it was unclear how a body came to be in the water. In 1816, a woman of approximately thirty years of age was found dead in the Grand Junction Canal near Paddington. A local paper reported that, "nearly naked" when found, she was "a complete picture of wretchedness, having but a part of a filthy flannel petticoat tied round her waist."[58] The verdict was an accidental drowning.[59] In 1826, Elizabeth Hanson, who was found by a boatman in a canal in Birmingham with no signs of violence, was categorized simply as "found drowned."[60] The association of suicide with insanity or mental aberrations was strong, as were popular understandings of suicide as an escape from sexual dishonor for women.[61] Yet, in the nineteenth century, the suicide rate among men was understood to be three to four times that for women (except for females during their late teen years).[62] Indeed, stories of suicide among male boat workers were prominent in the local press and may have reinforced a picture of suicide among men as "the quickly chosen escape of the proud, the weak, or the wicked from financial ruin, disgrace, or retribution."[63] In the Fall of 1809, a boatman named John Storey drowned himself in the River Blyth after recently experiencing the drowning deaths of his two daughters. Noted as an excellent swimmer, Storey had apparently tied a weight to himself as he entered the water.[64] The next year near Banbury, a boatman named Thomas Brown hung himself from an ash tree, the newspaper report indicating that he had been in "a low and desponding state for some time."[65] In 1818, the body of A. Wilkins was found by a boatman in the Paddington Canal [Basin], his feet sticking in the mud and his head just below the water's surface. An inquest revealed that he had been despondent since he had had a daughter drown in the canal.[66]

Although it is not possible to know the specific effect these reports may have had upon local readers, Olive Anderson has noted that while the circumstances of suicide were often treated sympathetically among the reading public of London, they could be much more harshly condemned in rural areas.[67] Broadly, those who survived suicide could be charged with *felo de se*, or suicide (although trial was rare), and be faced with restrictions on burial in churchyards.[68] However, Anderson notes that as concern regarding the "condition of England" became more prominent in the early nineteenth century, the circumstances of these deaths likely raised in the minds of some observers concerns regarding the tremendous gap between the struggling population of the poor and an uncaring middle class.[69]

In a society in which images of drinking, vice, and violence often followed all-male canal boat crews, and a reputation for squalor, ill-education, and debauched morals came to characterize boat families (and desperate circumstances were understood to lead to violence and suicide), missionary efforts soon emerged. In 1829, "The Paddington Society for Promoting Christian Knowledge among Canal Boatmen, and others" began publication of *The Canal Boatman's Magazine*, chronicling its efforts to morally reform canal boat workers and others in London. The journal largely focused upon the content of sermons, the distribution of tracts to boats in the Paddington area, and attendance at services. In the first issue, a letter inserted by one of those who preached for the organization, Joseph Wallis, recounted a recent experience of the wife of a boatman whose fortunes had significantly declined, indicating she had decided not to commit suicide after hearing one of his sermons.[70] However, the missionaries' early prognosis of the boatmen was not good.

> They are a class of men among whom ignorance prevails to a very great extent; . . . Brought up from childhood to manhood without education in many instances, and from the nature of their employment separated from society, they have naturally imbibed principles and habits in direct opposition to the Bible.[71]

However, for some, the effort to proselytize was a hopeful one. A letter to the magazine contended,

> I cannot but rejoice at the appearance of your unostentatious little Magazine as an addition to the number, in the hope, that under God it may extend far and wide, and draw the attention of benevolent Christians to that long neglected class of individuals—the Canal Boatmen. They have been, indeed, wonderfully neglected.

In adding that there was a "veil of darkness in which they [the canal boatmen] had been enveloped," the letter writer might well have been pointing not only to the spiritual challenges faced by missionaries but also to the uncertainties of later social reformers as well.[72]

Despite ongoing efforts to reach canal boat workers directly through the distribution of materials and services led by a variety of pastors, boatmen appear never to have dominated attendance at the Boatman's Chapel, established at Paddington Basin in 1827. The origins of the chapel appear somewhat unclear, either having been started by two officers connected with the Naval & Military Bible Society or members of the "West London Auxiliary" of the Solders' & Seamen's Bethel Union (which had already undertaken efforts to reach sailors, fishermen, and watermen on the River Thames).[73] On the morning of Sunday, April 5, 1829, 110 were present, of whom six were boatmen, and in the evening ninety people were present, including twelve boatmen, three of their wives, and "several persons" from the wharfs.[74] On

August 23, 1829, a morning service was conducted with forty-five people in attendance, including "a few canal-men;" the afternoon service with 300 present had "several canal-men," and an evening service was conducted with 150 people in attendance, nine of whom were canal boat workers.[75] By the end of the year, the organization reported that there were an average of 250 in attendance at its services on Sundays and that approximately 100 of those were associated with the canals.[76]

By the early 1830s, canal boatmen increasingly found themselves at odds with an emerging Victorian regulatory state. First, there was increasing attention to the hours and conditions of child labor. The Factory Act of 1833 restricted children's overall workday in textile factories and called for the provision of education to coincide with work. Most importantly, four factory inspectors were to be appointed to supervise adherence to the system.[77] Though it would not be until the 1870s that canal boat families felt the full effects of efforts to regulate child labor, attention in the 1830s signaled the difficulties the boat population and other itinerant groups were to face in coming years. Most immediately, it was the Poor Law, Britain's composite system of local poor relief practices, that raised the most concern within Parliament. Underemployed and sometimes itinerant classes were believed to be draining the strength of the nation (and its coffers) as they collected local relief through a patchwork set of parish supports. The Royal Commission for Inquiring into the Administration and Practical Operation of the Poor Laws—not directly addressing canals but giving vague indications of similarities between sailors and river boatmen—suggested a growing anxiety regarding mobile and intermittently employed populations making both settlement claims and demands for relief in parishes where they otherwise had few connections.[78] Concerns of the Commission ranged from steps employers might take to prevent "settlement" in individual parishes (in order to avoid paying the attendant rates), to efforts by recipients to game the system so they might receive relief in more than one location.[79] As passed, the 1834 Poor Law Amendment Act ("New Poor Law") established a three-member Poor Law Commission that had the power to determine practices of local boards of guardians including the classification of paupers and conditions under which significantly reduced outdoor relief would be distributed. Individual parishes were to be organized into poor law unions. The ability of migrant populations to establish claims of settlement, and thus demands upon poor relief, was significantly limited. This step likely lessened the concerns of those worried about the crime and violence associated with boatmen. Nonetheless, anxiety regarding inspection and monitoring of populations whose homes and work traveled with them and whose labor proved difficult to track ensured the challenges of regulating a working population were linked with questions of canals' economic impact in years ahead.

Concurrently with the Poor Law Commission, the Select Committee on Drunkenness convened by the House of Commons gathered testimony from religious, mercantile, military, medical, and legal authorities in an attempt to assess the effects of alcohol on the nation's fortunes.[80] The testimonies offered suggested alcohol was consuming the nation's working classes much like a wave, with problems of ill health, improvidence, crime, and degradation of women and families. Anxious questions regarding debt, violence, immorality, and incapacitation were leveled at witnesses with an eye toward reform that might limit the amount of drink available, the times and places of its availability, and foster moral uplift as well as a reduction in penury and crime. For the committee, crime came in a variety of forms including theft, fraud, prostitution, robberies, and "more hardened offenses." It argued such practices had now spread among those who would under "sober habits and moral training" be sources of gain to the nation.[81]

The reputation of canals as seedbeds of vice and criminality was exacerbated by the 1839 report of the Royal Commission Appointed to Inquire into the Best Means of Establishing an Effective Constabulary Force in the Counties of England and Wales.[82] The commissioners conducted their hearings amidst fear and uncertainty regarding the overall scope of crime in Britain. The first report included a wide range of topics from the security of travelers on highways, to piracy along the coastlines, to the dangers of violent protests by radicalized workers. Within the report, canals were carefully examined, and the connections made even earlier by newspapers between the boat population and crime and alcohol were further strengthened. Though noting that most goods transported on canals were "tolerably safe" from large-scale or violent robberies, the report concluded that portions of the country through which canals passed were subject to "petty depredation" and that canal boats traveling by night were often used by thieves for the "transmission and disposal of property stolen from the adjacent country."[83] Testifying before the commission, Mr. Dowling, Commissioner of Police in Liverpool, contended that there was a good deal of "plunder" of goods on canals and that,

> The bales are opened, boxes and cases are broken open and plundered, and it is so well done that the discovery is almost a matter of impossibility until they arrive at their destination, because they are addressed by order at Manchester to the parties abroad, . . . Till they arrive at their destination no discovery of the robbery is made, so that before the return of the loss arrives in England six or eight months may have elapsed.[84]

Dowling went on to describe the process by which some bulk goods could be unpackaged and pilfered at night, noting (in response to a specific question) that there would likely not be a reason for local police to focus on the goods' protection as no immediate or local pecuniary interest would be involved. Officials from carrying companies provided additional testimony, with one official noting that "cheap beer-shops" offered a common place for disposing of goods among the proprietors or parties gathered there.[85] Another witness, an official with a carrying company, offered a rationale for why boatmen engaged in ongoing theft,

> We find, also, that where men work their own horses they will work a boat for less than it can be done honestly for—lower than we can do it. . . . It can only be done by making it out from the barges.[86]

Another carrying official noted that, given the lack of police presence on rivers and canals, "A boat may stop anywhere on the river, and the men may do as they please with the cargo, without any interference from any one."[87] This same official, whose firm focused upon the River Severn and the Stourbridge Canal, added that he believed some transference of stolen goods was made at boatmen's homes bordering the canal.[88] At least two of those testifying agreed that prosecution of receivers of stolen goods should be a prime focus of policing efforts. In its initial report, the Commission also included the testimony of a major silk manufacturer at Leek who argued, "Robbery has become almost universal. Within the last eight years it has increased tenfold. There are as many as ten men in Leek who live in idleness and drunkenness on their gains as receivers."[89] The Commission's first report went on to include statements from men in jail, who indicated what they had done and how. One contended, "When not in honest employ I was maintained by poaching and stealing. We have many a time milked farmers' cows in the night. We had a borer for drawing sugar or dry goods: we slipped the hoop, made a small hole under it, and introduced the borer, and took what we liked. . . . Nothing would have altogether stopped us but the receiving-houses of the line being broken up; and if they had, there would be plenty others.[90]

Reports of the prevalence of theft and general insecurity along canals, and the active participation of some of those working on the canals, helped to once again mark out waterways as zones of vice and insecurity. Those detained for theft on the canals helped to tie the conditions of perceived lawlessness directly to the boat population, who reputedly perpetuated a cycle of theft as they drew new recruits into their ranks. As one imprisoned former boatman was recorded to have said, "I was taught to pilfer. . . . The whole crew were engaged in depredations, and I did as my companions did, and took goods of all sorts which they sold to the different receivers on the canal."[91] Another respondent offered,

> It is impossible to tell how many depredations I have committed altogether. . . . I don't think we ever went a trip without committing a depredation of some sort. Twenty-two years elapsed from my entering upon a course of crime to the period of my first apprehension.[92]

Indeed, the apparent propensity of canal thieves to usher young men into a life of crime indicated to many that a debased society alongside the waterways had indeed emerged. William Kershaw, a partner in the canal carrying firm of Buckley and Kershaw, remarked,

> They [youths] are seldom employed upon the canal until they get to be 12 or 14 years of age: . . . I have known instances where boys of that description have been participators, and had a full knowledge of all the circumstances: and I have supposed that being so young, I could get information from them, but I have found them quite as impervious as the oldest thieves in the trade.[93]

The commissioners expressed particular concern with incidents of murder in rural areas, taking stock of the ways in which murderers might avoid detection or prosecution through contacts or allegiances within the local community or an ability to take advantage of isolation and the lack of a constabulary force; however, it made no direct links between murder and the canals. The legislative outcome of its work was the County Police Act passed in August of 1839, which offered permissive powers to county magistrates to establish rural constabularies.[94] Only some proceeded to do so, and there is no evidence of a major impact upon canals.

However, in May 1839, parliamentary efforts to address the impact of drink and violence in relation to the canal world were profoundly affected by the case of Christina Collins, a young woman traveling from Liverpool to London on a Pickfords fly-boat who was believed to have been raped, murdered, and thrown into the Trent and Mersey Canal near Rugeley by a drunk crew.[95] Of the four-man crew, two were ultimately executed, one boy was not prosecuted, and one man, on the basis of a plea regarding his circumstances, was transported rather than executed.[96] Newspaper accounts of the case emphasized the circumstances of Collins, apparently a dress-maker of good reputation (both "modest and prudent") who had booked passage on the Pickfords fly-boat so that she could join her husband who had recently found a position in London.[97] Later, she would be found dead in the water at the edge of the canal as the boat continued on its journey. Her body was judged not to contain marks of physical violence, and whether she had died before entering the water (potentially as a result of suicide) or had died as a result of drowning would remain a point of dispute over the course of two impending trials. Charges of murder and felonious assault were brought against three of the four crew (with the boy supervising the horse towing the boat not charged).

Initially, the men faced trial on the charge of felonious assault. Numerous witnesses, many connected with the business of the canal, were brought forward who highlighted signs of Collins' distressed behavior early in the journey, cries of distress later on, and even an indication that she appeared to sharpen a small knife and stow it among her belongings.[98] Evidence of the crew's drunkenness and use of "coarse" language (not reprinted in news accounts though contextually indicated as of a sexually violent nature) also characterized the testimony.[99] Despite graphic testimony, a lack of evidence directly connecting the crew to violence against Collins (except an apparent earlier statement by one of the men that another had "raped and murdered" Collins) brought a verdict of "not guilty" on felonious assault.[100] Cleared of the first charge of assault, the three men were brought up for trial on murder early the next year with a significantly different result. Much the same evidence was offered, though notably it was reinforced with the testimony of a jailhouse witness who reported on confessional statements he said were offered to him by one of the men.[101] Condemned to death, the three men received a respite of one week after the intervention of a Rev. A. Talbot, a magistrate who was able to argue that doubt hung over the three men's actual participation in the act of drowning, which was the basis of the murder charge. After a stay of their execution, two of the convicted were hanged in mid-April while the third, William Ellis, was given a reprieve and faced transportation, it being concluded that he was not present at the point Collins had been murdered.[102]

Public outrage and condemnation of conditions on canals, along with condemnation of those who appeared to reflect canals' violence and moral depravity, swiftly emerged. In July 1839, the trustees of the River Weaver Navigation were reported to have begun considering petitions from flatmen (those working the "flats," or the primary barges of the River Weaver) to consider stopping all Sunday traffic on the river, a petition supported by a significant number of those in the "respectable trades." The same report indicated that the managers of the Mersey and Irwell Navigation had already closed their locks on Sundays and that Lord Francis Egerton (grandson of the third Duke of Bridgewater) had begun building churches for the poor alongside his canal holdings.[103] About one month later, another newspaper report indicated a group of churchwardens and clergymen in Staffordshire had begun preparing a statement encouraging carriers on the Trent and Mersey Canal to stop their Sunday trade.[104] In January 1840, a letter writer to the *Worcester Journal* (where the River Severn and the Birmingham and Worcester Canal joined) indicated that a memorial signed by 280 boatmen had been sent to the General Committee of the Worcester and Birmingham Canal calling for an end to operations on Sunday.[105] Reporting on the stay of execution allowed Collins' convicted killers in April of 1840, the *Staffordshire Advertiser* wrote,

> We have expressed our conviction, that whatever may be the fate of these unfortunate men, the shocking occurrence which has brought them into their present perilous circumstances, will have the effect of attracting the attention of philanthropists and Christians to the moral and spiritual destitution of the class to which they belonged.[106]

Nonetheless, a simple stop to Sunday trading might mean more difficulties in the eyes of some. The carrying firm of Robins & Co. contended that if only a few firms followed through in ending Sunday trade, canal boat workers would "not only spend them in idleness and drunkenness themselves, but would (if merely for want of occupation) lead others into the commission of the same vices, and compromise at the same time the interest and property of their employers."[107] Moreover, if no religious instruction or preparation of any other occupation for the workers was provided, "The want of it will inevitably increase the amount of immorality and profaneness, not limited to themselves, but spreading its contagion to all in their neighbourhood with whom they may happen to come into contact."[108] In 1840, an "Act to provide for keeping the Peace on Canals and Navigable Rivers" (Canals (Offences) Act) was passed, empowering canal companies, working in cooperation with justices of the peace or watch committees, to appoint constables with the right to patrol canal or river banks within approximately 0.25 miles of the water. The act specified conditions upon which arrests might be made and penalties levied for damage or interference with property.[109]

In the spring, the circumstances of Collins' murder girded both the questions and answers before a House of Lords Select Committee called to investigate whether canal traffic on Sundays ought to be prohibited. In proposing the creation of a Select Committee, Lord Normanby (Constantine Phipps) had argued it was "the first duty of the government to promote the moral and religious observance of the Sabbath-day."[110] Noting the men executed for Collins' murder had been deprived of both religious and secular instruction, he indicated that if appointed he believed the committee would prove "amongst other things, that there was no class of Her Majesty's subjects so illiterate as the persons engaged in this traffic."[111] In later testimony before the committee, representatives of canal companies and canal carrying companies generally indicated their support for an end to industrial traffic on Sundays, noting they thought the majority of boatmen would support it. Testifying before the committee, Sir George Chetwynd, a member of the managing Committee of the Trent and Mersey Canal, indicated he had spoken with Ellis, the reprieved member of the crew on Collins' boat.

> Since he has been in Stafford Gaol he has made such Disclosures as beggar all Description of the Abominations that are committed on Canals, and the Habits

> of Boatmen, their pilfering, and Habits of horrid Depravity. In the course of my communications I asked him if he could tell what he thought would better the Condition of Boatmen, and he made this Reply to me as near as I can recollect: By putting a Stop to the Sunday Trading and those Men on the Line of the Canal.[112]

In the course of his responses, Chetwynd indicated that although some boatmen married, it was not uncommon "for them to hire Females to accompany them on Voyages, say from London to Manchester and back again."[113] The lack of religious guidance was central to the understanding that Sunday trading needed to end.

> I may state with respect to Ellis, in the first place, that he is Twenty-five Years of Age; that he was never in any Place of Worship whatever until he was committed on a Capital Charge of Rape and Murder; he is quite uneducated, not having the least Idea of a God, or a Saviour, or a future State; in short, having no Idea of Religion whatever until he came to Stafford Gaol. I believe there are a number of Boatmen in a similar Situation.[114]

Ellis had received a reprieve and was transported for life, but others who went before the select committee testified to what they believed was the depraved nature of the canal boat workers. In addition to a lack of religious instruction, there were concerns regarding the disruptive effects of boatmen who might go ashore and concerns about the security of cargo. As Chetwynd warned,

> Wherever the Stoppages have been accidental they have been found trespassing and plundering, and the greatest possible Nuisance to the Neighbourhood. In short, when a Stoppage happened near Town we were obliged to swear in Special Constables and the Farmers to watch their Property, in consequence of the Boatmen being stopped, and not going to any Place of Worship.[115]

The testimony provided by Chetwynd and other observers indicated anecdotal observations regarding religion and boatmen's penchant for criminality could vary. For instance, Chetwynd, whose greatest experience lay along the Trent and Mersey, noted boatmen were generally paid by the voyage (and inadequately so). He said, "therefore they are induced to pilfer, and to many things they would not otherwise do."[116] Citing a great competition for employment, many of the boatmen and families who lived and worked on the canals were believed to have no homes or cottages ashore. Boatmen thought of themselves as a degraded "Class of Beings," and would likely "take directly to the Highway, or become Thieves" if expelled from the boating community, Chetwynd said.[117] In his testimony, Thomas Harrison, the superintendent of the shipping department of a concern making use of the Droitwich Canal whose experiences generally lay along the River Severn, indicated most boatmen had homes ashore, particularly in Upton-on-Severn, Tewkesbury, and Gloucester.[118] Those working the Severn, he said, were a more educated and respectable "race" of men than those on the canals, and he expected a day of rest would greatly benefit animals (horses) who were often driven harshly.[119] In his testimony, Francis Twemlow, Chairman of the Quarter Sessions of Staffordshire, who likely had knowledge of the Trent and Mersey as well as a wider range of Midlands' canals, argued he believed the captains of boats were generally of a better character, and those under them were "more doubtful."[120] In his observations, John Crowley, a partner in a carrying firm doing business between Birmingham, Wolverhampton, and London, argued that the men who operated fly-boats, and who had time to spend with families ashore, were of a "superior class" to boatmen who traveled with their families.[121] He also found them less prone to "plunder" cargoes, since such could not be done as easily if there were just one man and his wife.[122] In his testimony, John Davies, rector of St. Clement's, Worcester, offered several distinguishing characteristics of boatmen, indicating those who were able to take the Sabbath off were generally of a better character, and that Trent and Mersey Canal "flatmen" were superior to the "bargemen" on the River Severn.[123] Overall, he noted men working the slow boats were worse in their habits than those on the fly-boats, and that those of Staffordshire were worst of all.[124] He said:

> With regard to those who have not the Privileges of the Sabbath, I consider them in general to be a demoralized Class of Men in the extreme. . . . I consider them a Class of Men quite sui Generis. Their Habits and their whole Lives are detached as it were from those around them.[125]

Even if the men working the Trent and Mersey might be of a "better" condition than those elsewhere, Reverend Frederick Wade, who resided on the Trent and Mersey, argued that "Features, Dress, and every thing marks him as a distinct Being in Society. . . . In my Neighbourhood they intermarry almost universally with each other."[126]

Despite the enthusiasm raised between 1840 and 1842, no bill to prohibit Sunday trading was brought forth. There were some innovative efforts to extend missionary work to the canals, such as the "Episcopal Floating Chapel" to be based near Worcester, for which funds were raised in 1842 from the Society for Promoting Christian Knowledge, as well as from other organizations.[127] In addition, efforts to reach out to boatmen along the canals were continued by the Inland Navigation Society with a series of summer services held in the open air near the River Thames, or on board boats in the river or on canals.[128]

Pressure for legislative action focused on canals appears to have lapsed in the 1840s and 1850s, with little discussion of constabulary efforts, and pressure for a complete end to Sunday trading emerging only from missionaries themselves. Nonetheless, themes of violence, death, and the dangers associated with canals, particularly at night in industrialized regions near Birmingham and Manchester (and often connected with towpaths), continued to dominate stories involving boatmen. Reports of prosecutions for rape, perhaps given added attention due to the lingering memory of Collins' case, were a particular focus. In 1843, two men aged 18 and 20 were sentenced to death for, during the passage through Harecastle Tunnel, the rape of a woman whom they had taken on as an informal passenger.[129] Just a month later, a total of five boatmen working the Grand Junction Canal were charged in the suspected rape of an eighteen-year-old woman walking a short distance along the canal at night.[130]

At the same time that canals' reputations for violence and vice continued, their role in the commerce of the country was increasingly coming into question. They had once symbolized technological prowess. Yet, by the mid-nineteenth century, with the advent of steam railways, few would have spoken so loftily of them. In 1830, there were under 100 miles of railway in Britain, but by 1845 there were 2,200 miles, a good deal of it added in the early 1840s.[131] By the early 1850s, over 200 rail companies were in operation (though a significant amount of traffic was dominated by a handful of powerful companies).[132] New railways could see speeds of 25–30 miles an hour, vastly reducing the time for cross-country trips between London and the cities of Birmingham, Manchester, and Liverpool. Moreover, a number of the dilemmas of canal transport appeared to be overcome by the new railways. Rail advocates noted that stoppages due to winter ice and summer water shortages, delays associated with locks, and ongoing dangers of pilferage all could be ameliorated by a transition from canals to rails.[133] Railway expansion in the second half of the century included additions to mainline routes, as well as a host of rural and suburban lines. Canals, already maligned as spaces of danger, were overshadowed by the grand engineering works of railways and, in particular, the railway stations that dotted the landscape, often a focus of the efforts of town councils interested in reform.[134] While railways were constructed and powered by the new technologies of the age, canals and other river improvements, despite their own engineering significance and disruption of the landscape, might still be considered "artificial rivers" with the power upon them provided by human and animal muscle. Britain's mid-century culture was significantly affected by rail through an intensified intrusion upon property rights, massive dislocations of earth (prompting Victorians to grapple with mounting geologic evidence of the earth's past) and the emergence of an architectural infrastructure made manifest in railway stations and warehouses. Indeed, the transformations brought by rail appeared to overwhelm the descriptions of change brought by canals and would later inform the arguments of those who either yearned for, or sought to protect, the role of canals. As an account of rail travel in *Buck's Gazette* noted,

> The projections, or transits of the train through the tunnels or arches, are very electrifying. The deafening peal of thunder, the sudden immersion in gloom, and the dash of reverberated sounds in confined space, combined to produce a momentary shudder, or idea of destruction—a thrill of annihilation, which is instantly dispelled on emerging into cheerful light.[135]

As the age of the railway dawned, there were hints the story of canals and the populations living upon them would come to occupy a unique place within the wider, and rapidly changing, technological environment of modern Britain. Some of these hints would come with the emergence of yearnings for a world in which society was defined by local environs, and labor and industrial production driven by human and animal power. William Howitt's *The Rural Life of England* had in 1840 lovingly depicted the pleasant landscapes surrounding English country houses, the deeply entrenched traditions of sport, the meaningful rhythms of labor on farms, the pastimes of common people, and the traditions of the Romani ("Gypsies"/ "gypsies") who were "part and parcel of the landscape scenery of England."[136] A further powerful stroke proclaiming protection of the landscape came in October of 1844 with William Wordsworth's sonnet, "Kendal & Windemere Railway," published in the *Carlisle Journal* decrying plans for the extension of a railway to the Lake District, which Wordsworth believed would bring vast numbers of travelers from industrial centers that had been shaped by a "false utilitarian lure."[137] Yet, whereas Howitt had integrated changes in the countryside—particularly that of scientific farming—into his historical understanding of the ennobling qualities of agriculture, the arrival of rail represented a momentous shift. Wordsworth's sonnet had begun, "Is there no nook of English ground secure/From rash assault?"[138] His charge received a quick response from the *Morning Chronicle* decrying the elitism that seemed apparent in his writing and deriding the Poet Laureate's position, which seemed to have appropriated the Lake District as a reserve for the poet and ignored the degree to which railways were contributing to the nation's coffers and strengthening the prosperity of the empire. Wordsworth penned two further letters to the *Morning Post* in which he argued that the features of the Lake District, "cannot, in their finer relations to the human mind, be comprehended, or even very imperfectly conceived, without processes of culture or opportunities of observation in some degree habitual."[139] Wordsworth's opponents were ultimately successful in their plans for the railway, but his struggle foreshadowed the ways in which a democratization of travel brought by railways underlay controversies over access to, and preservation of, non-industrial areas. It was a struggle that was to become more intense in the late nineteenth and early twentieth centuries and see the first stages of canals' integration into the movement to preserve the countryside.

In the mid-nineteenth century, few understandings of canals as part of a cultural world being surpassed by a steady acceleration of technology had yet emerged, and there were perhaps fewer accounts that included boat people within an idyllic rural setting. However, in an 1860 compilation of essays published in *Odd Journeys In and Out of London*, the writer (and later theater manager) John Hollingshead caught first-hand the changing dynamics of local communities, technology, transport, and livelihood. His series of twenty-two mostly brief essays covered a myriad of travel experiences ranging from a brief visit to a military encampment near London to the building of a new telegraph system in London to a first-hand account of the journey of a postal package from the point of view of the package itself. Yet, among the most lengthy and defining of the essays were the first, entitled "On the Canal," chronicling a journey on a Grand Junction canal boat, and a second entitled, "Riding the Whirlwind," based upon a ride in the engine compartment of a steam train. In addition, the final essay, "The Last Stage-Coach," detailed Hollingshead's first-hand experiences of technological change by chronicling how his regular visits to a small village approximately 40 miles from London were altered by the advance of railway construction and the decline of a once-prominent local coaching business. Joining a Grand Junction Canal fly-boat on a journey from London in "On the Canal," Hollingshead emphasized what might have seemed to others a foreign journey, noting his London cabman appeared "astonished" by his requested destination of a barge wharf so that he could begin his journey. Later, on the first leg of his journey, he noted, there were those in London utterly ignorant of the "black and barge-loaded Styx" that flowed beneath them and on which he traveled.[140] He commented on the slow speed of the journey, which was governed by the horse pulling the boat, and marked by the frequent need to traverse locks. "Our pace is very slow . . . for we have gone at one bound a century back in the history of conveyance, and must be satisfied with an uniform and almost imperceptible rate of from two to two and a half miles an hour."[141] Yet,

> Melancholy pictures of drunken brawls, improper language, constant fights, danger to life and property, hordes of licensed ruffians beyond the pale of law and order, . . . all proved false before the experience of a few hours. . . . We were inmates of a new home and friends of a new family, whose members were

> honest, industrious, simple, and natural—too independent to stoop to the meanness of masquerading in adopted habits and manners, with a view of misleading the judgment of their guests.[142]

In the course of his journey, Hollingshead began to juxtapose images of a peaceful rural world with images of a rugged boat population. He wrote, "The further we went, the more did our long-cherished notions of the dry, utilitarian character of canals disappear, to give place to a feeling of admiration for the picturesque beauty of the country, and the artificial river."[143]

Hollingshead carefully delineated the workings of the four-man canal boat. Among the highlights were the role, characteristics, and routine of Captain Randle (likely selected by the company to captain the trip) who had been on the canals for 50 years, had once worked for Pickfords, had a rarely visited home on shore, and who had a contract for the voyage from the company but was in charge of paying his own crew. Hollingshead captured Randle's dialect, noted the distinctive (but not necessarily unique) clothing and accoutrement associated with canal travel, and found Randle to be very conscientious about the boat, though perhaps somewhat naïve about the affairs of the world, likely born of a lack of literacy, Hollingshead concluded. Randle's son was, "a light-eyed, full blooded, red-cheeked, good-tempered, clean-looking young man of twenty-three."[144] Most of the countryside through which they passed was peaceful, with Hollingshead remarking on fishermen he saw who may have come from London and been deposited near the canal for a day out.[145] "Many of the lock-houses are very pretty. All of them are neat and clean. . . . At all the lock-houses, coy little gardens peep out, and many of them are profusely decorated with flowers both inside and outside."[146] Though Hollingshead's experience on the water appeared peaceful, it was likely curated by the Grand Junction Company.[147] Conditions on the family boats they passed seemed different, and Hollingshead remarked on the danger of injury and death that seemed to be present upon them:

> There is the boatman, and his wife, a stout, sunburnt woman, and children, varying in number from two to ten, and in ages from three weeks to twelve years. The youngest of these helpless little ones, dirty, ragged, and stunted in growth, are confined to the close recesses of the cabin (the tarpaulin-covered part of the boat is inaccessible to children), stuck round the bed, like images upon a shelf; sitting upon the cabin-seat; standing in pans and tubs; rolling helplessly upon the floor, within a few inches of a fierce fire and a steaming kettle; leaning over the edge of the boat in the little passage between the cabin-doorway and the tiller-platform, with their bodies nearly in the water.[148]

In approximately the last one-third of his account, Hollingshead's narrative took a significantly darker turn as the boat made its way through the hub of boating life at Braunston (near the Grand Junction's intersection with the Oxford Canal) and entered the industrial regions of the Midlands on its way toward Birmingham. In an extended essay of approximately sixty-five pages, Hollingshead reserved only the final two pages for what may have been over 30 miles between Warwick and Birmingham. In this distance, he wrote that he and his companions left behind the vestiges of rural life where Shakespeare may once have walked, to travel:

> Past black, smoke-grimed, town-stamped boys, angling in the canal; past groups of ill-favoured, smoking, half-drunken men-youths—a mixture of the factory-hand, the dog-fancier, and the fighting-man; within sight of high viaducts, over which fly the hard-breathing engines—and clattering trains which have passed and repassed us on each side a hundred times during our slow journey; past a coal-dust-looking towing path, and under a sky of smoke; past tall chimneys and dingy gas-works; down another stair case of black locks, opened by a poor, active, grimy girl-child not more than ten years of age; past groups of demon-looking men who grin at us with white teeth, from coal-heaps, and the white, roaring mouths of furnaces; past backs of myriad-windowed factories, whose glass is broken, and under whose walls lie green sickly pools of stagnant water; past a dozen grimy boys with large set jaws and shrunken arms, and legs, bathing amongst the floating dead dogs and factory scum of the inky canal; past all this and more, we leave the Romance and beauty of our three days' journey far, in idea, behind us, to glide to our final destination.[149]

Hollingshead was shocked by the transition to the dirt and toil of a new industrial age as he passed Warwick; grappling with the transport technology of the era formed the theme of at least three more of the vignettes he sketched. His placement of the essay "Riding the Whirlwind" just after his recounting of the journey on the Grand Junction offered a quick shift from the age of canals to the age of rail. Opening this second essay with a brief look at what might surround each class of passenger elsewhere on the train, he indicated that his abode for the evening was the engine, where Tom Jones and John Jones, respectively, were driver and stoker. Although not mentioned, Hollingshead's journey was likely on the South Eastern Railway, as his trip appears to have been a nighttime mail route from London Bridge to Dover. Immediately, the contrast with the canal journey was central,

> We dart across the country—between high banks through valleys of chalk and sand—past trees—past roadside houses lighted up with the fires of a November night—starting away from twinkling villages like a skittish horse, or rushing madly across the quiet street with a roar and a whirlwind. . . . Before me is the

> shining brass, and steel, and iron of the engine, . . . and the partly opened doors of two glowing, ever-craving ovens—the bowels of our steed whose fiery hunger John Jones, the stoker, is constantly trying to satisfy with coke.[150]

The age of rail was the age of the masses, sparking a new relationship between the men of the engine and their charges in the cars that followed them. He wrote,

> The unwavering faith of the public in Tom Jones is something more than wonderful. . . . That where the swerving of a finger's breadth, or the carelessness of an instant, would send the whole precious freight to utter destruction, he will steer safely through all difficulties, and punctually deliver his charge at the appointed place at the appointed time.[151]

The exhilaration of the rail journey was intense, but the trip passed quickly and Hollinghead's narrative was not as long as that of the journey by canal. Much more extended, and the last essay in the collection, was "The Last Stage Coach." This contribution followed Hollingshead's changing impressions of repeated journeys to a town outside London where he had come to know not only the fishing nearby but the business of its principal coach operator—a man who found his way of life undone by the arrival of the rails.

> I continued my trips, as usual; . . . As if I had been the fine old English gentleman who lived in the olden time. . . . The horses were not so rampant; the turnpike keepers were not so watchful nor so obedient; the ostlers were not so numerous, and those who were left were not so admiring and so respectful; the guard had gone, and the coachman put on the drag himself, when we went down hill.[152]

Near the end of the 1860s, an anonymously published work chronicling the adventures of three travelers (and a cook) by canoe and skiff from Manchester to London also captured the contrast between the industrial Midlands, where canals saw greater continuing use, and the more sedate waterways of southern England, where decreases in industrial transport were more apparent. The travelers' journey began in Manchester, negotiated the rivers Mersey and Irwell, continued on the Ellesmere Canal (portions of which are today the Shropshire Union and Llangollen canals), navigated down the rivers Perry and Severn, and then later took the Thames and Severn eastward toward Oxford and then London. The travelers' narrator noted,

> Let those who have walked, ridden, and driven all through its beauties, start on a voyage, and they will find an endless succession of new scenes and landscapes on its [England's and Wales's] rivers; the most beautiful of which can only be reached in a canoe.[153]

An understanding that the railway was to be avoided was clear, as the narrator said,

> Having earned a holiday by a twelvemonth of hard work, we had resolved, instead of being whirled by steam, pistons, and wheels, on the Ironway to London at fifty miles an hour, to explore, with sail, paddle, and oar, the yet untried Waterway.[154]

The excursion proved to be less of a holiday than was expected as rain and mud accompanied the boats in their early stages. The narrator remarked that though the Ellesmere Canal was, "one of the most beautiful in the kingdom, . . . still it is a canal, and . . . rejoices in a number of locks."[155] The travelers did, however, find a number of boatmen accommodating and once turning on to the Perry and then joining the Severn, found something of the peaceful scenery they were looking for. "What a contrast to the junction of the Irwell and Mersey. One, the union of a dirty, sluggish stream with a crawling mass of ink; the other the confluence of a clear, rapid rivulet with a swift and bright river."[156] Later, as they traveled on the Thames and Severn Canal, which had already been affected by the loss of traffic to the rails, the narrator contended,

> The Thames and Severn canal is one of the most beautiful we ever were on, and the water is the purest I have seen. The canal-boats were conspicuous only by their absence, the traffic for all heavy boats being entirely stopped by the immense growth of the American canal weed, . . . We passed miles and miles of high banks of it that had been dragged out.[157]

Though the canal was bereft of traffic, the travelers still contended with a lockkeeper (who demanded to see a pass), and several locks, on their way to London.

The accounts of Hollingshed and the canoeists may have been among the earliest signals of change in cultural understandings of canals. Hollingshed's voyage began in the industrial heart of London, where he noted the first part of his journey began on a route that passed beneath the city's tumult of accelerating industrial development. The operations of the canal attracted little attention, though Hollingshead had clearly heard popular stories of a rough and violent world before departing. Soon, and particularly as the canal entered the countryside, the built environment of the waterway lost its menacing appearance and those who worked and lived near its waters largely came to embody a conscientious and independent labor force toiling in peaceful surroundings. Entering the industrial Midlands, the voyage once again turned dark, with Hollingshead's account providing little beyond a depiction of the grim visages of local inhabitants. The world of canals had thus been bifurcated, with rural waterways often associated with the traditions of the countryside and those making their way through the industrial regions often associated with the pollution and detritus of the industrial world. Eventually, it was upon the less industrialized southern canals that the canoe travelers found the peace they claimed to have been seeking. Such a peace, however, was not necessarily defined by an escape into nature. It was, rather, a peace discovered within re-naturalizing areas as industrial traffic departed on the Thames and Severn. Both the ambivalence regarding a system of canals that traversed both rural and urban areas and a renewal of peace in a post-industrial environment were to be themes significant to the story of canals in the late twentieth century.

It was the industrial utility of canals that was still the dominant theme of the late nineteenth century, however, and here economic fortunes were shifting. Although some already successful canal companies were prompted to lower tolls by the mid-nineteenth century in response to concerns regarding competition and financial returns, in many cases, loads of heavy materials such as coal and limestone continued to increase. The degree to which the structure of labor and the presence of families on canal boats was affected by the rapid expansion of railways has been the subject of debate. In 1875, a factory inspector, Charles R. Bowling, wrote that in the mid-1840s, few women or children lived and worked on canal boats, but that by the mid-1870s more did, as rail competition increased.[158] Among contemporary historians, Hanson has identified an overall decline in the number of boatmen (perhaps in response to wage concerns), an increase in the percentage of older workers, and a reduction in the number of fly-boats as a result of rail competition.[159] However, he has argued that, based upon testimony before the Select Committee on Sunday Trading, the presence of women on board canal boats was already well-established by the early 1840s and was not necessarily an effect of boatmen's efforts to reduce their costs.[160] Reviving Bowling's observations, Hadfield has argued that in the 1840s women and families were rare, increasing only later on in the century.[161] Despite these contrasting positions, numbers of women and children were a consistent source of concern in the late nineteenth century due to anxieties regarding health, sanitation, and morality in an increasingly regulatory and reform-minded environment. Census returns indicate a total of 12,562 people sleeping on board boats in England and Wales on census night in 1851, a number that appears to have diminished to under 7,000 by the end of the century. Among these were 10,059 males and 2,503 females in 1851, and 4,916 males and 1,953 females in 1901.[162] Among families afloat, life on the water may have alternated with some months living ashore with shifting seasonal and employment conditions. In this, Hanson has emphasized boat families made decisions regarding accommodations based on fluctuating costs of living ashore, rather than declining fortunes as a result of rail competition.[163]

The place of women and children aboard canal boats came to be of particular concern with the arrival of a new age of reform by the 1870s, which saw a significant expansion of the state not only in association with places of employment but in domestic housing as well. In 1873, a newly reformist Town Council in Birmingham (a city at the center of canal activity) selected one of the leading supporters of municipal authority, Joseph Chamberlain, as mayor. A successful businessman, Chamberlain turned to the power of the state to address the cramped and insanitary conditions of Birmingham's inner wards. First, a buyout of private gas companies allowed him to provide the city with a new source of revenue and a better-administered and cheaper gas supply; and support for the city's Sanitary Committee meant new systems for the treatment of sewage were initiated. Second, came a buyout of private water works and significant use of reformist parliamentary legislation—the Artisans' and Labourers' Dwellings Improvement Act of 1875 (Cross Act)—to begin buying and transforming an over-crowded section of the city.[164] In remarks before an 1875 conference of municipal officials, Chamberlain stated that not only did local governing bodies have a duty to care for the lives and health of the people entrusted to their charge, but they had a duty "to consider also their happiness and their morality."[165] For mid-to-late-Victorians, and in particular for those who had followed the reforming ideal of the "civic gospel," there was nothing so important as transforming the built environment of the city to improve both the physical health and moral habits of its inhabitants.[166] Crime, domestic violence, and intellectual stagnation were, just as much as disease, assumed to be the end result of polluted water, crowded and dirty homes, and poorly ventilated streets.[167] As Chamberlain pointed out:

> What folly it is to talk about the moral and intellectual elevation of the masses when the conditions of life are such as to render elevation impossible! What can the schoolmaster or the minister of religion do, when the influences of home undo all he does? We find bad air, polluted water, crowded and filthy homes, and ill-ventilated courts everywhere prevailing in the midst of our boasted wealth, luxury, and civilization . . . and then, when these people whom we have suffered to grow up like beasts behave like brutes, we rush to the Home Secretary, Mr. Cross, in a blind paroxysm of terror, and ask him to give us the humanizing influence of "the lash," in order to repress the instincts which our neglect and indifference have allowed to develop.[168]

Chamberlain soon helped to establish principles associated with the late nineteenth-century radical Liberal tradition at a national level. In this context, efforts to address living and working conditions on canal boats were spearheaded nearly single-handedly by the reformer George Smith ("of

Coalville"). Working-class in origin, Smith had taken on conditions for brickmakers among whom he labored and eventually gained a managerial position. By 1873, he turned to conditions on canals with a massive letter-writing campaign calling attention to what he believed was a debased population afloat and arguing, as he had in the case of the brickyards, for greater control of the conditions under which children labored. Already noting the degree to which the canals remained foreign to the daily experiences of many, his nineteenth-century biographer E .N. Hoare argued, "These poor boaters had, in great measure, dropped out of sight. They did not throng the railway stations and were not encountered on the highways. They had to be sought for and "looked up."[169] Smith's characterizations contained some of the most striking descriptions of immorality and filth that were attached to the boat population in the nineteenth century. On the surface of Britain's canal system, he wrote, there were over 25,000 boats carrying those,

> of all ages, together with filth, mud, manure, and the refuse of our large towns. Some of the [boat] cabins are models of neatness, and a man and two youths might pass a few nights in such very comfortably. Others are the most filthy holes imaginable, what with bugs and other vermin creeping up the sides, . . . Fathers, mothers, sisters and brothers, sleep in the same bed, and at the same time. In these places girls of 17 have given birth to children, the fathers of which are members of their own family.[170]

He added, "Let us take timely warning, and rescue these miserable women and children from the physical degradation, moral pollution, and unhealthy atmosphere they have to endure in those hot, damp, close, stuffy, buggy, filthy, and stinking holes, commonly called boat cabins."[171]

During the course of 1874, Smith embarked upon a campaign of letter-writing and a round of speaking engagements on behalf of the canal population that brought him both plaudits and criticism. He likely well overstated the number of boats in service, which likely barely topped 10,000, thus adding urgency to the crisis he had identified.[172] His perspective was challenged (as were his later assumptions regarding the unhealthiness of boat cabins) by those who emphasized the outdoor experiences of children on the canals. As one commentator wrote of boat people after listening to one of Smith's addresses, "Although we may regret this overcrowding on the score of morality and decency, yet we cannot help feeling that these children, as far as bodily health is concerned, are substantially far better off than the pale-faced denizens of city alleys."[173] Nonetheless, in coming years, Smith maintained what was virtually a one-man campaign regarding canals focused upon not only letter-writing but efforts to corner influential politicians. His goal was to bring the canal populations within the purview of one or more state regulatory efforts. Initially, he cited the Factories and Workshops or Sanitary acts, which he believed would prompt a significant change in the physical and moral condition of children on the boats. However, he soon turned to the Elementary Education Act of 1870 (amended in 1873), finding within its provisions for local school boards and the education of children between the ages of five and twelve, another route to compulsory state intervention in the lives of canal boat children. The education legislation was based upon a resident population, a situation confounded by the movement of the boat population and the reluctance of local authorities to make provisions for them. Yet, Smith argued, "I would suggest that the Government build the necessary schools and appoint masters to those districts where there are large numbers of boat children lying about waiting for the boats to be loaded, unloaded, repaired, or detained in other ways."[174]

The stakes were high, and Smith's characterization of the boat population stark. He wrote that canals running through large towns were nothing other than "cesspools of abomination and filth, black and muddy," adding, "Carcasses of dead animals float on the surface, from which a horrible stench proceeds, and bubbles of gas rise to the top from the decomposed animal and vegetable matter lying at the bottom. It is out of this water—if it can be called water—that these boat people have to wash and cook. Surely something needs to be done for this ignorant, neglected, but useful class."[175] As he wrote about pollution, Smith may well have had in mind canals where the saturation of the shoreline with polluting industries and dwellings, and an insufficient water supply to dissipate the pollutants, had created dire conditions. As the third report of the Royal Commission on Rivers Pollution noted in 1867:

> At the time of our inquiry Bradford Beck was the source of supply to the Bradford Canal, the fluid in which became so corrupt in summer that large volumes of inflammable gases were given off, and although it has usually been considered an impossible feat to "set the Thames on fire" it was found practicable to set the Bradford Canal on fire, as this at times formed part of the amusement of boys in the neighbourhood. They struck a match placed on the end of a stick, reached over and set the canal on fire, the flame rising six feet and running along the water for many yards, like a will-o-the-wisp; canal boats have been so enveloped in flame as to frighten persons on board.[176]

As Smith wrote, attempts to reduce sewage and other pollutants in canal waters took a step forward with the Public Health Act of 1875 prohibiting local authorities from discharging raw sewage or "filthy water" into either a canal or other body without treatment.[177] Canals were included in the provisions of the Rivers Pollution Prevention Act of 1876, which were designed to block new sources of pollution, guard against obstructions in water flow, and promote the best possible means of removing pollutants.[178]

Smith connected the boat people, suffering some of the worst effects of pollution, with Britain's Romani ("Gypsy"/"gypsy") population, some of whom maintained an itinerant existence and were regarded as a mysterious social caste whose morals and lifestyle remained degraded. In 1875, he quoted a newspaper article that said Britons had made "neighbors" of black South Africans, Malays, and even "cannibal islanders" in the South Pacific, but that

> those who live at our very doors, whom we pass and repass in our town and country walks, the inland navigators of our artificial rivers—our canal fellow-citizens—have been as the gipsey class, looked at by us in our pride, haughtiness and pharisaical self-content.[179]

He was at the very least skeptical of missionaries, noting he was shocked when he visited the Boatmen's Institution in Paddington Basin, London, and found only two to three children present as dozens of boats were tied up in the basin.[180] (Records of missionary activities in Paddington Basin are somewhat inconsistent, but this may have been a Ragged School established as early as 1848 in Paddington under the auspices of the London City Mission, a missionary society that had been founded in 1835.[181]) Smith's descriptions of canal boat life emphasized debauchery and yet seemed to express a mournful hope that the lives of those on canals could be reclaimed.

> I have frequently seen women in a half-nude state washing over the sides of the boat as it was moving along, out of the water of the canal, upon the top of which has been floating all manner of filth. . . . Swearing and drunkenness seem to be the two ruling passions. . . . Swearing is taught the children before anything else; . . . A respectable boater told me that there were boaters who read the Bible, and had family prayers daily. I have reason to believe it is true in some instances, and I could wish, with all my heart, that they were more numerous.[182]

The year 1875 was pivotal in establishing the context for the direct regulation of canal boats. First, the Public Health Act established rural and urban sanitary authorities, as well as local districts under their control. These local authorities were to provide clean water, regulate new construction in the interests of sanitation, and provide for inspection where "nuisances" threatened public health. The Local Government Board (LGB) was also given power to establish regulations that would prevent the spread of disease.[183] Central to the regulation of accommodations was inspection. Second, in 1875, the question of women and children on canal boats was taken up by the Factory and Workshops Acts Commission. Testimony before the Commission by inspectors and sub-inspectors who worked under the Factory Acts varied, with observations emerging on the significance that time outdoors played in balancing the close confinement of cabins to the limited education most canal boat children received as a result of migration. Effective instruction under the education acts, in the eyes of those who testified, did not seem possible.[184] The Commission's 1876 report indicated that it was "impossible" to place canal boat workers under the factory law and that it was "impossible" to adequately provide for education as long as women and children resided on the boats.[185]

> Nothing short of a prohibition of the residence of children in the boats appears to us adequate to the evil. Apart from the impossibility of the children obtaining education, there are mischiefs arising from the crowding together of whole families into a small cabin, and the withdrawal of the inmates from all religious and social influences, which appear to us sufficient to justify a strong measure.[186]

Citing the recent conviction of a canal-boat couple for manslaughter in the death of a seven-year-old girl living with them, the commissioners noted that such a dire experience was far from the norm, but that the current circumstances of canal boat life could help to bring about such events. It concluded,

> We recommend that the residence in canal boats of female young persons, and of children above the age of three years, should be forbidden. The operation of this law, we cannot doubt, will have the effect of withdrawing from migratory life the mother of the family, after the first year or two of married life; and will also rescue numbers of young girls from a very demoralizing condition.[187]

Likely seeking to highlight what he saw as the inadequacy of the missionary effort, Smith noted in a paper delivered before the Social Science Congress in 1876 his contention that there were perhaps 13,000 women on the canals living in an unmarried state and perhaps 40,000 illegitimate children.[188] He would later be taken to task for his estimates and the uncertainty which surrounded them, but they had their effect.[189] A sympathetic but skeptical nineteenth-century biographer noted,

> Life on the canals was a subject that made good 'copy,' . . . Everyone knew of the woman who, for twenty years, had never once slept in a house on land; and of another who had given birth to twenty-one children in the same cabin; of the little girl, wretched-looking, ill-clad, badly shod, who, while her drunken father smoked and steered, trotted along the towing-path, driving a couple of donkeys, until she staggered and fell, utterly tired out,—and a hundred other narratives of a similar type.[190]

The reinforcement of cultural stereotypes and marginalization soon became apparent in a campaign Smith undertook with regard to the Romani ("Gypsy"/

"gypsy") community. In August of 1879, he launched a campaign to combat dire conditions he believed he saw in mobile encampments.[191] The language in which he described their housing was similar to that in his campaign on the canals. In March 1880, he wrote,

> Not much regard is paid to the separation of the sexes in either the tents or vans. In the case of grown-up children, sleeping in the same van as father and mother, they generally sleep under the parents' bed at one end of the van. . . . Fully two-thirds of the gipsies living together as man and wife are unmarried.[192]

Critically, the campaigns affecting boat people and the Romani appear to have been drawn together. At the Social Science Congress in 1880, Smith offered two papers, one focusing on Romani children in mobile dwellings and one focusing on the canals.[193] In the minds of some, stories of the boat people and the Romani, already linked through concerns of criminality and mobility, were likely fused in racial identification. A clue to this may be found in the journal *The Field,* which in 1879 offered an article delineating differences between the Romani, a distinctive and foreign "race," and the boating population, which was composed of varying English "breeds."[194] The course and construction of the article in *The Field* suggest it was taking part in a wider debate that questioned the racial origins of the boat population. That same year, the moralizing writer L. T. Meade focused upon canal boat children in her work, *Water Gipsies, or the Adventures of Tag, Rag, and Bobtail*, highlighting the moral ignorance of the canal boat population, which (though noting what might be gained in the open air) condemned any romanticized view of them:

> The canal boat people are a distinct race, a race almost unknown. They have no parish, are reckoned in no "cure of souls." The overseer and the tax-collector are people they have never heard of; no board reports upon the sanitary condition of the boats, and no educational inspector takes the least interest in the children. Dwelling in their boats for generations, mingling exclusively with their own class, a more ignorant set of people can scarcely be imagined. Their ideas with regard to diet and disease are painfully old-fashioned; as to religion, few indeed of them have any. The children are unbaptized, and all good is literally shut away from them. In many respects they resemble the gipsies, being akin to them in their habits of thought and life, and like them enjoying the open air, the country scenes, the constant change; but they are decidedly more ignorant and less intelligent.[195]

Initially, Smith's efforts seemed crowned with success as the Canal Boats Act of 1877 assigned to the newly created LGB administrative responsibility for making regulations concerning canal boats used as dwellings. By the terms of the act, the numbers, ages, and sexes of those living on board the boats could be regulated, as well as the amount of space required in cabins, including provision for separation of the sexes. The LGB's regulations were also to address cleanliness and the "habitable condition" of the boats, as well as the prevention of the spread of infectious diseases by boats. The act allowed for the creation of local registration authorities which would issue certificates to boat owners indicating the provisions under which particular boats operated.[196] The act indicated that where a local sanitary authority was informed there was a person on a canal boat who was suffering from an infectious disorder, the local authority could prevent the disorder from spreading through actions such as removing an individual from that location and exercising all other powers of the Public Health Act of 1875. A boat could be detained long enough for cleansing and disinfection.[197] The act indicated that children aboard canal boats, for the purposes of the Elementary Education Acts of 1870, 1873, and 1876, were to be considered residents where their boat was registered. However, so long as a parent(s) satisfied the local school board that a child was under effective instruction outside the local authority, that local school board was to issue a waiver indicating the child was being accommodated by another school board.[198] The Canal Boats Act of 1877 was, however, permissive legislation that did not require local authorities to actually begin the process of boat registration (and thus enforce subsequent provisions of the act), and did little to force the LGB to move the process along. A lack of action by local authorities soon angered Smith. In his eyes, to a national problem, only a local and inadequate solution had been provided.[199] By 1880 at the Social Science Congress, he concluded of recent canal legislation, "It was manifest to the most careless observer that it was the deliberate intention of most of the authorities to allow the Act to become a dead letter."[200]

Smith finally saw the Canal Boats (Amendment) Act passed in 1884, which indicated that all registration and sanitary authorities through which a canal flowed had a duty to enforce the provisions of the act, as well as any ensuing regulations of the LGB.[201] Certificates of registration issued by local authorities recorded the owners of boats, their locations for purposes of the Elementary Education acts, the maximum number of those allowed on board (as well as statements regarding the separation of the sexes), and the amount of airspace assigned to each of the occupants.[202] Regular reports to the LGB were required under the act. John Brydone, who was appointed canal boats inspector—a position whose primary purpose was the compilation of boat registration and inspection records in sanitary districts along the canals—took up his responsibilities with enthusiasm.[203] Visiting nearly ninety-four districts in his first year on the job, he began prompting local inspectors to better fulfill and record their duties of inspection.[204] Those who had initially taken up the job of inspection after 1877 were often adding to duties they already carried and sometimes faltered and could provide little data. The challenges of fulfilling inspection obligations on the canals were many. Although all boats being used as dwellings were required to be registered in one of over 100 districts, few remained within the districts where they had been registered through their working lives. Outside of a registration district, there was no formal mechanism for removing them from the books if they became disused or were broken up. Thus, local inspections did little to clarify how many boats were on the waters, as enforcement did not necessarily mean the stopping and inspection of all boats, nor did a common database of all boats and their records of inspections exist. Enforcement in multiple jurisdictions meant that some boats might be stopped multiple times, while others not at all. From the 105 registration authorities providing information in 1886, there were reports of 9,068 boats on record as dwellings.[205] Brydone, however, felt compelled to provide an estimate of perhaps 7,000 to 8,000 boats in actual operation.[206] Enforcement of fines remained problematic as authorities often chose to give warnings before a fine was levied but had little control over a boat once it departed.

Yet, as reporting from local districts improved, a somewhat more nuanced view of the lives of the canal boat population did emerge in government reports, despite the crescendo in public organizing in the early 1880s against prevailing conditions for children in society.[207] This view in government reports contradicted the assumption that canal boat families were in either poor health or the natural transmitters of contagious diseases. As Brydone noted in his initial 1884–1885 report, "Exposed continually to all sorts of weather they grow up a hardy race, and many of them I have met with have not known a day's sickness in their lives."[208] Although he went on to report that he urged every canal boat inspector to use the utmost vigilance in reporting any outbreak of disease to local medical authorities, his reports over the next several years were notable for their only incidental need to chronicle outbreaks. As he wrote in his 1888–1889 report,

> The precautions taken generally throughout the country again enable me to record that in no case has a canal boat been found to be the medium of conveying infectious disease from one district to another, and the few cases of disease known to have existed on any boats were dealt with so promptly and satisfactorily as to confine the disease to the occupants of the particular boats in which it originated.[209]

Moreover, the basic construction of canal boats spoke to a level of efficiency and utility that was often masked by reports of the unsanitary conditions in which those aboard lived. For instance, as Brydone wrote in his report of 1885–1886,

> All the boats I have met with have been properly furnished or fitted with lockers, cupboards, and shelves of suitable construction and adequate capacity; and a stranger might well be amazed at the ingenuity displayed by the boat builder in these arrangements. The utmost is made of every available space, and where the occupants of the boat display the qualities of tidiness and order, even the initiated marvel at the quantity of various kinds of stores and utensils which can be stowed away in these little floating dwellings.[210]

This apparent organization and potential for tidiness could be violated in instances in which boats were carrying "foul" or "offensive" cargoes and violated the provision for a second bulkhead to be installed between the cabin and the cargo.[211] This type of vessel, often an older one which had been registered years before taking on "offensive" cargoes, was sometimes controlled by owners who resisted taking the step of providing double bulkheads until it was forced upon them by local officials.

In cataloging and summarizing the many reports of individual inspectors, Brydone emphasized both a steady improvement in the condition of the boating class and a strengthening of trust between inspectors and the boat population. What was often seen as the "indecent herding together of men, women, and children"[212] in narrowboat cabins (which struck at typical Victorian notions of privacy as well as raised the prospect of incest) was of particular concern and had been addressed in LGB regulations in response to the 1884 Act. In particular, girls above the age of twelve were prohibited from occupying boat cabins, unless married to a crewman. In many instances, local inspectors appear to have reported on the cooperation of the heads of canal boat families in either removing unmarried girls from the boats (perhaps sending them into domestic service) or re-organizing cabin space to meet LGB standards. But, the concern about incest was never far from the surface. It was sometimes referred to obliquely, as in the case of a fine levied upon a man from Derby who had been sleeping in the same cabin as his eighteen-year-old daughter, who herself was noted as having an infant aged one month.[213] Or, more directly, in an instance at Kidderminster as a twenty-two-year-old daughter and a twenty-year-old son were identified as sleeping in the same bed.[214] Dilemmas could arise as boat families attempted to grapple with the implications of the separation of the sexes. In one case noted at Beverley, a boat registered for a maximum occupancy of five in the aft cabin and two in the fore cabin encountered difficulty as the father slept in the aft cabin with two daughters aged eighteen and nine while two boys slept in the forward cabin. The father, a widower, kept a house onshore, but emphasized his fears of leaving his eighteen-year-old daughter ashore alone with the volume of passing canal traffic. Ultimately, he appeared to accept what Brydone identified as the "lesser of two evils" as he chose to put his daughters in the fore cabin.[215]

In Brydone's eyes, there was significant improvement in both the application of the Canal Boats acts and the responses of canal boat people by the end of the 1880s. In closing his report for 1887–1888, he noted an increase in the number of recorded infractions but attributed this to better record-keeping among inspectors. He argued that the "whole tone of their social life is being raised."[216] Moreover,

> Their regard for the sanitary condition of their cabin-homes, their greater respect for the observance of the ordinary decencies of habitation, their language, their general behaviour, their improved sobriety, and last, though not least, an increasing desire for the education of their children, all point to a substantial improvement.[217]

Alongside its major European counterparts, Britain had also begun to expand elementary education in the late nineteenth century, providing a framework for what would eventually be a national system of education through the Elementary Education Acts of 1870, 1873, and 1876. By the terms of the Canal Boats Act of 1877, canal boat children were initially identified as resident in the district within which a boat was registered. The Canal Boats Act of 1884 specified that the Education Department was required to report to Parliament on a yearly basis regarding the enforcement of the Education acts with regard to canal boat children.[218] Education Department actions regarding canal boat children were somewhat slow in coming, but by 1887, an agreement had been worked out for Brydone to work with the Education Department in establishing oversight. Here, he seems to have readily taken up the responsibility, establishing a "set of rules, regulations, certificates, attendance books, and general forms" that would be adaptable to children in canal boats or in migratory "gypsy" vans.[219] By the end of the 1880s, he began providing a special supplement to the annual reports of the Committee of Council which oversaw the Education acts. Yet, the rules and regulations he had designed were not implemented due to concern that they went beyond the purview of current legislation.[220] Indeed, education may well have been the most intractable problem in the regulation of canal boats. Parents complained about paying multiple school fees dependent on the districts through which they passed and found some teachers unwilling to accept their children for brief stays. Among teachers, canal boat children were reputed to be unruly. In contrast, there were reports from among the canal boat population that their children could be called names when at school.[221]

Brydone's otherwise generally positive view of conditions on canals was almost entirely contradicted by Smith, who in November of 1892 offered testimony before the Royal Commission on Labour. Smith revived once again his stories of grim conditions on canals and now argued that the continuing victimization of women and children was tied not simply to living conditions but (likely acknowledging the charge of the Royal Commission) to labor, itself. He contended that vast discrepancies existed between children's labor as it was regulated in the factory (and tied to education) and that as it was experienced by itinerant children. He argued emphatically that overall conditions had barely changed for canal children and that they were

> by the apathy of the inspectors and weakness at head-quarters . . . to all appearances to be kept in ignorance and slavery until doomsday. While the children of other working classes are brought under protective, educational and sanitary influences, our canal children, gipsy, and van children, and other little nomads are kept by our national unrighteousness in the gutter.[222]

Central to Smith's argument was that children performing tasks tied to the movement of their families' boats on the canals—driving horses, loading and unloading boats, and winding lock paddles—were in effect performing transport labor that ought to be prohibited. His call for an end to such labor was challenged by Sir Michael Hicks-Beach who both appeared reluctant to see children's involvement in the mechanics of their families' travels as labor which might be fairly regulated, and appeared skeptical of the possibilities of prohibition through inspection.[223] Nonetheless, Smith argued that the number of national inspectors ought to be increased and that it might be possible to periodically station inspectors at critical locks to enforce tightened standards. As his testimony before the commission continued, he seemed to turn to the elimination of women and children on canals altogether and pressed again the now several decades old argument that Sunday trading ought to be ended. He submitted a number of documents to the committee and painted a dire picture of the environment on or near canals, arguing that boatmen died at comparatively young ages—perhaps between forty and fifty. When pressed to provide statistics on the last point, he admitted he had none.[224] For some of his conclusions regarding health, Smith appeared to draw upon a *Joint Report by the Officers of the St. Marylebone, St. Pancras, and Paddington Vestries on the state of the Regent's and Grand Junction Canals in their Districts* which he had submitted as part of his evidence before the commission and which chronicled hazardous biological matter (composed partly of human and animal waste, as well as animal bodies) that was disposed of near the canals.[225] Smith's rhetoric was stark:

> I have no statistics from various parts of the country on which to base my opinion, but I give it from general observation and actual experience that many "dead ends," arms, wharves, or basins near large towns or in low-lying districts are little better than hotbeds of disease and putrefaction. The stench at times arising from these places is fearful, and the wonder to me is that our filthy canals have

> not been the means of spreading more disease and epidemics over the country than they have. . . . The canals round London, and those going through fashionable squares, are in a most dangerous condition, and an epidemic may break out at any moment. At the bottom of the canals in some places London sink or filth has accumulated to the depth of eight feet, and large bubbles of gas may be seen continually rising to the top of the black, thick, heavily laden waters, sending forth a stench almost unbearable. . . . And it is upon this London filth, and the filth of other towns, surrounded by dead cats and dogs and other rotting social pets, that our canal boatmen, with their wives and children, live and die.[226]

Brydone did not testify before the commission as a witness, but contributed his voice in a letter to Sir Hugh Owen, permanent secretary of the LGB, on July 1, 1893. Smith's evidence, he argued, was

> too inaccurate, exaggerated and contradictory to demand serious attention, the testimony of men actually engaged in the work is most encouraging, and forms an almost unanimous concensus [*sic*] of opinion that Boatmen and Boatwomen as a class, and the cabins of the Boats as homes, are in advance of many of the working classes and their homes on shore.[227]

He argued that there were in fact now few children of school age traveling with canal boats and that overall health conditions were good. By 1893, he indicated that there were just under 3,800 children on canal boats.[228] The irritation Smith presented to Brydone would eventually be brought to a close with Smith's death in 1895. For his part, Brydone had little time, as well, before his own death in 1899, but seems to have devoted significant time in his last few years to getting an accurate count of the number of boats actually in operation on the canals, which he estimated at 7,820 in 1897.[229]

As Smith's death brought to an end his ongoing campaign against canal boat conditions, so too did it mark an end to his perhaps more powerful association with Romani affairs. Like boat people, Romani and other landed "travelers" had weathered suspicions of crime and vice as well as evangelist efforts aimed at bringing them within the pale of settled, Christian society. While some missionaries had at first cast Britain's Romani as savage and degraded social parasites, others had been hopeful that missionary and educational efforts, as well as a focus on settlement, would fundamentally change their position. There were differences in the ways the populations were viewed. In the case of the Romani, mobility was understood to be associated with ethnicity and culture, much less so was it an effect of working conditions. Efforts at reform eventually focused on educating children to embrace a different set of labor practices, as well as limiting their own mobility. Among boatmen, the focus came to be on reforming the male working population and then reducing and removing the presence of women and families, not an end to the movement of the individual male worker. Nonetheless, David Mayall, assessing ways in which a perceived social and moral condition brought on by Romani "nomadism" was problematic for Victorians, has pointed to the ways in which a "nomadic" state (if not a similarly understood ethnic, racial, or cultural "instinct") likely helped define the boat population from the outside. This, despite the fact that an end to movement, integral to the operation of the boats, could not have been pressed by reformers.

> Travelling was the chief cause of the problem [for the Romani]. It resulted, according to some interpretations, from the nomadic instincts of the Gypsy tribe and was a cause of poverty, vice and degradation. By a process of re-education the nomadic instinct could be curbed and the evils resultant upon a travelling existence would be destroyed.[230]

Britain's canal boat population had not yet had any equivalent of the "Gypsy Lore Society," which, founded in 1888, sought to protect what it argued was a noble set of Romani traditions from the regimentation and oppression of contemporary industrial society and state supervision. Nonetheless, the "Gypsy Lore Society" did not disavow missionary activity or any attempted quashing of perceived problems brought on by itinerancy. Rather, it argued, distinctions ought to be made between those who remained of "pure-blood" Romani background (whose traditions ought to be preserved) and those of itinerants who maintained only partial and corrupted traditions.[231] For most boatmen, an endorsement of their traditions as part of a unique culture worthy of preservation would have to await the mid-twentieth century.

Yet, both Romani "travellers" and canal boat families had seen the regulatory contexts of their lives intertwined in the late nineteenth century. Smith's success with the Canal Boats Amendment Act of 1884 had brought the support of a "George Smith Fund" organized by the *Pall Mall Gazette*, and a total of £1480 from the Royal Bounty Fund.[232] With renewed confidence, he had returned to the Romani population in the late 1880s, despite his 1892 testimony regarding the boat population. In Romani vans (most often covered wagons) he identified the same degraded and unsanitary conditions he had seen aboard canal boats and which for him represented a sharp descent from any noble traditions cited by the "Gypsy Lore Society." His efforts to secure legislation sought to mirror the Canal Boats acts and focused on the registration of Romani vans with accompanying provisions for available cubic space and inspection. Early iterations of the Movable Dwellings bill, however, faced opposition from the Liberty and Property League, which decried the bill's emphasis on inspection and control, and from among others upset that local officials might be allowed to enter vans during a substantive portion of the day to ensure compliance. As Smith pressed his legislative case, it also drew antipathy from the Romani community.[233] Eventually, the bill died in Parliament at the time of Smith's death in 1895, failing to overcome opposition tied to its reach, the difficulties of enforcement, and the belief that existing legislation already controlled what were seen as the most pernicious activities of the Romani. Thus, at close of the nineteenth century, the Romani continued their efforts to navigate a variety of local controls focused on vagrancy (and by implication, mobility) as well as powers granted to local authorities, including those associated with health and sanitation. Britain's canal boat population, by contrast, found itself struggling to navigate specific national legislation aimed at controlling a mobile population, under the Canal Boats acts.

At the national level, monitoring and enforcement of conditions on canals continued to prove difficult. After Brydone's death in 1899, his place as canal boat inspector was taken by Owen J. Llewellyn, who was to continue as the LGB's canal boat inspector into the first decades of the twentieth century. After his first full year on the job in 1901, Llewellyn reported to the LGB that there had been a decrease in the overall number of inspections, but that these seemed to be traceable to just a few instances of local reductions in inspections.[234] He appeared to continue a rigorous inspection schedule with a total of 361 visits in 1900 and, at least in one instance, took the inspector at Gloucester to task. The inspector had reported a total of 205 inspections all without incident. Visiting Gloucester unannounced, Llewellyn indicated that he found "5 infringements out of eight boats in half-an-hour."[235] The records associated with one location indicate something of Llewellyn's early enthusiasm for bringing inspectors into line, but also his later acceptance of declines in the industry. Visiting the small market town Hinckley, located along the Ashby Canal, in 1901 and noting that there appeared to have been no inspections since 1899, he wrote to the local inspector in large red letters in the record book, "Please let me have a record of more and better work to inspect on my next visit."[236] On his next visit, he again wrote in large red letters,

> I hope the new inspector will put in more work than his predecessor. The only boats inspected by the late inspector since Nov 1899 were inspected when I was with him in March 1901, and as [there?] were 2 boats in the district [unreadable] that occasion, it is evident that boats are often [unreadable].[237]

Over the next few years, the inspection log book records a few trips to the canal side each year by the local inspector. However, there were few inspections and numerous notes regarding instances when no boats could be found. Llewellyn appears to have visited and left his signature in the log book on a near-annual basis through 1914; there are, however, no further notes of his frustration with the local inspector. Indeed, the absence of boats may have been predictable. In 1862, the canal had carried 138,117 tons, an amount that dropped to 74,604 tons in 1886 and 38,130 tons in 1906.[238]

In 1899, a Board of Education had replaced the Committee of Council on Education as the executive authority for education in Britain and three years later the Education Act of 1902 provided for a controversial overhaul of primary education in which some local rate support was directed to denominational voluntary schools.[239] By the terms of the act, supervision of most schools was turned over to county or county borough councils, which operated as local education authorities (LEAs). Controversy swirled as most voluntary schools were either Anglican or Roman Catholic; Nonconformist resistance spread. For canal boat children, these issues were remote from the basic controversy over whether they received a formal education at all. Overall, the Board's reports to Parliament regarding canal boat children were confident. In 1906, they reported with satisfaction that LEA reporting to them on canal boat children was increasing, that many more boats appeared to be without children and that the number of places with irregular attendance was lessening. They estimated there were about 1000 children living on canals.[240] However, such official reporting was underlain by extraordinarily messy accounting complicated by a variety issues related to brief and often inconsistent visits to canals. The Board attempted to refine the process of collecting data in the years ahead. However, basing its numbers on the return of surveys from individual districts, and making accommodation for some double counting, it arrived at an estimated 1,000 in 1910, 1,200 in 1911, 1,400 in 1912, and 1,500 in 1913, before dropping back down again to 1,200 in 1914.[241] As had been the case when Brydone first began to offer reports for the Committee of Council on Education, Llewellyn offered contributions for the Board of Education reports. However, despite frustrations with inconsistent inspection, he seems to have become more comfortable with ambiguity in the actual number of women and children on canal boats. As he noted in his report for the LGB in 1910–1911,

> I would remark that the numbers of women and children on the boats do not seem to alter much. The returns do not help to show how many of each actually inhabit them because they include the same craft repeatedly, but my experience agrees with that of many inspectors and is that, like a stage army, the same children appear over again and that the real number is much less than that popularly imagined.[242]

This apparent comfort with ambiguity in numbers likely reflected his overall belief that the boat children had not been enfeebled by their experiences on the waterways. In an uncharacteristically long assessment of canal boat children in his 1907–1908 LGB report just a few years earlier, he had emphasized a sense of overall health:

> It is perfectly true that canal boat children suffer from the want of facilities for education. If their parents have no other home than the canal boat (which is by no means the rule) they are obliged to live on the boat, and are more or less wanderers on the face of the earth. When the boat is tied up the children can go to the local school, and the managers cannot refuse to take them in. At the same time they fit in nowhere, as usually, whatever their size, they are capable mentally of instruction only in the infant class. They also lower the average attendances of the schools they thus occasionally attend, which naturally does not please the school authorities. They are of an independent spirit, and I am told look on slates and slate pencils as mediums for excruciating noises. I have heard that the natives resent their intrusions occasionally, and imply they may be verminous and use other uncomplimentary expressions. This does not make for order, so the departure of these families is looked on usually as a relief. In my view I do not exaggerate at all when I say that morally and physically the boat child is, for the most part, superior to the land child in a similar station of life. Its open air life inures it to all weathers, and its familiarity with its work from a very early age makes it reliant and strong and brings out its best qualities. . . . On the boats boys and girls (whose contemporaries ashore are in their hobbledehoyhood a source of worry and anxiety to their parents) can be found doing their duty in a quiet matter-of fact way, and taking a pride in its successful accomplishment. Undoubtedly, those ashore are more fully equipped in the matter of learning, but in the application of it I do not think most of them have any advantage over the boat children.[243]

Nonetheless, more widely, images of canal boat conditions oscillated between depictions of dank, cramped, and godless atmospheres within boat cabins and invigorating journeys in the fresh air of the countryside. This was evident in reports associated with the London City Mission, which had appointed its first missionary to the boatmen in 1877. [244] In 1895, it transferred a Mr. Bamber to west London, where he began what became the Canal Boatmen's Mission at Brentford, which was soon to include a day school.[245] (These Activities appear to have continued parallel to the Mission's activities near Paddington Basin, which also affected canal boatmen.) In extracts from the Fourth Annual Report of the Canal Boat Mission, which Bamber presented in 1900, he argued,

> In a sense our boat people are Outlanders. It is true several of them have their little cottages by the canal side in the country, and others in Birmingham and beyond, yet a great many are no one's parishioners, hence outside all parochial arrangements for visitation.[246]

Hugh B. Philpott, who in 1905 published an expanded history of the London School Board covering the years 1870–1904, published that same year an article entitled "Water Gipsies" in the *Methodist Magazine and Review* in which he noted that as nomads boat people had separated themselves from others since "the duties and the privileges of citizenship" were associated with fixity of dwelling. He wrote,

> The bargee pays neither rents nor rates nor taxes; no one solicits his vote or summons him on a jury; the Churches have hardly touched his life, and school boards—for all the use they have been to him—might as well have been non-existent. Like the gipsy, he belongs to a race apart; born on a canal boat, he has lived on a boat all his life and will probably die there.[247]

Philpott's review credited Smith as an "indefatigable worker on behalf of oppressed and neglected children" whose work on behalf of the canal boats acts was "a long struggle against difficulties which to a less earnest and persistent spirit would have proved insurmountable."[248] He indicated he had visited the Canal Boat Mission and had been introduced to those who were affected by Bamber's work.

In emphasizing the squalid nature of life on the canals, missionaries contrasted what seemed a delightful and picturesque vision from a canal bank with the reality of life afloat. That those upon the water rarely experienced the countryside in its majesty was a common theme for evangelistic writers. In describing the character of Ned in his novel of religious instruction *Ned the Barge Boy*, G. H. Sargent wrote,

> He was born in a barge, had spent the first eight years of his little life on board his father's boat, and then, when his father died . . . he had been hired by Bricky to drive his horses. . . . He had seen the beautiful green fields, the waving trees, the springing and the ripening corn as he drove the barge horses along the towing-paths, but he had taken little notice of these things.[249]

The missionary writer Annie Gray opened her novel *The Old Lock Farm*, a didactic and moralizing piece about evangelism on the canals, with the following:

> Threading its tortuous way through one of England's most picturesque counties, now glistening like a band of silver amidst broad acres of pasture-land, anon ascending, by means of numerous "locks," till it is lost to sight among the wooded hills, runs one of those great waterways upon whose placid bosom traverses by day and night so vast a portion of the traffic of our land.
>
> For many miles on either side of the quaint old farm which gives its name to our story, this canal is a very beautiful object. Parks and meadows, varied by an occasional tiny farmstead, run down to the water's edge, giving place here

> and there to pretty groves, particularly as you mount up beyond the "fourteen locks," groves made romantic by the multitude of little waterfalls which tumble tumultuously down the slope, and rush with a foam and a tiny dash into the quiet canal. The barges themselves, too, do much towards beautifying this picturesque spot. There are few prettier sights than, in the sunshine of a fair May day, or the twilight of a summer eve, to look down from the rising ground and watch one of these barges gliding calmly along beneath the shadowing trees. And when, as is often the case, you see the whole of its inmates seated on deck enjoying the sweet fresh air, you find it almost difficult to conceive that life on board can be otherwise than wholly delightful.
>
> Yet, were we to question the barge-folk themselves, how quickly should we be led to change our opinion![250]

Yet, by the 1890s, there was a cultural ambivalence emerging regarding canals. Though in industrial regions, they were still associated with the grit and dirt of industrialization, in primarily rural areas, they and those who worked on them were often seen as guardians of rustic traditions distinct from a rapidly changing world ashore. Dress, speech, rhythms of work, and knowledge and traditions associated with boat decoration increasingly came to be associated with a population that, if lacking in formal education, was hardworking, sincere, and independent. To enter the world of the canals for insight or pleasure, as was undertaken by a few in the late nineteenth century, came to be seen as entering a world wrought by industry but maintaining habits of a pre-modern era. For canal advocates in the late twentieth century, this was a unique culture that needed to be preserved alongside the engineering systems the boat people traveled. In recounting the life of her father William Buckland, an early geologist who later became Dean of Westminster, Elizabeth Oke Buckland Gordon offered in 1894 the following account of canal life, as she remembered it from her girlhood in Oxford. She wrote that

> several gaily-painted barges were often to be seen moored along the Canal Wharf, supplying the city with coals, salt, or pottery. However grimy their cargo might be, the owners contrived to keep fresh and bright the gay lines of colour on the sides of the little cabin at the end of the long black hull.[251]

She added that her father or, on occasion, "a good-natured bargee" would lift children into a boat so that they might "peep into the snug little abode, reeking with the savoury smell which issued from a black iron pot on its small hob, while from the tiny low chimney-pipe curled the prettiest possible wreath of blue-grey smoke."[252]

The precise degree to which this world—described in such varied ways by missionaries, recreational travelers, and nostalgic writers—was affected by the economic fortunes of the canals is inexact, but likely substantive. The most vibrant routes saw substantive efforts at maintenance and reconstruction as the century wore on, but no concerted effort at national reconstruction or administrative centralization to make them truly competitive with rail. Between the early 1840s and early 1850s, the opening of multiple rail lines forced the Oxford Canal to reduce tolls, cutting both gross revenue and the dividends paid to stockholders. Gross receipts were cut from £86,217 in 1842 to £26,312 in 1853; once attractive dividends of 30 percent in 1844 had dropped to 9 percent in 1855.[253] On the Wiltshire and Berkshire Canal, toll receipts dropped from £3,304 in 1837 to £2,205 in 1873, when a loss of £255 for the company was recorded.[254] Traffic ended in 1906 and an act of abandonment was passed by Parliament in 1914.[255] The Kennet and Avon had been taken over by the Great Western Railway and its tonnage had gone down from 360,610 in 1848 to 120,120 in 1888 and 83,373 in 1898.[256]

Yet, there were efforts to modernize, such as in the effort to introduce steam, and then petroleum motors. Early experiments at steam haulage had been carried out on Scotland's Forth and Clyde Canal between 1829 and 1830 and then later undertaken on a more limited basis on canals in England.[257] Steam power was most often associated with "fly" voyages and used in conjunction with a trailing "butty" boat that could extend capacity. Two notable fleets of steamers were employed by Fellows Morton & Clayton and the Leeds and Liverpool Canal Company.[258] Yet, particularly on narrow canals, because bank erosion from higher speeds remained a concern, and because machinery might occupy a significant amount of space which could be used for cargo aboard a boat, steam never induced a wave of transformation. Although some steam engines would continue into the twentieth century, the introduction of diesel motors soon eclipsed them, at least partially because diesels could be deployed in much smaller spaces of traditional family boats.[259] Despite a steady increase in diesels' presence in the early twentieth century and the ensuing decline of horse-drawn traffic, their advantages did not induce a revolution. Diesel boats could travel longer distances (not needing to rest as horses did) and could navigate their way through tunnels that might otherwise have required the assistance of "leggers" (men who lay on their backs aboard the boat and pushed it along by moving their feet against the tunnel wall). However, the pace of travel rarely rose above a few miles an hour, the advantages of increased speed being balanced by the costs of increased maintenance from the erosion caused by wave motion. Until the near end of industrial transport in the mid-twentieth century, horses—which now on land were associated with a bygone era of travel—would still be associated with everyday life on the canals.[260]

The overall transformations of the nineteenth century had been significant. Early on, there was little to distinguish canal boat workers from those who flooded into urban areas due to the growth of industry. Cramped living conditions, the impact of unemployment in times of bad weather and economic downturn, lack of education, sparse church attendance, and incidents of widely publicized crime, ensured formidable lines of division between canals and middle-class life. Boatmen were often associated with danger, vice, and violence specifically along canals, with stories of theft, accidental death, sexual violence, infanticide and suicide among the most significant in the press. Unlike highways, the presence of water added another level of fear and uncertainty. The ability to swim was limited in the early nineteenth century, and accidental deaths likely added to overall fears of canals.[261] Although boatmen often appear to have been among those able to offer assistance to those in distress or to be able to find bodies, suspicion of their world was significant. Unlike other working-class dwellings, canal boats might not be easily accessed, or escaped, as they passed through communities, thus adding to the complexity of inspection. Other marginalized populations such as the Romani ("Gypsies"/ "gypsies") could be subject to similar characterizations and associations, though unlike the Romani, the canal boat population may have been even more bereft of voices which could offer an independent narrative of their world.[262]

Boat people were subjects of missionary efforts to bring them within the pale of "Christian civilization" as early as 1829. Securing regular church attendance remained a difficulty, however. Moreover, state intervention to regulate living and working conditions proved a challenge due to the length of canal banks, the composite ownership and management of canal and transport companies, and practices in which individual captains might hire their own crews. The Royal Commission for Inquiring into the Administration and Practical Operation of the Poor Laws highlighted the problem of mobility in determining poor relief, the Select Committee on Drunkenness emphasized attachments between the boat population and alcohol, and the Royal Commission on constabulary forces reinforced links between the boat population and theft. The collection of testimonies regarding conditions on canals through governmental hearings provided some of the most substantive views of canal life, but offered no first-hand testimony from canal boatmen, themselves. The outrage generated by the murder of Christina Collins reinforced for many the significance of canals as spaces of danger and the need for moral reform. Nonetheless, characterizations of canal boatmen, as well as those working rivers with whom they overlapped, could be notoriously inconsistent. Anecdotal remarks varied based upon the waterway and type of boat referenced. Though graphic stories of violence and death continued at mid-century, pressure for legislative reform lapsed by the early 1840s.

At this same time, investment and engineering initiatives were transferred to the rails, whose impact upon the landscape suggested not so much a blending as a triumph, cutting through hillsides and making use of coal extracted from the earth. What had once been the pulling power of the horse was transformed into simple horsepower. Railways introduced new sights and sounds into the landscape and, perhaps most significantly, demolished previous limits of speed, which had generally been measured by a simple walking pace. They offered speed and an ability to distribute products over wide swaths of territory, their construction facilitated by modern methods of engineering that had already made significant progress in the era of canals. They also offered the ability to better secure freight, which no longer spent several days in the unmonitored custody of a canal boat captain and his crew. A standard railway gauge, established with the assistance of the government in the mid-1840s, reinforced business practices, making the cross-country transport of goods easier and more efficient. Railways were able to overcome some of the difficulties of canal infrastructure, avoiding water shortages and water routes that might freeze. The income generated by passenger traffic helped to ensure steady returns, drawing investment monies away from the waterways and prompting agreements that saw some canals purchased by railways. In this environment, there was no nineteenth-century movement for an overhaul of canals. Some boats, despite the introduction of steam and later diesel power, continued to be horse-drawn until the closing days of industrial traffic.

Though many canals remained active and profitable in the mid to late nineteenth century, others lapsed in both traffic and maintenance, the diminishing ranks of boatmen left to bear the mixed legacies of more active days. Hollinghead's account of at least three different modes of travel in 1860 spoke to contrasting images of canals. The beginnings of his canal journey were a peaceful transit of the countryside in the company of a competent crew. However, his observations included images of an unkempt family boat and, as he entered the industrial Midlands, dark and threatening aspects of a polluted waterway and debauched population. No such memories made it into his account of rail travel, a trip which passed speedily—akin to a galloping horse—with no contact with the surrounding population.

Even upon waters with decreasing traffic, some found a population in desperate need and want. George Smith ("of Coalville") took up the baton of reform in the 1870s, painting conditions on canal boats in the most dire terms and pressing for state intervention. Smith drew implicit connections between the Romani and boatmen in interlinked campaigns in the late nineteenth century. He, like some others, addressed both groups using the language of race, though no assumption that the latter were descendants of the former appears to have emerged explicitly among other writers until the early twentieth century. Canal boatmen did not share in the substantive linguistic, cultural, or ethnological examinations associated with the Romani population. Rather, their legacies were largely determined by the conflicting assessments of their living conditions that arose from governmental inspections and missionary impulses. The Canal Boats acts of 1877 and 1884 only partially brought the canal boat population within the pale of state regulation. Appointed as a canal boat inspector for the LGB, Brydone approached his work with enthusiasm, though his commitment was tested by the challenges of attempting to enforce diligence among local inspectors who had other responsibilities and a boat population which remained difficult to track and suspicious of outsiders. Nonetheless, he soon rejected the most dire pronouncements regarding the health, sanitation, and morality of the boat population and instead pointed to what he believed might be the invigorating effects of canal boat life. His observations were rejected by Smith, who remained steadfast in his sketching of the horrendous conditions of health and sanitation on the canals. Similarly, missionary works of the late nineteenth century also pointed to the dire position of children. Yet, by the turn of the century, anxiety decreased. Llewellyn continued the system of inspection of canal boats in the early twentieth century with empathy for the dilemmas faced by the steadily diminishing numbers of canal boat workers and their families.

Martin Wiener has argued that the vision of a "southern" England began to establish its ultimate power over the imagination of the middle and upper classes by the late nineteenth century, reaching into an imagined idyllic and rural past to grasp a perceived soul of England.[263] Indeed, particularly in the south, one could find canals (many now bereft of traffic) very much parts of the natural landscape, or at least spaces in which rural values persisted. Yet, their cuttings, reservoirs, towpaths, and bridges spoke of the boldness of eighteenth-century capitalism and the exploits of the early to mid-nineteenth century, when individual entrepreneurs had founded businesses which became local engines of development and urban civilization. Arguments that canals represented unique spaces bringing together the natural and man-made worlds, that restraints imposed upon their motive power reinforced valued labor traditions, and that divergences in their construction represented regional geographic variations, had emerged in sporadic fashion and would come into their own after World War II. Before World War I, British canals were still far too immersed in the world of functional transport to draw interest from groups associated with the protection of the countryside. As industrial traffic continued to depart in the decades ahead, they would offer opportunities for embracing nature and simplified engineering—chances for peace amidst a fraught industrial present. In the immediate term, however, there was anxiety regarding the canals' future. Despite the opening of the Manchester Ship Canal in the 1890s (to be discussed in the next chapter), the waterways were, in the minds of most observers, a reminder of the nation's industrial weaknesses, as nations on continental Europe undertook major canal improvements. Depictions of quiet canals captivated a few but sobered many others concerned about the nation and its competitiveness in the twentieth century.

NOTES

1. As it is used here, the word "environment" has come to reference, broadly, the natural systems influencing, and influenced by, human activity. This should be distinguished from its nineteenth-century usage. As James Winter has written, "In the nineteenth century, connotations would have been closer to the etymological roots of *environment*: the country around, the neighborhood, the environs, the stretch of topography that gave definition to a place, one's own surroundings." Winter, *Secure from Rash Assault* (Berkeley: University of California Press, 19).

2. Harry Hanson, *The Canal Boatmen* (Manchester: Manchester University Press, 1975),13.

3. "Friday's Post," *Ipswich Journal*, January 31, 1795, 2. British Newspaper Archive (BNA).

4. Hanson, *The Canal Boatmen*, 48, 50–52. Hanson suggests that "fly-boats" became more common as the canal system expanded and potential routes were lengthened. (Hanson, *The Canal Boatmen*, 33–45.)

5. Hanson, *The Canal Boatmen*, 17, 21.

6. "Droitwich Navigation," *Aris's Birmingham Gazette*, August 19, 1771, 1. BNA.

7. "Droitwich Navigation," 1.

8. "Droitwich Navigation," 1.

9. See Wendy Freer, "Canal Boat People, 1840–1970," Vol. 1 (PhD Dissertation, University of Nottingham, Nottingham, 1991), 78.

10. Watermen, or boatmen, have been a substantive occupational group on the Thames, as well as on some other rivers, since the medieval era. Lightermen were specifically those operating a type of light barge with a flat bottom on the Thames.

11. Hanson, *The Canal Boatmen*, 201–202.

12. This should likely be treated as an upper limit. For a discussion of the uncertainties in estimating female labor on the canals, see Freer, "Canal Boat People, 1840–1970," Vol. I, 85–88.

13. Victor Bailey, ed., *Nineteenth-Century Crime and Punishment* (London: Routledge, 2022), 1:21. Before 1805, historians must rely upon numbers of committals for trials on indictments presented to the courts of quarter sessions and assizes. Many minor offenses, such as those for vagrancy, drunkenness, and breaches of the peace, were tried at petty sessions for which records have not survived. Parliament ordered the collection of information concerning those committed to trial as well as the outcomes of their proceedings beginning in 1805. Though the data collected were significant enough to raise alarm over rising numbers of indictments, questions remain regarding what may have been significantly varying levels in the reporting,

investigation, and prosecution of crime. This is reinforced in Rob Sindall, *Street Violence in the Nineteenth Century: Media Panic or Real Danger?* (Leicester: Leicester University Press, 1990).

14. Bailey, *Nineteenth-Century Crime and Punishment*, 1:30.

15. See Great Britain. Royal Commission into the Condition of the Poorer Classes in Ireland. Appendix G: *Report on the State of the Irish Poor in Great Britain* (London: W. Clowes and Sons, 1835), 142–143, 163.

16. David Mayall, *Gypsy-Travellers in Nineteenth-Century Society* (Cambridge: Cambridge University Press, 1988), 15. Here, Romani here refers to a diverse community (also commonly identified as Roma) whose genetic ancestry has been linked to South Asia. Though the term "gypsy" is today often identified as pejorative, it was in frequent use in the nineteenth century. The census of the United Kingdom has continued to identify a "Gypsy, Roma and Traveller" group into the early twenty-first century, more recently establishing separate categories for "Gypsy or Irish Traveller" and "Roma." The latter identifier is meant to encompass more recent migrants from Central and Eastern Europe. See United Kingdom "Gypsy, Roma and Irish Traveller Ethnicity Summary," accessed May 23, 2024, https://www.ethnicity-facts-figures.service.gov.uk/summaries/gypsy-roma-irish-traveller/.

17. In the early to mid-nineteenth century, emphasis was placed on cultural and historical understandings of human difference in addressing race rather than biologically determined categories. The latter of these was emphasized in the late nineteenth century.

18. Bailey, *Nineteenth-Century Crime and Punishment*, 1:30–32.

19. Bailey, *Nineteenth-Century Crime and Punishment*, 1:30–37.

20. For a discussion of the significance of accessing the press in an attempt to better understand the nature of middle-class fears of crime, see Sindall, *Street Violence in the Nineteenth Century*, 29. The use of newspaper archives, as has been the case with the use of the BNA in this book, should be undertaken with great care given the incomplete but evolving content of electronic resources and access to them. See Yann Ryan, *A Short Guide to Historical Newspaper Data, Using R* (Bookdown), accessed December 13, 2023, https://bookdown.org/yann_ryan/r-for-newspaper-data/.

21. Although, as Sindall notes, the role of provincial newspapers in exercising an independent editorial bent and influencing the attitudes of the nation may have been limited, the inclusion of stories covering local crime and violence is significant in establishing how canals may have been viewed by those in the areas through which they ran. See Sindall, *Street Violence in the Nineteenth Century*, 31.

22. As an example of an early, rarer incident, in early 1775 a boat belonging to William Crosby was broken open and robbed of a variety of goods. Two men were detained and a search was put out on a third, "Birmingham Wharf, May 16 or 17, 1775," *Aris's Birmingham Gazette*, July 31, 1775, 1. BNA.

23. "Salt," *Gloucester Journal*, August 4, 1800, 2. BNA.

24. "Chester, Tuesday, March 17," *Chester Courant*, March 17, 1801, 3. BNA.

25. "Derby, Wednesday, Jan. 14," *Derby Mercury*, January 15, 1807, 3, and "Warwickshire Sessions," *Aris's Birmingham Gazette*, July 25, 1808, 3. BNA.

26. "Market—Harborough," *Northampton Mercury*, December 17, 1808, 1. BNA.

27. "Lost," *Staffordshire Advertiser*, April 22, 1809, 4. BNA.

28. “Pilfering on Canals,” *Hull Advertiser*, October 5, 1811, 4. BNA.
29. “Sessions,” *Aris’s Birmingham Gazette*, October 31, 1814, 1. BNA.
30. “Fifty Guineas Reward,” *Staffordshire Advertiser*, June 5, 1813, 1. BNA.
31. “Northampton,” *Northampton Mercury*, February 28, 1801, 3. BNA.
32. “Old Bailey, Sept. 19,” *Saint James’s Chronicle*, September 10, 1810, 3. BNA.
33. “Old Bailey . . . Murder,” *True Briton*, January 15, 1803, 3. BNA.
34. “London, Friday, Sept. 8,” *Hampshire Chronicle*, September 11, 1809, 3. BNA.
35. “Bow Street,” *English Chronicle and Whitehall Evening Post*, November 15, 1821, 1. BNA.
36. “Trial and Execution of Abel Hill,” *Staffordshire Advertiser*, July 29, 1820, 4. BNA.
37. “Windsor and Eton,” *Windsor and Eton Express*, February 2, 1817, 4. BNA.
38. “Shocking Outrage,” *Cork Constitution*, October 17, 1826, 4. BNA.
39. “Leeds, March 29,” *Leeds Intelligencer*, March 29, 1802, 3. BNA.
40. “Derby, Wednesday, September 7,” *Derby Mercury*, September 8, 1803, 3. BNA.
41. “Stafford, Saturday, July 9,” *Staffordshire Advertiser*, July 9, 1803, 4. BNA.
42. [Untitled], *Staffordshire Advertiser*, August 2, 1806, 3. BNA.
43. “Northampton, Saturday Evening, Dec. 19,” *Northampton Mercury*, December 29, 1810, 3. BNA.
44. “Lancaster, December 5, 1812,” *Lancaster Gazette*, December 5, 1812, 3. BNA.
45. “Accidents,” *General Evening Post*, November 3, 1818, 4. BNA.
46. “Local Intelligence,” *Leeds Intelligencer*, November 18, 1822, 3. BNA.
47. “Local Intelligence,” *Yorkshire Gazette*, March 6, 1824, 2. BNA.
48. [Untitled] *Johnson’s Sunday Monitor*, September 20, 1812, 4. BNA.
49. [Untitled] *Staffordshire Advertiser*, October 1, 1825, 4. BNA.
50. [Untitled] *Staffordshire Advertiser*, September 6, 1828, 4. BNA.
51. “Leicester, Wednesday Morning, October 8,” *Leicester Herald*, October 8, 1828, 2. BNA.
52. “Quarter Sessions,” *Aris’s Birmingham Gazette*, January 19, 1801, 3. BNA.
53. “Saturday’s Post. London, Friday, August 9,” *Aris’s Birmingham Gazette*, August 12, 1816, 2. BNA.
54. “Miscellaneous,” *General Evening Post*, December 7, 1819, 4. BNA.
55. [Untitled] *The Examiner*, July 29, 1827, 13. BNA.
56. [Untitled] *Northampton Mercury*, November 8, 1823, 3. BNA.
57. “Crown Court . . . ” *Staffordshire Advertiser*, March 22, 1828, 3. BNA.
58. “Accidents,” *Johnson’s Sunday Monitor*, November 17, 1816, 4. BNA.
59. “Accidents,” *Johnson’s Sunday Monitor*, 4.
60. [Untitled] *Birmingham Journal*, December 16, 1826, 3. BNA.
61. Olive Anderson, *Suicide in Victorian and Edwardian England* (Oxford: Clarendon Press, 1987), 196.
62. Anderson, *Suicide in Victorian and Edwardian England*, 45.
63. Anderson, *Suicide in Victorian and Edwardian England*, 196.

64. "Provincials," *The National Register*, October 29, 1809, 16. BNA.

65. "Northampton, Saturday Evening, July 28," *Northampton Mercury*, July 28, 1810, 3. BNA.

66. "Inquests," *General Evening Post*, October 8, 1818, 4. BNA.

67. Anderson, *Suicide in Victorian and Edwardian England*, 241.

68. Anderson, *Suicide in Victorian and Edwardian England*, 284, 270, 291.

69. Anderson, *Suicide in Victorian and Edwardian England*, 213.

70. "Suicide Prevented," *The Canal Boatman's Magazine* 1, no. 1 (April 1829): 6–7.

71. "Report," *A Supplementary Number of the Paddington Canal Boatman's Magazine* 1 (May 20, 1829): 35.

72. "To the Editor of the Canal Boatman's Magazine," *The Canal Boatman's Magazine* 1, no. 2 (May 1829): 17.

73. Wendy Freer and Gill Foster, *Canal Boatmen's Missions* (Oxford: Railway and Canal Historical Society, 2004), 13.

74. "Journal of Attendance," *The Canal Boatman's Magazine* 1, no. 2 (May 1829): 28–29.

75. "Journal of Attendance," *The Canal Boatman's Magazine* 1, no. 7 (October 1829): 141.

76. "Announcement of Meeting," *The Canal Boatman's Magazine* 1, no. 9 (December 1829): 162.

77. An Act to regulate the Labour of Children and Young Persons in the Mills and Factories of the United Kingdom [August 29, 1833], 3 & 4 William IV, *c.* 103.

78. Great Britain. Parliament, *Report From His Majesty's Commissioners for Inquiring into the Administration and Practical Operation of the Poor Laws* (London: B. Fellowes, 1834), 1:160.

79. Great Britain. Parliament, *Report From His Majesty's Commissioner . . . Poor Laws*, 1:156–158.

80. Buckingham, J. S., *Evidence on Drunkenness, Presented to the House of Commons, by the Select Committee Appointed by the House to Inquire Into This Subject, and Report the Minutes of Evidence, with Their Opinions Thereupon* ("J.S. Buckingham Esq., M.P., in the Chair"). London: Samuel Bagster, 1834.

81. Buckingham, *Evidence on Drunkenness, Presented to the House of Commons*, 585.

82. Great Britain, *First Report of the Commissioners Appointed to Inquire as to the Best Means of Establishing an Efficient Constabulary Force in the Counties of England and Wales* (London: HMSO, 1839).

83. Great Britain, *First Report . . . Efficient Constabulary Force*, 92.

84. Great Britain, *First Report . . . Efficient Constabulary Force*, 92–93.

85. Great Britain, *First Report . . . Efficient Constabulary Force*, 95.

86. Great Britain, *First Report . . . Efficient Constabulary Force*, 96.

87. Great Britain, *First Report . . . Efficient Constabulary Force*, 96.

88. Great Britain, *First Report . . . Efficient Constabulary Force*, 97.

89. Great Britain, *First Report . . . Efficient Constabulary Force*, 99.

90. Great Britain, *First Report . . . Efficient Constabulary Force*, 101.

91. Great Britain, *First Report . . . Efficient Constabulary Force*, 100.

92. Great Britain, *First Report . . . Efficient Constabulary Force*, 101.

93. Great Britain, *First Report . . . Efficient Constabulary Force*, 103.

94. An Act for the Establishment of County and District Constables by the Authority of the Justices of the Peace, 2 & 3 Vict., *c*. 93.

95. The case continued to generate interest well into the modern era, even prompting its incorporation into Britain's popular series of crime stories revolving around the fictional "Inspector Morse." See "Inspector Morse Writer Visits Scene of a Crime," *Staffordshire Sentinel*, April 15, 1991, 5. BNA.

96. "Execution of Owen and Thomas for the Murder of Christina Collins," *Staffordshire Gazette and County Standard*, April 18, 1840, 2. BNA.

97. See "The Murder near Rugeley," *Staffordshire Gazette and County Standard*, July 6, 1839, 3, and "Summer Assizes. Oxford Circuit," *English Chronicle and Whitehall Evening Post*, July 25, 1839, 4. BNA. Pickfords appears to have initially donated £30 toward the prosecution of the men who were to be charged in her murder. ("The Rugeley Murder," *Liverpool Standard and General Commercial Advertiser*, July 16, 1839, 2. BNA.)

98. "Summer Assizes. Oxford Circuit," 4. BNA.

99. "Summer Assizes. Oxford Circuit," 4. BNA, and "Alleged Violation and Murder of a Female on the Canal by Boatmen," *Staffordshire Advertiser*, July 27, 1839, 2. BNA.

100. "Alleged Violation and Murder of a Female on the Canal by Boatmen," 2. BNA.

101. "Assize, Intelligence," *Sun*, March 18, 1840, 3; and "The Murder of Christina Collins on the Trent and Mersey Canal, Near Rugeley," *Staffordshire Gazette and County Standard*, March 21, 1840, 2. BNA.

102. "Local Intelligence. Staffordshire," *Coventry Herald and Observer*, April 17, 1840, 2; and "Execution of Owen and Thomas for the Murder of Christina Collins," *Staffordshire Gazette and County Standard*, April 18, 1840, 2. BNA.

103. From the *Chester Gazette*, as reported by the *Liverpool Standard and General Commercial Advertiser*, July 20, 1839, 2. BNA. It was also reported that Egerton and his wife walked to church in order to allow more time for domestic servants to attend church. See "Lord F. Egerton," *Blackburn Standard*, June 12, 1839, 3. BNA.

104. "Sunday Trading," *Staffordshire Gazette and County Standard*, August 17, 1839, 2. BNA.

105. "The Watermen's Memorial," *Worcester Journal*, January 16, 1840, 3. BNA.

106. "The Condemned Boatmen Respited for One Week," *Staffordshire Advertiser*, April 4, 1840, 3. BNA.

107. "The Condemned Boatmen Respited for One Week," 3.

108. "The Condemned Boatmen Respited for One Week," 3.

109. An Act to provide for keeping the Peace on Canals and Navigable Rivers, 3 & 4 Vict., *c*. 50. See also, John Frederick Archbold, *Snowden's Magistrates Assistant*. 4th ed. (London: Shaw and Sons, 1859), 58–63.

110. "Imperial Parliament. House of Lords," *Derby Mercury*, May 5, 1841, 4. BNA.

111. "Imperial Parliament. House of Lords."

112. Great Britain, *Report from the Select Committee of the House of Lords Appointed to Inquire Into the Expediency of Restraining the Practice of carrying Goods and Merchandize on Canals, Navigable rivers and Railways on Sundays; . . . with the Minutes of Evidence*, 4, in House of Lords Sessional Papers, 1841: Reports From Select Committees of the House of Lords and Evidence, 21:4. Chetwynd also added parenthetically, "(by 'those Men' he meant the Receivers of stolen Goods)," 4.

113. Great Britain, *Report from the Select Committee . . . Goods and Merchandise . . . on Sundays*, 10.

114. Great Britain, *Report from the Select Committee . . . Goods and Merchandise . . . on Sundays*, 4.

115. Great Britain, *Report from the Select Committee . . . Goods and Merchandise . . . on Sundays*, 7.

116. Great Britain, *Report from the Select Committee . . . Goods and Merchandise . . . on Sundays*, 4.

117. Great Britain, *Report from the Select Committee . . . Goods and Merchandise . . . on Sundays*, 10.

118. Great Britain, *Report from the Select Committee . . . Goods and Merchandise . . . on Sundays*, 31.

119. Great Britain, *Report from the Select Committee . . . Goods and Merchandise . . . on Sundays*, 31.

120. Great Britain, *Report from the Select Committee . . . Goods and Merchandise . . . on Sundays*, 45.

121. Great Britain, *Report from the Select Committee . . . Goods and Merchandise . . . on Sundays*, 49.

122. Great Britain, *Report from the Select Committee . . . Goods and Merchandise . . . on Sundays*, 51.

123. Great Britain, *Report from the Select Committee . . . Goods and Merchandise . . . on Sundays*, 67.

124. Great Britain, *Report from the Select Committee . . . Goods and Merchandise . . . on Sundays*, 69.

125. Great Britain, *Report from the Select Committee . . . Goods and Merchandise . . . on Sundays*, 64. William August Miles, an observer of crime and policing in the nineteenth century, used the phrase "Race 'sui generis," generally meaning a unique group of its own kind, in providing evidence on thieves to the 1835 House of Lords Select Committee on Gaols. Bailey, *Nineteenth-Century Crime and Punishment*, 1:119.

126. Great Britain, *Report from the Select Committee . . . Goods and Merchandise . . . on Sundays*, 89.

127. "Local Intelligence. Worcester," *Cheltenham Journal and Gloucestershire Fashionable Weekly Gazette*, August 29, 1842, 2. BNA.

128. "Inland Navigation Society," *Patriot*, April 27, 1843, 1. BNA.

129. "Summer Assizes. Oxford Circuit," *Morning Advertiser*, August 15, 1843, 4. BNA.

130. "Hemel Hempstead Police," *Buckinghamshire Advertiser and Aylesbury News*, September 23, 1843, 7. BNA.

131. Michael Freeman, *Railways and the Victorian Imagination* (New Haven: Yale University Press, 1999), 1.

132. Freeman, *Railways and the Victorian Imagination*, 2.

133. For references to the impact of ice and "droughts," see Thomas Gray, *Observations on A General Iron Rail-Way: (With Plates and Map Ilustrative* [*sic*] *of the Plan,) Showing Its Great Superiority, By the General Introduction of Mechanic Power, Over All the Present Methods of Conveyance by Turnpike Roads and Canals; And Claiming the Particular Attention of Merchants, Manufacturers, Farmers, and Indeed, Every Class of Society*. 4th ed. (London: Baldwin, Cradock, and Joy, 1823), 4; and John Francis, *A History of the English Railway; Its Social Relations and Revelations, 1820–1845* (London: Longman, Brown, Green, and Longmans, 1851), 1:77. See also Freeman, *Railways and the Victorian Imagination*, 27.

134. Freeman, *Railways and the Victorian Imagination*, 4.

135. Quoted in Freeman, *Railways and the Victorian Imagination*, 40. As James Winter has written,

> Extranatural power, iron machinery, artifice, railway speed, innovation, precision, synchronization, professionalism, rationalization, calculation, utility, abstractness, efficiency—these phenomena, values, and processes were what Victorians had in mind when they described their age, as they frequently did, as the age of steam. (*Secure from Rash Assault*, 83)

136. William Howitt, *The Rural Life of England.* 3rd ed. (Philadelphia, PA: Carey and Hart, 1841 (1840)), 154.

137. "Kendal & Windemere Railway," *Carlisle Journal*, October 26, 1844, 3. BNA.

138. "Poetry Against the World," *Carlisle Journal*, October 26, 1844, 3. BNA.

139. "Kendal and Windemere Railway" (Letter), *The Morning Post*, December 11, 1844, 4–5. BNA. See also, "The Windemere Railway" (Letter), *The Morning Post*, December 20, 1844, 5; and, for a critical summation, "Poetry and Railways," *Carlisle Journal*, December 28, 1844, 3. BNA.

140. John Hollingshead, *Odd Journeys In and Out of London* (London: Groombridge and Sons, 1860), 5, 11.

141. Hollingshead, *Odd Journeys In and Out of London*, 9.

142. Hollingshead, *Odd Journeys In and Out of London*, 25–26.

143. Hollingshead, *Odd Journeys In and Out of London*, 26.

144. Hollingshead, *Odd Journeys In and Out of London*, 19.

145. Hollingshead, *Odd Journeys In and Out of London*, 27–28.

146. Hollingshead, *Odd Journeys In and Out of London*, 29.

147. Hollingshead identifies the Grand Junction Canal Company as his "hosts." (Hollingshead, *Odd Journeys In and Out of London*, 7.)

148. Hollingshead, *Odd Journeys In and Out of London*, 37–38.

149. Hollingshead, *Odd Journeys In and Out of London*, 64–65.

150. Hollingshead, *Odd Journeys In and Out of London*, 70.

151. Hollingshead, *Odd Journeys In and Out of London*, 72–73.

152. Hollingshead, *Odd Journeys In and Out of London*, 375.

153. Anonymous, *The Waterway to London, as Explored in the "Wanderer" and "Ranger," with Sail, Paddle, and Oar, in a Voyage on the Mersey, Perry, Severn, and Thames, and Several Canals* (London: Simpkin, Marshall, & Co., 1869), Preface.

154. Anonymous, *The Waterway to London*, 9–10.

155. Anonymous, *The Waterway to London*, 24.

156. Anonymous, *The Waterway to London*, 40–41. Jodie Matthews offers an analysis of this work, along with others, drawing attention to the links between nineteenth-century journeys of exploration on canals and accounts of European voyages into regions of an expanding British imperial world. She endorses the finding of Richard Fairhurst ("Pioneer Paddlers," *Waterways World* (December 2016)) that the two travelers are Alfred Taylor Schofield, author, his brother R. Harold A. Schofield, and their friend, Will. (See Matthews, *The British Industrial Canal: Reading the Waterways from the Eighteenth Century to the Anthropocene* (Cardiff: University of Wales Press, 2023), 72.)

157. Anonymous, *The Waterway to London*, 72. The American canal weed appears to have been a particularly successful invasive species. According to a late nineteenth-century canoe traveler, Malcolm Black,

> In many places it [the Thames and Severn Canal] is almost choked up by the American canal weed, a long stringy kind of plant said to have been introduced into England on a piece of wood brought over by some American sailors and thoughtlessly thrown into the canal without proper attention being paid to the certain result. Miles and miles of it that had been pulled out to facilitate navigation lay upon the banks. The presence of this weed may account for the striking purity of the water for its filtering properties are well known. (Malcolm Black, *"Our Canoe Voyage," How We Launched at Hereford and Landed at Reading* [Manchester: Thomas and Jones, 1876], 176)

158. Factory and Workshops Acts Commission, *Report of the Commissioners Appointed to Inquire into the Working of the Factory and Workshops Acts, with a View to their Consolidation and Amendment* (London: Eyre and Spottiswoode, 1876), 1:124. See also, Hanson, *The Canal Boatmen*, 54.

159. Hanson, *The Canal Boatmen*, 90–91.

160. Hanson, *The Canal Boatmen*, 54–55.

161. Hanson, *The Canal Boatmen*, 54.

162. Hanson, *The Canal Boatmen*, 205.

163. Hanson, *The Canal Boatmen*, 54–55, 90–91.

164. The Artizans' and Labourers' Dwellings Improvement Act [1875], 38 & 39 Vict., *c*. 36. See Alexander Glen, *The Artizans' and Labourers' Dwellings Improvement Act, 1875 . . . Re-issued with the Artizans' and Labourers' Dwellings Improvement Act, 1879* (London: Knight and Co., 1880), 18–22.

165. Joseph Chamberlain, "On the Sanitary Conditions of Large Towns," Birmingham, January 13, 1875, in *Mr. Chamberlain's Speeches*, vol. 1, ed. Charles W. Boyd (London: Constable and Company, 1914), 63.

166. See Jules Gehrke, "Municipal Anti-Socialism and the Growth of the Anti-Socialist Critique in Britain, 1873–1914" (PhD Dissertation, University of Minnesota, Minneapolis, MN, 2006), 53.

167. Gehrke, "Municipal Anti-Socialism and the Growth of the Anti-Socialist Critique . . . ," 53.

168. Chamberlain, "On the Sanitary Conditions of Large Towns," 63–64.

169. Hoare, *Notable Workers in Humble Life* (London: Thomas Nelson and Sons, 1889), 201.

170. George Smith ("of Coalville"), *Our Canal Population: the Sad Condition of the Women and Children,—with Remedy* (London: Haughton and Co., 1875), 8.

171. Smith, *Our Canal Population,* 2. The general structure and outline of Smith's campaign may be reviewed alongside the influential model for the emergence and empowerment of "moral panics" introduced by Stanley Cohen in *Folk Devils and Moral Panics*. 3rd ed. (London: Routledge, 2002 (1973)). Sindall uses Cohen's model as the basis for his look at street violence in the late nineteenth century (Sindall, *Street Violence in the Nineteenth Century.*).

172. Wendy Jane Freer, "Canal Boat People, 1840–1870" Vol. I, 43. Freer's work offers one of the most substantive assessments of the numbers of boats and male and female workers on canals. Women likely made up between 20 and 30 percent of those who slept aboard boats. See pp. 77–78 and 84 of Freer in particular.

173. "Canal Boat Children," *The Graphic*, May 16, 1874, 462–463. BNA.

174. "Mr. George Smith and 'Our Floating Population,'" *Leicester Journal*, August 21, 1874, 8. BNA.

175. "Life on Canal Boats," *The Manchester Evening News*, May 29, 1875, 2. BNA.

176. *Third Report of the Commissioners Appointed to Inquire into the Best Means of Preventing the Pollution of Rivers* (Rivers Aire and Calder.) (Report, Appendix, Plans) (London: HMSO, 1867), 1:xxxviii.

177. 38 & 39 Vict., *c*. 55, s. 17. See the discussion in Alexander Glen, *The Rivers Pollution Prevention Act, 1876, 39 & 40 Vict. c. 75; with Introduction, Notes, and Index* (London: Knight & Co. 1876), 14.

178. See the text of the act in Alexander Glen, *The Rivers Pollution Prevention Act*, 1876, vi–vii.

179. Smith, *Our Canal Population*, 90. See also, George Smith, *Gypsy Life: Being an Account of Our Gipsies and Their Children, with Suggestions for Their Improvement* (London: Haughton, 1880).

180. "Coalville," *The Leicester Chronicle and Leicestershire Mercury*, August 7, 1875, 11. BNA. See also, Freer and Foster, *Canal Boatmen's Missions*, 42–44.

181. Freer and Foster, *Canal Boatmen's Missions*, 42

182. "Our Canal Population," *The Staffordshire Daily Sentinel*, August 23, 1875, 3. BNA.

183. "Public Health Act," 1875, 38 & 39 Vict., *c*. 55.

184. Factory and Workshops Acts Commission, *Report of the Commissioners*, 118–126.

185. Factory and Workshops Acts Commission, *Report of the Commissioners*, lxxxvi.

186. Factory and Workshops Acts Commission, *Report of the Commissioners*, lxxxvi.

187. Factory and Workshops Acts Commission, *Report of the Commissioners*, lxxxvi.

188. "Canal Boats and Brickyards," *The Staffordshire Sentinel*, October 21, 1876, 8. BNA.

189. "The Canal Boat Population," *The Worcestershire Chronicle*, July 12, 1884, 2. BNA.

190. Edwin Hodder, *George Smith (of Coalville): The Story of an Enthusiast* (London: James Nisbet and Co., Ltd., 1896), 77.

191. [Untitled] *The Newcastle Daily Chronicle*, August 14, 1879, 2. BNA.

192. "Among the Gipsies," *The Graphic*, March 13, 1880, 13. BNA.

193. "The Social Science Congress," *The Evening Standard*, October 9, 1880, 2. BNA.

194. "What Shall We Do With the Gypsy?" *The Field, The Country Gentleman's Newspaper*, August 23, 1879, 251. BNA.

195. L. T. (Elizabeth Thomasina) Meade, *Water Gipsies, or the Adventures of Tag, Rag, and Bobtail* (London: John F. Shaw and Co., 1883?), 20. Originally published in 1878, the novel contains the following footnote in this slightly later edition (perhaps 1883): "In 1877, the "Canal Boats Act" passed through Parliament, by which boatmen's children are obliged to attend school. It commenced its operation in July 1877. But many years must pass before its good effects will be reaped," 20.

196. "An Act to provide for the Registration and Regulation of Canal Boats used as Dwellings," 40&41 Vict., *c*. 60. See Robert Webster, *The Law Relating to Canals: Comprising a Treatise on Navigable Rivers and Canals; and Including all Legislation to the Close of the Last Session of Parliament, Together with the Procedure and Practice in Private Bill Legislation; with a Map of the Existing Canals in England and Wales* (London: Stevens and Sons, 1885), 311.

197. Webster, *The Law Relating to Canals,* 311.

198. Webster, *The Law Relating to Canals*, 312.

199. Overall national attention to working conditions was intensified in 1878 with the passing of an additional Factory and Workshop Act, which consolidated much of the legislation on working conditions passed since the 1830s. In it, children under ten were not to be employed, and most children between the ages of eleven and fourteen were to be going to school half time before employment became legal. In this and other provisions affecting the employment of youths and women, no mention was made of employment on rivers or canals. ["An Act to Consolidate and Amend the Law Relating to Factories and Workshops," 41 Vict., *c*. 16. See George Jarvis Notcutt, *The Law Relating to Factories and Workshops with Introduction and Explanatory Notes.* 2nd ed. (London: Stevens and Sons, 1879), x–xi.]

200. "Mr. Geo. Smith on the Inspection of Canal Boats: How the Act is Worked in Leicestershire," *Leicester Chronicle* [*Supplement to the Leicester Chronicle and Leicestershire Mercury*], October 16, 1880, 11. BNA.

201. Canal Boats Act [1884], 47 & 48 Vict., *c*. 75. See Webster, *The Law Relating to Canals*, 330–331.

202. See sample certificates of registration under the Canal Boats Act, 1877, BW 162/10/5/12, Waterways Archive, Canal & River Trust, Ellesmere Port, United Kingdom. The 1884 act was passed amidst what may have been the peak of late Victorian anxieties regarding housing for the working classes, marked by Andrew Mearns's *The Bitter Cry of Outcast London* (1883) and the Royal Commission on Housing (1884). The Housing of the Working Classes Act of 1885 offered significant new power to the LGB to close homes considered unhealthy.

203. Roy M. MacLeod, "Social Policy and the "Floating Population: The Administration of the Canal Boats Acts, 1877–1899," *Past & Present* 35 (December 1966): 105.

204. Local Government Board (LGB), *Fourteenth Annual Report of the Local Government Board, 1884–85* (London: Eyre and Spottiswoode, 1885), ciii.

205. LGB. *Sixteenth Annual Report*, cxxii.

206. LGB, *Sixteenth Annual Report*, cxxii.

207. In particular, see the increasing organization and activism of the National Society for the Prevention of Cruelty to Children (NSPCC), which became active in many major cities in the early 1880s. George K. Behlmer, *Child Abuse and Moral Reform in England, 1870–1908* (Stanford, CA: Stanford University Press, 1982), 71.

208. LGB, *Fourteenth Annual Report*, 74.

209. LGB, *Eighteenth Annual Report*, 208–209.

210. LGB, *Fifteenth Annual Report*, 68. In his review of Brydone's correspondence with the LGB, MacLeod has emphasized the degree to which administrators sought to erase more extended and emotive evaluations of living conditions on the canal boats with perfunctory language that simply responded to the demands of administrative record-keeping. "The assumption that government reports must be uncritical of their terms of reference, and that inspectors must be content to work precisely within a defined legal context, gradually reduced central canal boat administration to routine, unquestioning acquiescence." MacLeod, "Social Policy and the 'Floating Population,'" 123.

211. "Foul" or "offensive" cargoes were often biological waste. LGB, *Fifteenth Annual Report*, 18, 68–69.

212. LGB, *Fifteenth Annual Report*, civ.

213. LGB, *Fifteenth Annual Report*, cv.

214. LGB, *Sixteenth Annual Report*, 168.

215. LGB, *Seventeenth Annual Report*, 119–120.

216. LGB, *Seventeenth Annual Report*, 124.

217. LGB, *Seventeenth Annual Report*, 124.

218. Webster, *The Law Relating to Canals*, 331.

219. LGB, *Sixteenth Annual Report*, 171.

220. LGB, *Sixteenth Annual Report*, 171.

221. LGB, *Eighteenth Annual Report*, 213.

222. Royal Commission on Labour, *Minutes of Evidence, With Appendices, Taken Before Group "B." Transport and Agriculture . . . of the Royal Commission on Labour* (London: HMSO, 1893), 3:455.

223. Royal Commission on , *Minutes of Evidence, . . . Taken Before Group "B,"* 3:455–456.

224. Royal Commission on Labour, *Minutes of Evidence, . . . aken Before Group "B,"* 3:455.

225. "Appendix CLXXXV. The Regents and Grand Junction Canals. Canals Conference. Extracts from Report of the Conference of Delegates from the Vestries of St. Marylebone, St. Pancras, and Paddington, upon the state of the Canals in those Parishes, together with recommendations submitted by the Conference for adoption by the several Vestries with a view to ensure the future sanitary condition thereof" (Royal Commission on Labour, *Minutes of Evidence, . . . Taken Before Group "B,"* 3:610–611).

226. Royal Commission on Labour, *Minutes of Evidence, . . . Taken Before Group "B,"* 3:457.

227. John Brydone to Sir Hugh Owen, July 1, 1893, MT 32–94 (John Gordon Brydone, 1883–1889) [Personal correspondence]. National Archives, United Kingdom.

228. Committee of Council on Education, *Report of the Committee of Council on Education* (England and Wales), 1892–1893, 528.

229. LGB, *Twenty-Sixth Annual Report*, cxxxvii.

230. Mayall, *Gypsy-Travellers in Nineteenth-Century Society*, 126.

231. Mayall, *Gypsy-Travellers in Nineteenth-Century Society*, 131.

232. Mayall, *Gypsy-Travelers in Nineteenth-Century Society*, 134.

233. Mayall, *Gypsy-Travelers in Nineteenth-Century Society*, 142–149.

234. LGB, *Thirtieth Annual Report*, 210.

235. LGB, *Thirtieth Annual Report*, 215.

236. Register of Inspections [Hinckley Urban Sanitary Department], BW 117/2/2, Waterways Archive, Ellesmere Port, United Kingdom. There appears to be no catalog information indicating that this is an inspection book from Hinckley. However, incidental geographic references, as well as the number of boats indicating Hinckley for their registration, make the case a strong one.

237. Register of Inspections [Hinckley Urban Sanitary Department], BW 117/2/2.

238. Charles Hadfield, *The Canals of the East Midlands.* 2nd ed. (Newton Abbot: David & Charles, 1970), 212–213.

239. Education Act [1902], 2 Edw. 7, *c*. 42. See Montague Barlow and H. Macan, *The Education Act, 1902, with Notes*. 2nd rev. ed. (London: Butterworth and Co., 1903).

240. Board of Education, *Report of the Board of Education for the Year 1905–1906* (London: HMSO, 1906), 35.

241. Board of Education, *Reports of the Board of Education*, 1909–1910, 1910–1911, 1911–1912, 1912–1913, and 1913–1914 (London: HMSO, 1910, 1911, 1912, 1913, and 1914). See pp. 54, 88, 50, 83, and 71, respectively. Nonetheless, specific concerns regarding women on canal boats continued to manifest themselves. In 1906, a Wolverhampton coroner investigating the death of a small child indicated his anger at the situation in which the child's mother had reported that she had had fifteen children, nine of whom died in infancy, and that she had worked on a canal boat with her husband until with a short period of giving birth. It was reported that he had adjourned the inquest in order to bring greater attention to the situation. See

"Life on Canal Boats: Coroner's Strong Comments," *Nottingham Post*, May 4, 1906, 5. BNA.

242. LGB, *Fortieth Annual Report*, 67.

243. LGB, *Thirty-Fifth Annual Report*, 82–83.

244. H. C. Moore, "Among the Water Gipsies," *The Churchman* 16, no. 160 (January 1902): 214.

245. Moore, "Among the Water Gipsies," 214, 218.

246. "Canal Boatmen's Mission," *The County of Middlesex Independent*, March 31, 1900, 3. BNA.

247. Hugh B. Philpott, "Water Gipsies," *Methodist Magazine and Review* 61 (January to June 1905): 509.

248. Philpott, "Water Gipsies," 509.

249. G. H. Sargent, *Ned the Barge Boy* (London: The Religious Tract Society, 1885), 10.

250. Annie Gray, *The Old Lock Farm: A Story of Canal Life* (London: Sunday School Union, 1883), 9–10.

251. Mrs. Gordon, *The Life and Correspondence of William Buckland, D.D., F.R.S., Sometime Dean of Westminster, Twice President of the Geological Society, and First President of the British Association* (London: John Murray, 1894), 109.

252. Mrs. Gordon, *The Life and Correspondence of William Buckland*, 109.

253. Hadfield, *The Canals of the East Midlands*, 214.

254. Charles Hadfield, *The Canals of South and South East England* (New York: Augustus M. Kelley, 1967), 291.

255. Hadfield, *The Canals of South and South East England*, 294.

256. Hadfield, *The Canals of South and South East England*, 261.

257. L.T.C. Rolt, *The Inland Waterways of England* (London: George Allen & Unwin, 1979 (1950)), 154.

258. Rolt, *The Inland Waterways of England*, 155–157.

259. Winter, in analyzing the ongoing restraint in introducing new machinery in the Victorian era, has also emphasized the practical limitations of steam engines, which he indicates helped to lessen the overall disruptive impact of such technology and maintain an agricultural system that did not overwhelm the agricultural landscape. "Stationary alternating steam engines were costly to construct, service, and house; heavy power users like textiles and other enterprises whose operations allowed for a concentration of machinery around the power source were the first to employ them." (Winter, *Secure from Rash Assault*, 13).

260. See, for instance, David Blagrove's account of seeing horse-drawn craft in Birmingham in the early 1960s. (David Blagrove, *Bread Upon the Waters* (Burton-on-Trent: J. M. Pearson, 1984), 89.) Thus, while the opening of the carbon era has been commonly associated with freeing societies from the limitations of immediately available solar and biological fuel, such sources were not displaced in the case of canals. For a discussion of the revolution in fuel sources, see Sara B. Pritchard and Thomas Zeller, "The Nature of Industrialization," in *The Illusory Boundary*, eds. Martin Reuss and Stephen H. Cutcliffe (Charlottesville, VA: University of Virginia Press, 2010), 92.

261. Knowledge of swimming increased in the course of the century, however. See Vybarr Cregan-Reid, "Water Defences: 'The Arts of Swimming in Nineteenth-Century Culture,'" *Critical Survey* 16, no. 3 (2004): 33–47.

262. The canal boat population lacked a prominent and influential voice such as that of Rodney "Gipsy" Smith, who was both an evangelist and someone able to write and speak sympathetically of his background. See [Rodney] "Gypsy" Smith, *His Life and Work* (New York: Fleming H. Revell Co., 1909 (1901)).

263. See Martin Weiner, *English Culture and the Decline of the Industrial Spirit* (Cambridge: Cambridge U.P., 1981), 42–43.

Scenes of domesticity and labor in canal boat life, as documented by *The Illustrated London News* in the late nineteenth century. *Source:* Courtesy of National Waterways Archive, Canal & River Trust, Ellesmere Port, UK.

Boatmen in traditional dress aboard a fellows Morton & Clayton boat about 1910.
Source: Courtesy of National Waterways Archive, Canal & River Trust, Ellesmere Port, UK.

Canal boat family on a boat in the late nineteenth century. Reverse of photo print indicates the "Lapworth family" about 1887. *Source:* Courtesy of National Waterways Archive, Canal & River Trust, Ellesmere Port, UK.

Canal boat family posing with horses, likely used to pull canal boats, in the late nineteenth century. Reverse of photo print indicates the "Lapworth family" about 1887. *Source:* Courtesy of National Waterways Archive, Canal & River Trust, Ellesmere Port, UK.

A peaceful image on the Thames and Severn Canal in approximately 1910, at a time when traffic had significantly slowed. The canal would be largely abandoned by the mid-twentieth century, though restoration efforts began in the early twenty-first century. *Source:* Courtesy of National Waterways Archive, Canal & River Trust, Ellesmere Port, UK.

The Boatmen's Institute at Brentford was the outcome of missionary efforts directed at the canal boat population, which continued into the early twentieth century. This image dates from approximately 1905. *Source:* Courtesy of National Waterways Archive, Canal & River Trust, Ellesmere Port, UK.

Children, including canal boat children, at school at Brentford. *Source:* Courtesy of National Waterways Archive, Canal & River Trust, Ellesmere Port, UK.

Chapter 3

An Era of Decline

The Debate Over Canals in the Early Twentieth Century, 1906–1945

Between 1906 and 1945, many canals reached the nadir of their social and economic fortunes. Many owned by rail companies continued to see little investment, and some, owned by independent companies, faced minimal returns and the prospect of closure. Deteriorating conditions and intensifying international competition across a variety of industrial fields suggested to some observers that a nationally guided policy for canals was finally due. In 1906, a Royal Commission initiated an investigation of canals' finances and engineering, eventually identifying a series of steps for modernization and reform. Little action was forthcoming. And, despite the upheavals of World War I, many canals became known for both stagnant water and stagnant policy, a judgment little changed a generation later despite the mobilization of World War II. Canals continued to fade into obscurity. In 1945, many stood on the cusp of nationalization, not as a result of strategic economic and industrial investment, but as a result of the nationalization of the railways to which they were attached. Amidst rapid changes in technology, transport, and overall patterns of living, the stage was set for reconsiderations of canals' places as embedded features within Britain's natural and cultural landscapes. The reputation of canal boat people, so often maligned in the nineteenth century, had begun to be reframed as that of a stolid community struggling amidst economic, industrial, and social forces it could not control. More widely, under-utilized canals offered social and economic counterpoints for those concerned about what they believed was a twentieth-century fixation on speed and competition. The transformation of many canals from symbols of beleaguered industry to routes of recreation and cultural renewal would await the period after 1945. Yet, the early twentieth century was essential in setting the backdrop against which such a change would take place.

The visible decline of the canals' infrastructure intersected with national self-reflection regarding the nation's *laissez-faire* traditions, which had often kept national oversight and coordination in both industry and transport to a minimum. Constructed by a myriad of companies for local and regional purposes, canals often differed in their widths and lock standards, and their ability to accommodate boats of varying sizes. There were often separations between the companies that owned and managed canals and the companies that supervised transport. Railways, however, had largely adopted standardized technology due to the desire to take advantage of speed and affirm national links for both passengers and cargo. In the words of one observer comparing canals and railways as early as the 1840s, in contrast to "an air of assumption and parade about a railway," canals seemed to be "the embodiment of quiet, plodding undisguised sluggishness."[1] Between 1888 and 1898, the weight of goods carried by canal rose 8.5 percent to just under 40 million tons, due in part to the opening of the Manchester Ship Canal in 1894 (one of the few navigations that could accommodate larger loads with boats up to 600 feet in length).[2] The weight of goods carried by rail rose over 34 percent to above 378 million tons in the same period.[3] The fate of individual canals was rarely determined by those devoted to their interests. In the mid-nineteenth century, rail companies had rapidly submitted bills to Parliament that called for the unification of their holdings with those of local canal companies, and Parliament had generally acquiesced. By 1883, more than one-third of the navigable mileage of canals had passed into rail control.[4]

Between 1880 and 1920, Britain's economic and industrial position was threatened by industrial and urban transformations in Europe and the United States.[5] British officials were unnerved by the role canals still played in those other nations. In France, among the most significant projects was the *Canal de la Marne au Rhin*, built at mid-century linking the two great rivers across northeastern France and negotiating significant changes in height. French water routes could offer greater capacity and early on introduced steam and diesel propulsion as well as innovative systems of mechanized towing.[6] Beginning in the late 1870s in France, over 2,000 miles of canals and over 2,000 miles of rivers were upgraded and provision made for ships carrying 300 tons.[7] Britain's riposte to this expansion was the Manchester Ship Canal, first proposed in 1877 and then given concrete form in 1885 with the formation of the Manchester Ship Canal Company and an initial capitalization of £10 million.[8] The company sought to create a direct route between the inland city and the sea. Work on the canal began in late 1887, and despite a financial shortfall and the resumption of work on the basis of a loan from the Manchester Corporation in 1891, it was finally completed at the end of 1893 and opened by Queen Victoria on May 21, 1894. With a final price tag of £15 million, as well as a 170-feet width and 26-feet depth, it was just over 35 miles in length but transcended in scope the nation's other canals.[9] In 1904, the waterway offered access to ships 600 feet in length and up to 65 feet in width.[10] It made Manchester the third busiest port in Britain, though it was 40 miles from the seacoast. The drastic differences between the Manchester Ship Canal and the rest of the canal system of England meant that it was often assessed separately and apart from the rest of the nation's canals. Though some questioned its financial returns, carriage on it did not reach its peak until 20 million tons in 1958, as loads on many other canals lapsed.[11]

In 1901, H. Gordon Thompson (later to be famous for his medical work in China) was awarded a medal by the Cobden Club, an organization focused on classical liberal economics, for an essay he prepared on canals. In it, he found Britain's canal system, outside of the Manchester Ship Canal, in a "deplorable" state, noting the problems created by canals' differing dimensions and lock sizes. Lack of investment had been perennial, in contrast to France, where significant investments in waterways were now paying off. This lack of investment was, in some instances preventing British canals from taking advantage of the cost savings provided by steam engines, which offered greater efficiency but could not be used where canal banks had already fallen into disrepair.[12] Thompson went on to note that on the route between Liverpool and London, one might have to negotiate nine to ten different canal companies, and between London and Hull, eight to nine different companies; he argued for an amalgamation of companies controlling through routes.[13]

By 1906, Britain's new Liberal government, having sought to protect free trade in its most recent election battle, engaged with a host of issues regarding national efficiency and competitiveness, most notably how the nation might sustain its commercial and financial supremacy in the midst of competition from nations in Europe, as well as the United States.[14] Though canals were only one part of a wider debate, the government soon endorsed a Royal Commission on Canals and Inland Waterways to enquire into their status.[15] Over five years, the vast material gathered by the Royal Commission suggested canals' situation was both complex and uncertain. In particular, the Royal Commission was directed to report on five aspects of the nation's canals, including: (1) their current condition and financial position; (2) the reasons for which improvements on them had not been made by private enterprise, and "whether such causes are removable by legislation;" (3) improvements "desirable" in seeking to construct a system of through communication "by water between centers of commercial, industrial, or agricultural importance, and between such centres and the sea;" (4) "the prospect of benefit to the trade of the country compatible with a reasonable return on the probable cost;" and (5) "the expediency of canals being made or acquired by public bodies or trusts, and the methods by which funds for the purpose could be obtained and secured; and what should be the system of control and management of such bodies or trusts."[16]

Between March 1906 and April 1909, the Royal Commission interviewed 266 witnesses including canal and railway officials, merchants, manufacturers, and mine owners, and representatives of chambers of commerce and county and city councils. The report, as completed by 1911, contained eleven volumes covering minutes of evidence, several appendices, periodic reports covering Britain and Ireland, and a report on the commissioners' visit to the Continent. The Commission offered a sobering picture of the participation of canals in Britain's industrial growth in the mid to late nineteenth century.

> Considered as a whole, the waterways have had no share in the enormous increase of internal transport business which has taken place. . . . Their position . . . has been at best one of a stationary character . . . while the whole transport business of the country, including that taken by railways and that taken by coasting vessels, has multiplied itself several times over.[17]

At least part of this seemed due to the ways in which the rush of railway investment in the early nineteenth century had been allowed to overwhelm the canal industry. In some cases, canal owners, seeking to preserve the earnings of their shareholders, had sold out to railways.[18] The dysfunctional intermingling of canal and railway enterprises made a fractured canal system more difficult to manage. The Commission reported,

> The multiplicity and diversity of character and of interest of the various canal companies, public authorities and railway companies, which divide the control of varying lengths of this chaotic system of waterways have made it difficult to organize through rates and other arrangements for facilitating long-distance traffic.[19]

It concluded that the industry seemed strongest where there was a high population density and where significant numbers of collieries and wharves were located on or near the canal banks.[20] Overall, long-distance travel was limited, and transport was based upon the ownership of small groups of boats. One of the largest canal carriers, Fellows Morton & Clayton, had 219 boats in 1906.[21] (John Brydone had estimated there were approximately 7,800 boats in operation in 1897. See chapter 2.) As C. J. Aubertin, who traveled the canals in the immediate pre-war period, wrote in a 1916 account:

> Imagine, for instance, what a railway journey from London to Penzance would be like if the original order of things had subsisted . . . as it has on canals. The route would have been in the hands of five companies. . . . Imagine these of various gauges, necessitating a change and a fresh booking at the beginning of each

> new line, and you have a very fair parallel of the way canal traffic is conducted to-day. The railways saw the error of their ways and amalgamated; canals are largely as they were built. The railways, too, more by luck than judgment, agreed on the same size of gauge—except the original Great Western—but in consequence of the fatuous irregularity on the part of canals the whole system can only be navigated by the smallest unit. The smallest lock rules the whole, just as the strength of the chain is that of its weakest link.[22]

The Royal Commission concluded that there seemed to be few incentives or resources available for major investment in canals. The railway companies that controlled many of them were focused upon the development of rail, and overall access to capital seemed limited. Moreover,

> A company whose canal forms only a section in a through route would be wasting its money if it improved its own section . . . while any other section of the route remained without corresponding improvements. . . . The non-improvement of a single lock, indeed—might block improvement all along the line.[23]

Railway companies, it was noted, found little reason to direct investment money to canals, and if confronted by an improved canal under a competitor's ownership, would lower their own rates, undermining the incentive for investment in canals in the first place.[24] A lack of investment in canals had important implications in international trade as the twentieth century opened. Areas around Birmingham, the Commission concluded, "lose part of the natural advantage which they once possessed through the vicinity of coal and iron in days when foreign mines were less developed and sea transport was more costly."[25] Some manufacturers, it continued, had chosen to move their works to places where coal could be found close to seaports that also handled the import of raw materials and the export of finished products. Inland manufacturers were now

> in danger of being under-sold by competitors established on the coast, and even by manufacturers on the banks of the Rhine or the Scheldt, who may be able to obtain their raw materials at cheaper prices, and to export their manufactured articles, even to London, at lower rates than those possible, as means of transport now stand, for English manufacturers whose works are situated 70 or 80 miles from the nearest port.[26]

The Commission argued:

> Waterways are in all countries at a disadvantage with respect to certain kinds of business in competing with railways. But waterways in this country are also at a disadvantage, due not to the nature of things, but to a state of things which, in our opinion, has been to some extent brought about by errors of legislation,

> and by neglect on the part of Government and the Legislature. With a want of foresight not unusual in the affairs of this country, the control of the waterways was, in the first instance, allowed to fall into the hands of a multitude of small authorities; and, in the next place, a third part of these waterways was allowed to pass into the hands of railway companies, who had no interest in promoting trade upon them. The first evil might have been cured by a process of amalgamation, but this process was arrested and rendered almost impracticable by the second of these errors. Both errors have been followed by a period during which great things have been done for waterways on the Continent and nothing at home. Hence, our canals have been and still are in a position of arrested development.[27]

By 1909, the Commission's foreign and domestic inquiries had brought a majority to two major conclusions: (1) the capacity of four major canal routes connecting the industrial Midlands in England with major river systems ought to be significantly increased; and (2) supervision of England's and Wales's major canal systems ought to be placed under the authority of a Waterway Board that would have the authority to amalgamate, refurbish, and supervise traffic in the interests of providing a viable alternative to rail transportation. The four routes, often referred to as "the Cross" because of the way they connected with each other in the center of the country, were to be upgraded to better handle heavy traffic between the Birmingham and South Staffordshire region of England and the four major river estuaries of the Thames, Mersey, Severn, and Humber. The plan reflected, in many respects, that which had been set forth by the engineer James Brindley in the early stages of the canal era (see chapter 1). The results would both ensure that industrial areas were connected with major seaports and allow for the transport of goods that could be moved efficiently by water or that were unsuited to roads or rail. Goods envisioned as being particularly suitable for canals were dangerous items such as acids and explosives, and bulk items including coal and building materials.[28] Among the Commission's concerns were the limits on the size of boats that could be accommodated on Britain's canals. On the Continent, the size and capacity of boats had increased rapidly in the late nineteenth century. In Germany, the average tonnage of boats in 1877 had been eighty, while by 1902, it had increased to 220.[29] After the final volume of the Royal Commission's record of minutes and evidence, as well as reports, was published in 1911, no parliamentary action was forthcoming.[30] The roots of governmental inaction on canals aren't immediately clear. The nation's political attention appeared to turn to labor disputes, the struggle for women's suffrage, and debates over Irish Home Rule, all of which dominated headlines between 1912 and 1914. Nonetheless, the one-paragraph assessment the noted transportation writer Edwin A. Pratt provided in 1912 may have reflected the assumptions of many politicians:

> There is no reasonable ground for expecting much from the projected scheme of canal revival. When the country comes fully to realise (1) the natural unsuitability of England's undulatory surfaces for transport by artificial waterways; (2) the enormous cost which the carrying out of any general scheme of canal revival would involve; (3) the practical impossibility of canal-widening in the Birmingham and Black Country districts; and (4) the comparatively small proportion of traders in the United Kingdom who could hope to benefit from a scheme for which all alike might have to pay;—it is hardly probable that public opinion will sanction the carrying out of a project at once so costly and so unsatisfactory in its prospective results.[31]

War in 1914 fundamentally reshaped Britain's society and economy, as well as its standing in the world and ability to project power. Recruiting offices were full in late 1914 and 1915, as nearly 2.5 million men volunteered for the military effort, with consequent labor shortages felt in a variety of industries across the nation. Governmental power was extended through various iterations of the Defence of the Realm Act (DORA) and new limitations on both wartime industrial profits and the ability of workers to move between industries. Over 1,000 miles of railway-owned canals came under state supervision with the establishment of governmental control over the rail industry under the Regulation of the Forces Act of 1871, which had provided for government takeover of the railways in a declared emergency (though it also indicated owners were to be compensated for loss and injuries).[32] As part of the compensation agreement reached with the railways, the rail companies (as well as the canals associated with some) were guaranteed their net receipts of 1913 for the period they were under government control.[33] For the remainder of the war, portions of the system would be supervised by the Railway Executive Committee, made up of managers of the railway companies, to ensure military needs were met. Many day-to-day operations of the major railways, as well as many smaller lines, remained largely in the hands of non-governmental administrators. From an early point, many railway companies struggled with reduced staffs and the deferred maintenance that ensued. Many were prompted to give their workers increased compensation in the form of bonuses, due to an early tightening of the labor market, and so too were independently owned canals.[34] Men employed on railway-owned canals were subject to recruiting restrictions that had been put in place to protect the staffs of the railway companies, and received wartime bonuses. However, those working for independent canal companies, as well as carriers, could be actively recruited for the military.

The overall costs of operation increased on the waterways, and many independent canal companies struggled to maintain themselves. As they raised their tolls, some traffic was driven to already congested railways.[35] Eventually, the chair of the Railway Executive Committee called for government intervention to direct more traffic to canals, which might relieve pressure on strained rail infrastructure. For their part, the leaders of independent canal companies and traders complained of a lack of workers. Because workers were "born and bred on canals," they said it was difficult to find replacements for those who chose to leave the water.[36] As a result, just over half of the independent canal system was transferred to a Canal Control Committee supervised by the Board of Trade in early 1917, through powers granted by the DORA, with additional provisions made for the control of carriers. The tonnage on these canals had diminished from about 27 million just before the war to under 20 million when control began.[37] Of approximately 8,273 men employed in 1914 on these non-railway canals, only about 5,452 men were employed by early 1917.[38] The government soon reached compensation agreements with these companies, similar to those that had been established for railway-owned canals.[39] (A minority of canal mileage remained under the authority of independent companies and outside of government control through the end of the war.) The government also began paying subsidies to some carriers.[40] Helping to sustain the industry was that employees over twenty-five years of age were now protected in their positions under provisions set forth by the Ministry of National Service. Moreover, additional labor was directed to the waterways through the Transport Workers' Battalions. As part of an effort to combat the labor shortage, a training school for boatmen was created at Devizes on the Kennet and Avon Canal, where just over 200 boatmen were trained in the remainder of the war.[41]

Despite the Canal Control Committee's efforts, many industries resisted turning to the waterways, reticent to give up their access to rail facilities and reluctant to take on added costs associated with transport from plants to canals, as well as loading and unloading. In addition, as later noted by I. T. Williams before the Select Committee on Transport in October 1918, railways offered faster transit, the ability to order and maintain small stocks of material, the convenience of a single financial transaction using publicly accessible rate schedules, uniformity of gauge across an entire network, the flexibility of wagon cars which could be moved directly to a coal mine pit or positioned at an appropriate factory door, and fewer stoppages. Indeed, such a roster of complaints was to follow industrial canals through much of the twentieth century. In the late years of the war, tonnage on major non-railway canals continued to decline. A total of 1,724,207 tons were carried in August 1916 and 1,491,952 tons in August 1918. Compensation payments for non-railway canals were £433,000 between March and December 1917, £829,000 for the whole of 1918, and £1,324,000 for the 1919–1920 financial year, as wartime controls continued.[42] While the war had represented the first steps in shaping a unified national approach to canals, coordination within the industry had not been achieved. Railway-controlled canals were to follow the rail industry into supervision by the Ministry of Transport in 1919, but independent canal mileage was to be released from government control in late August, 1920.[43]

In the context of efforts to cut spending as the war closed, government officials noted the potential savings of ending compensation agreements with independent canals, as well as the future risks to those companies as the agreements ended.[44] On October 15, 1919, the Minister of Transport Eric Geddes and members of his staff received a visit from Neville Chamberlain, who was then a backbencher from the newly created Birmingham constituency of Ladywood, which surrounded the Gas Street Basin where the Worcester and Birmingham Canal met the network of waterways controlled by the Birmingham Canal Navigations company (BCN). Annual tonnage on the BCN had declined from 3,598,000 to 2,095,000 between 1913 and 1916.[45] Neville Chamberlain, whose father was Joseph Chamberlain, had demonstrated an early interest in social affairs and town improvement and had briefly become Lord Mayor of Birmingham before serving a short time as Director of National Service under Prime Minister David Lloyd George. He would soon take up chairmanship of the "Slums Committee" (Unhealthy Areas Committee).[46] Both Geddes and Chamberlain agreed that the position of canals was a bad one. Canals faced increasing costs and there was uncertainty about the effect on them of changes in the rail rates which might be introduced under government direction. Geddes, with pressure mounting to cut government spending, noted that canal subsidization seemed impossible, while Chamberlain made the case for eventual canal nationalization and the appointment of an executive body within the Ministry of Transport to encourage an increase in their use at least partially, he argued, to relieve pressure on a strained rail infrastructure. This executive committee eventually came to focus on the impact upon canals of any increase in rail rates.[47] Most important, however, Chamberlain secured Geddes' support for a more wide-ranging "advisory panel" that would consider the canals' future.[48] Chamberlain's steps were part of wider-ranging efforts to bring pressure to bear on canal issues. Early in February 1920, a deputation from the Canal Association, a group formed of many of the non-rail-owned companies that had been absorbed under government control in 1917, had met with the Canal Control Committee and outlined dilemmas facing the industry ranging from pressure for the eight-hour working day, to restrictions on the export of coal, to the departure of boatmen during wartime. The loss of government subsidies and support would mean a virtual collapse of the industry, its members argued. W. W. Sheepshanks, chair of the Aire and Calder Navigation, concluded that the Ministry of Transport ought to exercise its powers to take over the industry using the powers embedded in the Transport Act of 1919, at least until the future could be evaluated.[49] In July of 1920, Geddes recommended to the Cabinet that the Ministry of Transport take control (particularly for the purpose of establishing new rates) of any canal currently covered by the DORA, whose leadership sought it.[50]

At the same time Geddes made his recommendation to the Cabinet, Chamberlain became chair of the Committee on Inland Waterways, which had emerged as a result of his initial conversation with Geddes. Among the members were the railway economist W. M. Acworth, R. B. Dunwoody (who had served on the Royal Commission on Canals a decade and a half earlier), and the geographer and M. P. Halford Mackinder.[51] Later Ernest Bevin, then a leader of the Dock, Wharf, Riverside and General Labourers' Union, was added as a labor representative. The committee's directives were to consider what portion of the inland waterway system should be acquired by the Ministry of Transport and on what terms, what the form of succeeding ownership ought to be, what types of improvements might be carried out (and how an ensuing increase in the value of canal frontages might be recouped), and what type of cooperation could be established between water, rail, and road transport across the country.[52] In a memorandum of April 1921, Chamberlain concluded that thus far it was not possible to indicate that returns from tolls and increased traffic would justify the acquisition and improvement of canals by the Ministry of Transport. At least a portion of this uncertainty was related to the unsettled financial and economic conditions of the time. In comparing the questions confronting the Royal Commission just after the turn of the century and those being confronted presently, he contended that the recommendations of the Royal Commission had been bold but that the

> very size of their ideas frightened the Government and the public and proved their undoing. To-day, in so far as circumstances have changed, they have changed for the worse. The capital required for the Royal Commission's schemes would be multiplied several times and the country is in no mood to listen to any proposals of any expenditure by the State which is [*sic*] not absolutely necessary.[53]

Chamberlain's early assessment revealed the present committee's turn toward a series of regional trusts, each of which might encompass at least one main through route.[54] Chamberlain tentatively laid out a plan for six regional canal groupings, suggesting that recommendations on them be developed on a step-by-step basis.[55] Later, an interim report in 1921 from his committee argued for improvements to the River Trent that would make it possible for 300-ton boats to reach Nottingham. Afterward, the committee's main report spoke against nationalization but called for the creation of a Waterways Board that would re-organize Britain's canals into seven regional groups, each with its own Board of Management.[56]

Despite Chamberlain's efforts, canals received little attention or governmental action in the course of the 1920s; no regional groupings or Waterways Board were established. The nation's attention was focused on the rails and, increasingly, the roads. In 1924, Ministry of Transport statistics indicated 2,183,933 tons were carried by railway-owned canals and 14,272,016 tons were carried by other companies for a total of 16,455,949 tons. In 1929, those same statistics indicated a total of 1,772,663 tons on railway-owned canals and 12,589,517 tons on non-railway-owned canals for a total of 14,362,180 tons.[57] In the course of the previous decade, two major infrastructure/business operations had been undertaken. The Corporation of Nottingham and the Trent Navigation Company, supported with government grants, had undertaken a series of major improvements which allowed for greater capacity on the River Trent.[58] In addition, five independent operations, each owning a portion of the overall route from London to Birmingham—the Regent's Canal and Dock Company, the Grand Junction Canal Company, the Warwick and Napton Canal Company, the Warwick and Birmingham Canal Company, and the Birmingham and Warwick Junction Canal Company—amalgamated in 1929 to form the Grand Union Canal Company.[59] Yet, in making a final report in 1931, a Royal Commission on Transport appointed in 1928 paid relatively little attention to canals, offering only a sober overall assessment of the failures of canal companies, emphasizing the continuing problems of fractured ownership and the routing of through traffic.[60]

As the nation grappled with postwar structural and economic conditions on the canals, so too did it with the experiences of the remaining canal boat population. The loss of canal boat workers to other industries and the armed services had been significant in wartime, a trend which continued into the 1920s. Yet, public understandings of the canal boat population were shifting. In a 1917 account remarking upon its efforts at the Boatmen's Institute in Brentford, the London City Mission (LCM) argued that "lurid pictures of their moral degradation, of over-crowded cabins, and of neglected children belong almost entirely to the past." It added,

> Nor is the world of the bargee and his confrere on the monkey boat [narrowboat] any longer bounded by the canal banks. Education and labour fellowships altered that. The ability to read, now fairly general, has widened their horizon, increased their interests, making them less than formerly a race apart.[61]

Rather than a den of immorality and filth, boat cabins were quaint and neatly kept, in-step with a patriotic, national endeavor. "Cleverly arranged, compact as a watch, and models of neatness and cleanliness, the cabins reflect great credit on these 'water gipsies,' as they are called."[62] Women's dedication to maintaining a home on a canal boat mirrored the dedication of wider domestic efforts in wartime.

> The missionaries cannot speak too highly of the women, whose work never seems done. . . . Since the war started hundreds of boatmen have entered the Navy and Mercantile Marine. Instead of sailing through peaceful, rural England, they are now mine-sweeping, or helping to defend our shores. In many cases this has necessitated the 'missus' acting technically as the 'mate,' and right nobly is she fulfilling the extra task.[63]

Thus, in the context of war, the boat people had apparently become not a foreign race, but a domestic one. As the LCM contended,

> Monkey-boat people are the aristocracy among boat dwellers. "Born on a boat, they have lived on a boat all their lives, and will probably die afloat," writes a missionary who has close on one thousand such craft under visitation. "Inclined to be clannish, they have high moral ideas and invariably live up to them. They are bad 'mixers,' and prefer to stand aloof from aliens—a term used for canal workers who have drifted thither from necessity, or, as is frequently the case, from misfortune."[64]

In November 1919, the President of the Board of Education, along with representatives of the newly formed ministries of Health and Transport, received a deputation (including Ernest Bevin representing the National Transport Workers' Federation) which drew attention to the living conditions aboard canal boats and in particular to the problems of education.[65] At the same time that Chamberlain had become chair of the Committee on Inland Waterways, the groundwork had also been laid for a Departmental Committee on Living-In on Canal Boats (also to be chaired by Chamberlain) under the authority of the Ministry of Health.[66] At the outset, the committee was hampered by a lack of definitive information from both the Board of Education and the canal boat inspectors, which would allow them to estimate the size of their problem. Moreover, vital statistics by which to measure the health of the canal boat population seemed non-existent.[67] Testimony before the committee generally built upon the responses submitted by witnesses ahead of time and focused on education and the prospective steps in restricting living-in that might assist in children's education, followed by concerns regarding morality. Committee members represented a wider field of progressive political and social thought than bodies convened before the war.[68] Among the witnesses providing new perspectives on canal boat life was Clara Jones, a "certificated" nurse and midwife in Brentford. With seven years' experience in attending approximately thirty births associated with the canal boat population, Jones said that she had had few difficult births aboard canal boats, though she believed it would be better for women not to work up to the time of birth, and that rooms ashore would be much better than the cramped accommodations of boat cabins.[69] Her overall testimony was brief, but suggested the difficulties faced by canal boat women during birth were little different than those faced by other poor women. She had known only one mother to die during birth and there had been only one instance of a baby being born dead.[70] Startingly for members of the committee (as they queried her on it more than once), she had apparently seen no illegitimate births on the canals. She contended emphatically:

> Never; many on shore among those round about the boats, but never once on the boats. In seven years I have never been called to a case. I find them a clean, honest, hard-working race of people, and very good to pay up. When your work is finished, there is your money for you; and very often they are confined to-day, and when you go to look for them the next day they are gone up the river. That is the only thing I do not like about them: they are gone before you really know how they are getting on. I never heard of any wrong come to them; they just come up in another 12 or 15 months and want me again. They seem to get on all right.[71]

As she departed, Chamberlain thanked her for her "quite unique" testimony.[72] And yet, the committee completed its inquiry with perhaps an even more "unique" perspective—that of a boatman, himself. Called to offer testimony, William Bagnall, whose family, he said, had been in boating for generations, reported in relatively few words that he was not sixty and had been on the canals since the age of eight, noting that he was the youngest of fourteen children. His parental family had maintained a house ashore (and their children were born ashore), but all had been involved in boating. Bagnall said he had been on the canals his entire life, adding, "Manchester, Liverpool, Ellesmere, Chester, all round Staffordshire and that way—but continually, most of my life, up and down the Grand Junction."[73] Bagnall said that boats had improved "wonderfully" in his time, contending that, "There was very low life on the boat I was on when I first came about. Everything has improved."[74] From his perspective, Bagnall said it seemed as if there were more women now on canals, and he did not think they would easily leave if prompted to. Children, he said, were healthier aboard the boats, though education remained a problem. To put children ashore for education, however, would ruin their chances of taking on the trade. "If I had put my children on shore for education, they would simply have lost their work in the canal boats. If they are not brought up to it from their cradle, they cannot follow the employment."[75] Overall, Bagnall said, most adults on the canals now could read and write. In response to the overall question of whether children were more liable to accidents onboard the boats or in the streets, he argued,

> They are under our control every minute on the boats, and they are not in the streets; . . . Our boat-children are under better control than any class of children going; . . . They are better controlled by the parents, simply because if they touch anything, they are stopped at it there and then; and if they are in any kind of mischief, we are there to see it; because the place is so small and such a little compass that they are not out of sight.[76]

The conclusions of the Departmental Committee's report in May of 1921 made it clear that condemnation of the health, sanitation, and morality of the boat population, so common in the late nineteenth century, had been replaced by a more nuanced understanding of the conditions they faced and a belief that their living conditions were not significantly worse (or better) than similar working-class populations. Educational efforts were, however, significantly behind those of children ashore and required intervention, it contended—most likely the removing of children from the canal boats to attend traditional schools. The committee emphasized the lack of statistical information available on the numbers of people living on canal boats as well as their vital medical statistics. Nonetheless, it indicated that "the majority of the witnesses have agreed that, so far as health, cleanliness, morality, feeding and clothing are concerned, they are fully equal, if not superior, to town-dwellers of a similar class."[77] It went on to say that standards of morality and decency generally seemed to have been followed even though boats were crowded, but that "sanitary conveniences" were needed in a number of areas.[78] Information concerning accidents seemed to be somewhat contradictory, but that regarding education was significant. The committee wrote that,

> here the evidence is overwhelming and practically unanimous that under present circumstances canal boat children are scandalously under-educated. When their manner of life is considered, it is not surprising; their only opportunities for schooling occur when the boats are tied up . . . and the fact that many of the adult boat population are themselves unable to read or write has a tendency to make them lax in seeing that their children take full advantage of their opportunities.[79]

The committee concluded that there could be significant concerns with the separation of families if children were removed from canal boats, but that this seemed to be outweighed by the benefits: "We therefore recommend that children of school age, that is, at present, between the ages of five and fourteen years, should be prohibited from living-in on canal boats during the term times of the schools which they are attending."[80] In implementation, the committee recommended that a grace period of twelve months be provided from the passing of any act of enforcement.[81] No action was taken as a result of the committee's recommendations, though tension regarding education continued through the 1920s.

The immediate post-World War I era had drawn attention to the struggles of canal companies, carriers, and the canal boat population. Each facet of canal life had highlighted the continuing struggles of an industry, even as its workers had shed the nineteenth-century cloak of moral condemnation. In the coming years, there would be little to attract new workers to the waterways, and the relative insulation of remaining workers and their families would continue, if not intensify. These concurrent transformations intensified the association of canals with a bygone industrial era in Britain and the association of workers with traditions of labor and living that were at once anachronistic but also independent. Indeed, determining the number of children onboard boats remained extraordinarily difficult, at least partly because of inconsistent inspection. In Hinckley on the Ashby Canal, a total of seven boats were inspected in 1915 and nearly twenty in 1916, apparently under a different inspector.[82] Under yet another inspector, only two seem to have been inspected in 1917 and two in 1919.[83] (No record of inspection in 1918 is apparent, though this was likely due to a major breach of the canal that was not fully repaired until 1919.) Trade on the canal, over 138,000 tons in 1862, had declined to just over 33,000 tons by the early 1890s.[84] In response to a circular of December 1919, the inspector J. S. Featherston, reported,

> During the year frequent visits have been paid to the Canal, but only two boats were found. . . . No infringements of the Acts and Regulations have been noticed, and no cases of infectious disease found. No boats have been detained for cleansing and disinfection. There is only one boat on the Register which can be traced as being in use. There are no motor propelled boats on the Register. No boats have been registered in 1919.[85]

Yet, the next year, while noting the same general information on infringements, infectious diseases, lack of new registrations, and lack of motor-propelled craft, a different inspector reported visiting twenty-five boats. The inspector reported seeing no boats in 1922 or 1923. Perhaps a portion of this was due to the lack of a visit by Owen J. Llewellyn, as his initials for 1922 were not indicated until 1924. After an apparent lapse of two years, inspections returned in a much more significant way beginning in 1927 and appear to have continued under a single and much more consistent inspector until the mid-1930s, when they again began to dwindle.[86]

Some efforts to have children removed entirely from canal boats did emerge in the late 1920s, most notably when Labour MP Harry Gosling introduced a bill in 1929 to ban children under fourteen from traveling or living on boats.[87] Gosling's roots in both waterborne trade and education ran deep. The son of a waterman and schoolteacher, Gosling was an early member of the Amalgamated Society of Watermen, Lightermen, and Bargemen and later served on the Port of London Authority. In addition, he became a member and president of the Transport and General Workers' Union in 1922.[88] He was a member of the London County Council between 1898 and 1925 and later became MP for Whitechapel and St. George's and, briefly, Minister of Transport and Paymaster General in Ramsay MacDonald's first government. While General Secretary of the Amalgamated Society of Watermen and Lightermen, Gosling campaigned against the cabin accommodations of boys and men on the River Thames.

Gosling's bill faced opposition from a number of sources. Although arguing children should not be on canal boats carrying refuse or "foul" cargo, J. H. Clayton, director of Fellows Morton & Clayton, argued that it would "ruin" canal boat traffic, bringing an increase in the expenses of working the boats, and that he believed women would follow children in leaving the canals, leading to potentially unhealthy conditions ashore.[89] Moreover, he indicated that there were proposals for a new floating school currently being considered.[90] G. H. Collier, head of water transport for Messrs. J. Lyons & Co., contended that, "Bargees' are bred and born to the work for generations, and would not be happy in houses ashore, even if such accommodation *could* be found."[91] The Canal Carriers Association and the Canal Association, in a combined statement, indicated that narrowboats could likely not be worked economically by just a crew of two men, that the custom for 150 years had been for families to live on the boats, that men would be separated from their children, and that new homes would have to be provided for wives and children. Referring to what they believed to be the much more limited proposals of the Departmental Committee, the associations argued that the bill should only apply to children of school age during term times and that a grace period should be provided before its provisions came into operation.[92] A memorandum from the Paddington Boatman's Institute also indicated,

> The fact is that the people of the barges are particularly attached to their children, (the story which is put about that they are bad and unnatural parents being a wholly unfounded aspersion), so much so that they will not put up with the compulsory ration of the family which the Bill proposes to enact, and the effect of such an enactment will be to destroy the hereditary trade of an inoffensive and useful class of people.[93]

In supporting a Second Reading of the bill, Gosling argued that it was supported by many men who worked on the canal boats as well as the National

Society for the Prevention of Cruelty to Children (NSPCC). However, Conservative Lord Eustace Percy, who had served as President of the Board of Education between 1924 and 1929, argued that there was an unsettling level of state compulsion attached to the bill,

> The proposal of the Bill is that the population, as a matter of principle, should live invariably in places where they can be properly inspected by a public authority. . . . That is a very dangerous principle, and one which we have never yet quite adopted, that the locus of the family habitation should be determined by its immediate inspectability by a public authority, and that is what you are saying in the first place in this Bill.[94]

Sent to a Standing Committee, the Canal Boats bill was amended to prohibit children over the age of five and younger than school-leaving age from being on board boats except after the end of the last school hour of one week and the beginning of the next school week, including the period between terms. It also prohibited children from being on boats carrying manure or other foul refuse. Enforcement was to begin in 1932.[95] The parliamentary session ended, however, with the bill failing to become a law. In subsequent days, interest in the fate of canal boat workers continued. Labour MP Susan Lawrence, parliamentary secretary to the Ministry of Health and chair of the Labour Party, pressed for a medical examination of young children living on boats.[96] Subsequently, George Newman, chief medical officer of the Ministry of Health, recommended that Lilian Wilson, a medical officer of the Board of Education, be dispatched to Brentford in London and Wolverhampton.[97] As part of her study of Brentford that year, Wilson visited nineteen boats, directly observed seventeen children on the boats, and a further nineteen at the Brentford school. (Also included in her study were forty-two boat children examined by the School Medical Officer for Chiswick and forty-two children who had accepted dental care.)[98] Wilson noted in her report that there was suspicion among families and difficulty in getting reliable data, but that no active resistance was encountered from those on board. She wrote:

> The barge people are unique in many ways. They have a strong clan and family sense and are even patriarchal in type. For instance, no information would be given to a stranger unless approach was made through the wife of the Captain. In one or two cases where the mistress of the barge was temporarily absent, nothing could be gained from adults until her return. The occupants denied the presence of children on board, and did not even know their names until it became safe to do so. (It seems possible that the barge folk have a rather large number of casual inquirers into their mode of life and they are sturdy independent people who dislike interference).[99]

Wilson found that the principal causes of infant mortality were "convulsions, pneumonia, burns, drowning or overlaying," adding that she felt that the last three causes were potentially due to the cramped conditions of narrowboats.[100] Nonetheless, "deaths from drowning were probably fewer than would be attributed to street accidents in a corresponding slum on land."[101] Wilson noted that infant mortality at first appeared high but that Llewellyn had indicated some estimates on family size were high and "a few persons may tend to overestimate their losses."[102] The feeding of infants, she remarked, was frequently "not on modern lines," due at least partly to "the strong influence of the grandmother—a powerful force with antiquated views—and also because cupboard accommodation on the boats makes fresh milk an impossibility."[103] Wilson went on to remark that women's "confinements" appeared normal, though work soon after birth may have meant poor-quality breast milk.[104] She added,

> Barge etiquette apparently demands that mother and new-born child should not issue from the cabin until the mother has been "churched." This may not occur for 4 weeks or even longer, so it is somewhat surprising that postpartum recovery is so normal and that more infants do not succumb, as fresh air is considered detrimental at this period.[105]

Despite its comprehensiveness, Wilson's work did not affect the progress of Gosling's bill as he died in October 1930 and the bill was not re-introduced. Moreover, the Labour government collapsed on August 24, 1931, to be replaced by Ramsay MacDonald's National Government.

While there was a significant blunting of anxiety regarding morality and health on canal boats in the 1930s, the more acute concern for the status of the system's physical infrastructure led to only limited changes. These included the creation of the Grand Union Canal Company stretching from London to Birmingham primarily on the routes held by the Regent's Canal and the Grand Junction Canal, and improvements to navigation on the River Trent. Wilfrid Henry Curtis, Chairman of the Grand Union Company, offered a hopeful presentation at the Royal Society of Arts in May 1935. Rather than an industry in decline, Curtis saw one in a period of revival, a "priceless" asset for the nation.[106] After its formation, and supported by a government loan of £500,000, the Grand Union began an over £800,000 project of deepening channels and updating locks that facilitated increased tonnage from London to Birmingham. The company opted for craft of 12 feet, six inches in width and the reconstruction of fourteen bridges while carrying out an update of locks that would ultimately accommodate 14-foot wide barges.[107] A major dredging project accompanied this effort and was completed in late 1934. The reconstruction efforts meant that it was eventually possible to travel the distance from Limehouse in London to Birmingham (151 miles) and Brentford to Birmingham (135.5 miles) in approximately fifty hours, traveling day and night, and seventy-two hours if traveling according to a normal schedule. Curtis emphasized the advantages drawn from the company's independently controlled dock in London, which allowed goods to be loaded and unloaded directly to/from ships.[108] In addition to the efforts undertaken by the Grand Union, Curtis also remarked upon concurrent improvement efforts being undertaken on the River Trent. The efforts of the Grand Union, he noted, had been hampered by the drought experienced over the past eighteen months.[109] Nearly all the boats now being manufactured for the Grand Union were motorized, he added. Yet, Curtis's evaluation could not speak for much of the rest of the system, where routes remained too narrow for large loads, lock systems were of differing sizes, no efficient means of collecting tolls had yet been implemented, and little capital was available for investment. Some previously important canals, such as the Grantham and Nottingham canals, had now begun to be closed to traffic.

Curtis's emphasis on the technical updating of the Grand Union ran counter to his views of the canal boat population, which he saw as a cultural anachronism, even if they were an essential part of the continued operation of his route. The Grand Union, he said, "passes through some of the most beautiful country we possess, and in view of the healthy open air life that the canal community lead, it would be strange indeed if the general level of health were not correspondingly high."[110] Most canal "folk" looked the "picture of health," he said, adding that they were an "interesting and worthy body of men and women" and that their children "escape even the minor ailments" that children from families in other walks of life encounter.[111] "Epidemic diseases are virtually unknown among these children of the open air," he wrote. The population, he said, had changed little over the last 100 years, and it was only in the last twenty to thirty years that "steps have been taken to assist them socially and spiritually."[112] Indeed, Curtis closed his talk before the Royal Society with a sense of pride that the Grand Union Company had provided a school afloat that was ready for use after the defeat of the Canal Boats bill of 1930, an "ill-fated" effort that "would have ended family life on our canals."[113]

The number of lives which might be lost as a result of canals, however, ranged far beyond just those of the canal boat population. Throughout the 1920s and 1930s, drownings were a consistent part of news stories focused on canals. Often, the stories were conveyed through brief notices of coroners' inquests or verdicts, likely heightening concerns regarding the links between canals and death. Victims, often working men walking along the canal bank or young boys playing near a canal bank, died accidentally as they slipped into deep water near the shore. For instance, in the early summer of 1922, a man walking along the St. Helens Canal in the company of two other men fell in and was found later that same evening, apparently drowned.[114] In the spring of 1923, a child aged six, whose parents worked a boat that traveled between Paddington Basin and West Drayton on the Grand Junction Canal, died apparently after leaving her house near the canal bank to play with other children. Her body was found in the water near a plank from the shore to her family's boat.[115] In the spring of 1925, a man walking along the Stourbridge Extension Canal at Pensnett drowned after apparently slipping and falling into the water, though there had been an immediate attempt to rescue him.[116] Sometimes, a coroner's inquest could reveal prevailing conditions and prompt some action. For instance, in the late summer of 1926, an inquest on a boy who had fallen into a canal while accompanying others fishing prompted the coroner to request the posting of warning signs for children at particularly dangerous points on the waterway.[117] On January 6, 1931, a chemical worker's body was recovered from the St. Helens Canal. He had gone to work in the fog at night, with a search of the canal the next day revealing his body.[118] A few months later, on April 9, 1931, a woman who had had fits of fainting was found drowned in the Leeds and Liverpool Canal, the circumstances indicating she had likely fallen in.[119] On June 21, 1935, an eight-year-old boy drowned in the Coventry Canal while out playing with friends.[120] On November 4, 1938, a railwayman returning from work on foot along the Grand Union Canal in London was recorded as having fallen in and drowned, perhaps as a result of a narrowed towpath due to railway work.[121] Continuing stories of deaths along canal banks in the interwar period would soon help to drive pressure to close little-used or unused canals. By the mid-1930s, these conditions affected the delicate position of canals in the British mind. Crime no longer appearing a pressing issue, danger and death were now tied closely to drowning.

Beginning in 1939, in calling for the commitment of the nation's resources to war, Britain's government directed new attention to canals. Though such attention would pale in comparison to the intensified demands placed upon road, rail, and international shipping, it initiated one of the most targeted, if brief, investigations of canals' then present physical and economic statuses, as well as future potential. At the beginning of the war, railway-owned canals followed their rail counterparts into government control. Independently owned canals faced no nationalization, partly because of concerns regarding the costly expenditures of World War I.[122] They were organized into six Regional Canal Committees to better respond to the needs of industrial transport, but faced escalating costs.[123] By the summer of 1940, the Government began granting subsidies for canal tolls, with the Ministry of Transport required to give approval to all increases in tolls.[124] Recruitment for labor on the canals was difficult, particularly since the practice of carrying companies engaging captains who then hired their crews had made no provision for specific wage apportionment to crew members. Efforts to reach agreements on wages later in the war had only limited success.[125] Efforts to steady the labor supply on canals with women workers proved difficult. Though women had been an enduring part of the canal labor force as members of family boats, a Grand Union Canal official said in April of 1941 that he doubted women could take charge of canal boats themselves, primarily due to lack of strength; he also indicated that thus far the company had been turning down applications from a large number of women who had applied.[126] However, by 1942, women had been trained and were at work in Birmingham.[127] In addition, Sir Osborne Mance, the Director of Canals at the Ministry of War Transport, noting that women were now being trained as part of an experimental program on the Grand Union Canal, made a plea for labor from both young men and women. He cited canals as essential to the war effort.[128] By March of 1944, eighteen women were working on the Grand Union taking cargo between London and Birmingham, with the call for more women workers being repeated.[129] In 1945, Eily Tolley Gayford, a trainer of women's crews for canal boats, was made a Member of the Order of the British Empire for her work.[130] Despite government action in taking over a further group of canals in 1942, and women's work on canal boats in the absence of men, the overall decline in canal tonnage continued, with the 13 million tons carried in 1938 dropping to 10 million by 1946.[131]

This decline and the forces behind it were critical for Frank Pick, a man who had already established a brilliant career in rail administration, serving as chief executive officer of the London Passenger Transport Board between 1933 and 1940 (during which the famous "bulls-eye" branding of the London Underground system was established), and was now appointed to study further utilization of canals. Compiling his report over the span of just eight to nine weeks with a focused and practical eye, Pick offered an accessible and concise assessment of the overall system, including an examination of the state of its finances, tonnage, engineering works, and materials available for boat repair. It was designed to provide a series of proposals which the Ministry of War Transport might act upon to increase the capacity of the system given wartime exigencies. There were many complexities that were ignored. The report included no testimony from experts engaged in engineering, business, or the social welfare of canal workers. Detailed assessments of individual canals, representing unique combinations of engineering and business dilemmas, were limited. Moreover, Pick himself noted what might be the unreliability of some numbers included in the report.[132] Yet, in its content, structure, and conclusions it delineated, perhaps for the first time, critical logistical dysfunctions which if acted upon might ensure a place for canals as a consolidated transport system providing service to the nation.

Pick wrote that the financial position of the waterways was mixed, with only some now able to pay a return on capital invested. The most financially successful had been able to secure income streams from other sources, such as proceeds from land investments or water rights purchased by other industries along canal banks. The most financially successful of the canals Pick studied was the Oxford Canal, which before the war was able to pay an eight percent return on its ordinary stock. Overall, across the canals, the average return on all sources was less than two percent, and less than a third of this return came from the operation of canal traffic. Pick wrote, "Most were just carrying on, while some were already hard up."[133] According to the report, the wartime regional Canal Committees should be strengthened and a centralized administrative body established that could effectively coordinate with other wartime ministries.[134] In Pick's assessment, such a step should precede more effective centralized control of canals and an end to direct railway control over the canals they owned. Old management systems that had failed both financially and in maintaining infrastructure should no longer prevail in the postwar era. He argued for upgrading transshipment facilities to improve efficiency and highlighted the minimal and uncoordinated standards to which canals had been maintained. Dredging, prevention of bank erosion, and lock standardization, all needed to be upgraded. The introduction of motors had increased the rate of bank erosion perhaps four to five times in some areas. [135] Pick wrote,

> Canals appear to have always lived from hand to mouth. . . . When they have made an improvement it has had relation to the requirements of the day and has not been related to any consistent scheme for the redevelopment of the whole property. There is too much patchwork.[136]

In the context of government attempts to rationalize their operations, complexities and inefficiencies had been revealed in canals' carrying operations and labor policies. Pick described carrying as "really disorganized."[137] Rather than reporting based on the compilations of returns from local administrators, an inspecting officer should be tasked with visiting each canal, perhaps on a three-year cycle, so that the circumstances of individual undertakings could be assessed and investigations carried out on proposals for the abandonment or closure of routes; serious accidents could be evaluated; and approval given to extensions and improvements. Pick's frustration with the lack of information available on individual watercraft was palpable, as he argued in support of registration forms for vessels, which would be renewed at regular intervals. He wrote:

> The only registration of canal craft that now takes place is for purposes of the Ministry of Health. This registration relates only to craft used for human

> habitation and is used primarily for sanitary and hygienic purposes, secondly, for education purposes when there are children on board. The local authorities carry out a spasmodic inspection of inhabited craft, some being more diligent in the matter than others.[138]

Retaining wartime workers and ensuring standardized practices in a system where no single authority had yet managed finances, logistics, maintenance, or labor practices was a formidable challenge, according to Pick. He noted that the boat building and repairing business was broken up across many yards and that as independent workmen, builders were often men "of force of character" who had advanced themselves and become independent but were not prepared to grapple with the new demands of a coordinated industry. He wrote that they lacked that "quickness and agility of action and thought which government intervention in so many phases of their business renders necessary, if success is to be obtained or even if equity is to be secured."[139] Furthermore,

> They have no association or organisation to watch their interests or look after their requirements. As a result, they appear not to have been consulted or considered in any of the instructions or regulations issued by the Ministries of Supply and Labour and have been left to make the best they can of a worsening situation.[140]

Pick cited the adverse impact upon the canal industry of those who left for more attractive occupations and called for canal carriers to apply to the government for protection of their personnel.[141] Pick's reflections on the unique experiences of families who worked the boats were limited, though he noted:

> At points of regular frequency it might be worth while to build a bath house and laundry with drying room, to improve the lavatory accommodation and generally supplement the present inadequate facilities. It may be said that a war is the wrong time to embark on such an innovation but the effect of such a venture might be considerable. An expenditure of £5,000 of government money might be worth while. The first experiment is sure to be abused but by perseverance a bit better standard of living will be reached. Such a course may be unavoidable if outside labour is to be brought in. The canal folk are a class apart used to their peculiar life, inter marrying and continuing from generation to generation. On one day a grandfather, father and son were met with on separate craft. These folk are only just beginning to realise that there is the possibility of a better life before them. The young folk are drifting away to munition factories and high pay. Some measure to arrest a sudden withdrawal may have to be contemplated. An interest in the canal life and a suggestion of amenity might serve the turn.[142]

In the latter stages of his report, Pick discussed several issues associated with logistical changes meant to improve the overall operation of canals. There was a good deal that spoke against some elements of the system. Narrowboats were inefficient, and locks, particularly where narrowboats were concerned, intensified the problems of canals in comparison to roads.[143] Pick argued that what the canal industry needed most was "encouragement."[144] It must be offered assurance that a "progressive" management was in place and that it would have "a stable and secure future" within a coordinated national transport system.[145] "Only the Ministry of Transport by discharging its responsibilities generously and frankly, can create this. Such must be the conclusion of this Report," he wrote.[146] After Pick's report had been received by the Ministry of War Transport, it was circulated on a very limited basis, with copies being sent to the Ministry of Supply and the Mines Department, but only a digest of recommendations to the Inland Transport War Council, the Central Transport Committee, and the Canal (Defence) Advisory Committee.[147] The restriction of the report prompted numerous inquiries, though in February of 1942, Philip Noel-Baker, the parliamentary secretary for the Ministry of War Transport, cited the problems of security and the potential precedent of publishing an internal departmental report, and noted that the hurried nature of the report and the need for further consideration argued against publication, particularly in light of potential postwar policy.[148] Though Pick began a series of follow-up reports on the status of individual canals, after his death in November 1941, much of his specific work on canals lay dormant. Beginning in June 1942, eighteen canal companies and what would eventually be twenty-three carrying companies were put under the authority of the Ministry of War Transport.[149] By 1945, government intervention in the canal industry may have cost approximately £2.6 million.[150]

Across Britain, government intervention in the economy was unprecedented, and new calls emerged for wartime planning to be applied to a postwar world in which the benefits of victory would be widely shared. Collectivism gained strength, marked by Sir William Beveridge's 1942 study (the "Beveridge Report") calling for a full plan of national insurance, as well as a national health program.[151] Although an increasing share of Britain's population would continue to reside in towns and cities, the war profoundly affected the countryside politically, socially, and economically. In addition to the placing of some 20 percent of England and Wales under military control, "ploughing up" of new areas to increase agricultural production meant an increase in land under cultivation from 8.3 million acres to 13 million acres between 1939 and 1945.[152] Production gains by 1944 were significant, ranging from 90 percent for wheat to 19 percent for sugar beets, though there were losses in pastoral farming.[153] The fertility of the soil was strained and thus greater amounts of fertilizer were applied including potash, phosphate, and sulfate of ammonia.[154] The number of tractors in use in Britain increased from 52,000 in 1939 to 125,000 in the spring of 1943 (which the postwar planner Sir Dudley Stamp later claimed as the addition of 2 million horsepower to the land).[155] The increase in tractors was accompanied by significant increases in the tools pulled by tractors, as well as threshing machines and gyrotillers. The number of combine harvesters rose from 150 to 1500 between 1939 and 1943.[156]

Planned use of the countryside became a theme of both war and postwar visions of a transformed Britain, including land nationalization and government intervention to secure continued production gains as well as stable prices and incomes for farmers. In what Alun Howkins has described as a rural Beveridge Report, the 1942 *Report of the Committee on Land Utilisation in Rural Areas* by the Ministry of Works and Planning (Scott Report) offered a picture of an idyllic rural England but called for reconstruction that would include new towns, the re-building of villages to include social centers and schools, and the proper supply of electricity, gas, and water.[157] It recognized the countryside of Britain as essentially man-made, arguing that if landowners, farmers, and foresters were removed from their functions as the landscape gardeners of the nation, the landscape would return first to brambles and brush, and then forest.[158] The nineteenth century had privileged the interests of urban commercialism and disrupted the countryside with incongruous town villas and bungalows. Now, planning with the understanding of equalizing the advantages of urban and rural life was necessary. Those visiting rural areas from the city would find national parks and reserves, clear access to land with well-marked footpaths, and hostels, according to the report.[159]

By 1945, the attitudes of the postwar Labour government could only build upon assumptions regarding the potential of the canals which had become entrenched over the previous half century. In 1900, they were a noted source of anxiety for politicians and industrialists who watched Britain falter in its global race against the United States and other European powers. Declining financial returns, continuing skepticism regarding the decisions which had led to their initial construction, distrust of new investment schemes, and a belief that in both their engineering works and patterns of labor they represented an outmoded and uncompetitive artifact of the early industrial era, were all prominent. Though discussions of decline at the turn of the twentieth century have continued to be highlighted in the historiography of Britain, there has been little focus on canals, which offered one of the most conspicuous examples of declining infrastructure. The Manchester Ship Canal in the 1890s had signaled that a new era of water transport would be one of heavy investment to support increasingly large loads, dwarfing those for which most of Britain's canals had been designed. Between 1906 and 1911, the Royal

Commission on Canals and Inland Waterways took the testimony of hundreds whose analyses and recommendations were recorded over several volumes. Canals had been built by small groups of investors seeking economy in transport, primarily for bulk goods, and many represented innovative use of water management techniques. But, in emphasizing the generation of income on a local and regional level, companies fostered disconnected schemes that later undercut the savings and efficiency of the industry as a whole. The creation of a more integrated system capable of carrying heavy loads was possible but difficult to achieve within a fractured system of ownership, a significant proportion of which had been placed under rail control. The proposal for a Waterway Board suggested the possibility of a reconceptualization and rebuilding of the waterways, as did the proposal for an upgrade of four major routes linking England and Wales. But, a lack of political commitment and the arrival of World War I meant a continuance of *ad hoc* efforts to address canals' difficulties.

World War I was a transformative era for both industry and transport in Britain, notably in the creation of large new administrative bodies to meet the demands of wartime production. Canals experienced no overall nationalization, though some were taken under government authority due to their links with the rails, and others by the later formation of the Canal Control Committee. However, canals entered the postwar era with tonnages that continued to drop and no overall plan for renewal. In some areas, they became relics hardly noticed by local populations. But it was hard to miss them in Birmingham, the home of Neville Chamberlain, who had only recently entered Parliament and national politics. Under his direction, the Committee on Inland Waterways recommended a scaled-down reform proposal based upon the creation of regional waterways trusts and a Waterways Board. Yet, financial constraints meant no major new governmental initiatives to restructure canals were undertaken. The creation of the Grand Union Canal Company meant administrative coordination and an upgrade of tonnage on one of the system's most important throughways beginning in 1929. But many other routes continued to lose traffic, with some seeing virtually none by the end of the 1930s. As was evident in the scant attention given to canals in the report of the Royal Commission on Transport in 1931, attention was now focused on rails and roads.

Nonetheless, Chamberlain also responded to the ongoing and integral relationship between canals and those who worked their waters as he took charge of a Departmental Committee under the authority of the Ministry of Health that focused on the canals and the practice of "living in." Declining infrastructure continued to make canals spaces of danger, but their population was becoming one marked as beleaguered rather than a moral or sanitary threat, as revealed through the committee's efforts. This understanding that boat people had withstood a maligning of their character born of the Victorian era was significant. Rather than purveyors of crime and moral corruption, they were seen as anachronistic but honest guardians of a family and economic life that had thus far withstood the twentieth century's increasing emphasis upon speed and homogenization. Reports of visits to canals, such as that of Lilian Wilson, were now largely sympathetic. Support for governmental intervention to end the presence of children on boats between 1929 and 1930 gained relatively little traction. Missionary groups sympathized with the domestic impact the removal of children might have on some family boats; overall political interest in pressing the issue was short, with an overall decline in numbers of boatmen and their families. Nonetheless, while the boat population was seen as less of a threat, the danger canal waters represented in drowning would stir interest among some in closing them as an overall public health threat. The postwar period would be marked by these two often dueling visions of the waterways—one which saw them as the home of one of Britain's unique cultural groups, and one that saw them as none other than a source of danger.

The era of World War II was marked by seismic events in which the dedicated contributions of new workers to the nation's defense would be celebrated, and the possibilities that a collective state might bring to the postwar era would emerge. As in World War I, governmental intervention to utilize the canals came late and with relatively minimal planning for the industry as a whole. During wartime, Pick offered a hurried but succinct and insightful look at why the system was increasingly marginalized. Though the report's contents were never published in time to influence postwar public discussions of nationalization, the report would likely have met a skeptical public. Most in Britain saw canals as an outmoded system of transport representing an expanding problem. Standing and unattended water brought drownings and other accidents on top of the eyesores of refuse and collapsing infrastructure. The financial collapse of canal companies raised the question of whether some routes should be filled in and whether governmental intervention could, or should, attend to the system's difficulties.

Thus, canals would enter the postwar era characterized by danger and dilapidation. The lack of centralized coordination or administration meant they would face nationalization in a disjointed fashion. There was only superficial communication and coordination among many leaders and no effective plan to solve the system's problems. The grand designs of the Royal Commission were hopelessly out of reach, and there was little vision for an overhaul. As the canal system became an administrative adjunct to transport officials' wider attention to rails and roads, the initial revisioning of the system would have to come from an entirely different quarter.

NOTES

1. Philip S. Bagwell, *The Transport Revolution* (London: Routledge, 1988), 144.
2. Bagwell, *The Transport Revolution*, 145.
3. Bagwell, *The Transport Revolution*, 145.
4. Bagwell, *The Transport Revolution*, 146.
5. For a discussion of major eras of concern for British "decline," see Andrew Gamble, "Theories and Explanations of British Decline," in *Rethinking British Decline*, eds. Richard English and Michael Kenny (London: Macmillan, 2000), 5–6. Significant historical analyses have questioned whether Britain actually continued to prosper as it grappled with evolving economic challenges in the nineteenth and twentieth centuries, or whether leaders' choices had prompted the nation's abdication of its economic and political position. In the first category are the optimists, who argue that the nation continued to experience overall economic growth in the twentieth century and transformed its peoples' living standards. In the second category are the pessimists, for whom Britain's declining rate of economic growth in comparison to its competitors was an indication that the nation had failed to grapple with its problems. Among the two most well-known voices in the pessimist versus optimist debate have been Martin J. Wiener, who in *English Culture and the Decline of the Industrial Spirit* (rev. ed., Cambridge: Cambridge U.P. 2004) suggests that Britain's social elites shied away from challenging social and economic decisions, instead seeking domestic political tranquility and comfort in the countryside; and W.D. Rubinstein, who in *Capitalism, Culture, and Decline in Britain, 1750–1990* (London: Routledge, 1993), contends that rather than stepping away from critical decisions, the nation's elite made a logical transition from focusing upon heavy industry to the world of finance.
6. Charles Hadfield, *The Canal Age,* 2nd. ed. (London: David & Charles, 1981), 187–189.
7. Bagwell, *The Transport Revolution*, 154.
8. H. Gordon Thompson, *The Canal System of England* (London: T. Fisher Unwin, 1902), 15.
9. Thompson, *The Canal System of England*, 16.
10. Henry Rodolph de Salis, *Bradshaw's Canals and Navigable Rivers of England and Wales* (London: Henry Blacklock & Co., Ltd., 1904), 228.
11. Ben Johnson, "The Manchester Ship Canal," *History Magazine*, accessed August 29, 2023, http://www.historic-uk.com/HistoryMagazine/DestinationsUK/The-Manchester-Ship-Canal/.
12. Thompson, *The Canal System of England*, 18, 26, 31, 34.
13. Thompson, *The Canal System of England*, 46, 49.
14. See G. R. Searle, *The Quest for National Efficiency: A Study in Politics and Political Thought* (Oxford: Oxford University Press, 1971). See also Gamble, "Theories and Explanations of British Decline," 5.
15. The sense that trade in Britain was hobbled by the dominance of specific railway companies prompted the Associated Chambers of Commerce to sponsor bills in Parliament between 1904 and 1906 containing provisions they believed would strengthen canals as a component of Britain's industrial base. In a memorandum

attached to the bill of 1906, they argued for the creation of a "Central Canals Board" that would take over the operation of specific canals in Britain and potentially acquire a host of other canals later on with the objective of improving "the facilities for water carriage" and establishing "a complete system of intercommunication." United Kingdom, Royal Commission on Canals and Inland Waterways, *Report of the Royal Commission Appointed to Enquire into and to Report on The Canals and Inland Navigations of the United Kingdom, Fourth & Final Report*, Vol. 7 (London: Printed for H. M. Stationery Office by Wyman and Sons, Ltd., 1909), 7.

16. United Kingdom, Royal Commission on Canals and Inland Waterways, *First Report of the Royal Commission Appointed to Enquire into and to Report on The Canals and Inland Navigations of the United Kingdom*, Vol. 1 (London: Printed for H. M. Stationery Office by Wyman and Sons, Ltd., 1906), ii.

17. United Kingdom, *Royal Commission on Canals and Inland Waterways, Fourth & Final Report*, Vol. 7, 5.

18. United Kingdom, *Royal Commission on Canals and Inland Waterways, Fourth & Final Report*, Vol. 7, 16.

19. United Kingdom, *Royal Commission on Canals and Inland Waterways, Fourth & Final Report*, Vol. 7, 17.

20. United Kingdom, *Royal Commission on Canals and Inland Waterways, Fourth & Final Report*, Vol. 7, 57.

21. United Kingdom, *Royal Commission on Canals and Inland Waterways, Fourth & Final Report*, Vol. 7, 57.

22. C. J. Aubertin, *A Caravan Afloat* (London: Simpkin, Marshall, Hamilton Kent & Co., Ltd., 1916), 126. The Great Western Railway had been designed by Isambard Kingdom Brunel, who originally chose a rail gauge approximately 7 feet in width, as opposed to what became the standard gauge of between 4 and 5 feet. Some trains using the broad gauge ran through the nineteenth century, though use of the standard gauge became dominant.

23. United Kingdom, *Royal Commission on Canals and Inland Waterways, Fourth & Final Report*, Vol. 7, 70.

24. United Kingdom, *Royal Commission on Canals and Inland Waterways, Fourth & Final Report*, Vol. 7, 81.

25. United Kingdom, *Royal Commission on Canals and Inland Waterways, Fourth & Final Report*, Vol. 7, 6.

26. United Kingdom, *Royal Commission on Canals and Inland Waterways, Fourth & Final Report*, Vol. 7, 6.

27. United Kingdom, *Royal Commission on Canals and Inland Waterways, Fourth & Final Report*, Vol. 7, 82.

28. United Kingdom, *Royal Commission on Canals and Inland Waterways, Fourth & Final Report*, Vol. 7, 94.

29. United Kingdom, *Royal Commission on Canals and Inland Waterways, Fourth & Final Report*, Vol. 7, 119.

30. Local communities did, however, attempt to emphasize what they saw as the Commission's significance. As a resolution passed by the Sheffield Chamber of Commerce stated: "That this Chamber having in view the cheaper transit facilities due to

Canal development in Germany, France and Belgium, the countries chiefly competing with our manufacturers in our own and foreign markets, the congested conditions of transit of heavy and bulky goods in this country and the subsequent hindrance to trade, urges the Government to take immediate action upon the report and recommendations of the Royal Commission on Canals appointed upon their advice, by introducing a Bill to give immediate effect to Paragraph 973 sub-sections 1 and 2 of the Report, which recommends the appointment of a Waterways Board." "Appendix B—With Paragraph 13. Public Meeting at Sheffield," November 13, 1913, Committee on Inland Waterways, 1920–1921. Minutes and Notes of Evidence, MT 52/13, National Archives, United Kingdom.

31. Edwin Pratt, *A History of Inland Transport and Communication in England* (London: Kegan Paul, Trench, Trübner & Co., Ltd., 1912), 495. Moreover, the writer J. B. Firth remarked,

> Another Royal Commission has expended time and labour on a forlorn hope, and has only succeeded in recommending—and that in a half-hearted way—to the public a project for the expenditure of millions which no responsible Government will ever dare to lay before the House of Commons.

See J. B. Firth, "Water Transport or Rail," *Fortnightly Review* 87, New Series (January to June, 1910): 755. The American economist and canal researcher H. G. Moulton was also critical. "The future of British waterways appears . . . utterly devoid of promise." H. G. Moulton, "England's Waterway Revival," *Journal of Political Economy* 19, no. 1 (January 1911): 16.

32. An Act for the better Regulation of the Regular and Auxiliary Land Forces of the Crown; and for other purposes relating thereto [August 17, 1871], 34 & 35 Vict., c. 86. See *The Law Reports. The Public General Statutes . . . 1871*, Vol. VI (London: Eyre and Spottiswoode, 1871): 438–446.

33. See Edwin A. Pratt, *British Railways and the Great War: Organisation, Efforts, Difficulties and Achievements*, Vol. I (London: Selwyn and Blount, 1921), 100; and "Memorandum for the Cabinet by the Minister of Transport," July 9, 1920, 1. Future Policy with Reference to Canals, MT 52/7, National Archives, United Kingdom.

34. R. B. Dunwoody, "The Economic Requirements for Inland Navigation Transport in the British Isles," *The Vernon Harcourt Lectures, 1920* (Institution of Civil Engineers, 1920), 32. In Memoranda Circulated, MT 52/14, National Archives, United Kingdom.

35. Edwin A. Pratt, *British Railways and the Great War: Organisation, Efforts, Difficulties and Achievements*, Vol. II (London: Selwyn and Blount, 1921), 743.

36. Pratt, *British Railways and the Great War*, Vol. II, 743–744.

37. Dunwoody, "The Economic Requirements for Inland Navigation Transport in the British Isles," 34.

38. Dunwoody, "The Economic Requirements for Inland Navigation Transport in the British Isles," 33–34.

39. "Position of Controlled Canals. Memorandum for the Cabinet by the Minister of Transport," July 9, 1920, 1. Future Policy with Reference to Canals, MT 52/7, National Archives, United Kingdom.

40. "Position of Controlled Canals. Memorandum for the Cabinet by the Minister of Transport," July 9, 1920, 2.

41. Dunwoody, "The Economic Requirements for Inland Navigation Transport in the British Isles," 34. See also, The War Cabinet, *Report for the Year 1918* (London: HMSO, 1919), 195.

42. "Position of Controlled Canals. Memorandum for the Cabinet by the Minister of Transport," July 9, 1920, 2.

43. "Position of Controlled Canals. Memorandum for the Cabinet by the Minister of Transport," July 9, 1920, 2.

44. [Undated Memorandum], 1. Future Policy with Reference to Canals, MT 52/7, National Archives, United Kingdom. See also "Report of the Committee appointed by the Minister of Transport to consider the effect of increased Railway rates upon the Controlled Canals and Private Carriers," January 20, 1920. Future Policy with Reference to Canals, MT 52/7, National Archives, United Kingdom.

45. Pratt, *British Railways and the Great War*, Vol. II, 752.

46. "Canals . . . Memorandum of Meeting," October 15, 1919, Future Policy with Reference to Canals, MT 52/7, National Archives, United Kingdom, and Simon Pepper and Peter Richmond, "Homes Unfit for Heroes: The Slum Problem in London and Neville Chamberlain's Unhealthy Areas Committee, 1919–21," *TPR: Town Planning Review* 80, no. 2 (2009): 143–171.

47. "Railway Rates in Competition with Canal Rates and Coastwise Traffic," March 30, 1920, 1. Canals-Policy. Mr. Neville Chamberlain's Committee, MT 52/11, National Archives, United Kingdom.

48. "Railway Rates in Competition with Canal Rates and Coastwise Traffic," March 30, 1920, 2.

49. "Proceedings of a Deputation from the Canal Association, held at noon in Conference Room C.," February 24, 1920, 5. Future Policy with Reference to Canals, MT 52/7, National Archives, United Kingdom.

50. "Position of Controlled Canals. Memorandum for the Cabinet by the Minister of Transport," July 9, 1920, 3–4. Future Policy with Reference to Canals, MT 52/7, National Archives, United Kingdom.

51. "Committee on Inland Waterways," June 2, 1920, 1. Canals-Policy. Mr. Neville Chamberlain's Committee, MT 52/11, National Archives, United Kingdom.

52. "Committee on Inland Waterways," June 2, 1920, 1.

53. "Memorandum by the Chairman Upon Past and Future Work of the Committee," April 15, 1921, 3, Committee on Inland Waterways, 1920–1921. Minutes and Notes of Evidence, MT 52/13, National Archives, United Kingdom.

54. "Memorandum by the Chairman Upon Past and Future Work of the Committee," April 15, 1921, 3–4.

55. "Memorandum by the Chairman Upon Past and Future Work of the Committee," April 15, 1921, 4–5.

56. Reginald Hill, "The Future of Our Inland Waterways," *The Journal of the Institute of Transport* 25, no. 9 (March 1954): 343. For further reference, see Ministry of Transport, *Second Interim Report of the Committee on Inland Waterways*. Cmnd. 1410. (London: HMSO, 1921).

57. Royal Commission on Transport, *Final Report: The Co-ordination and Development of Transport* (London: HMSO, 1931), 117.

58. Royal Commission on Transport, *Final Report*, 117.

59. Royal Commission on Transport, *Final Report*, 117.

60. Royal Commission on Transport, *Final Report*, 111.

61. "Unfamiliar Ministries. IV.—On Barge and Monkey Boat," *The London City Mission Magazine* 82 (1917): 87. Here, while "bargee" would have referred to those who managed a particular type of flat-bottomed boat known as a barge, those onboard "monkey boats" would have been those managing "narrowboats."

62. "Unfamiliar Ministries. IV.—On Barge and Monkey Boat," 88.

63. "Unfamiliar Ministries. IV.—On Barge and Monkey Boat," 89–90.

64. "Unfamiliar Ministries. IV.—On Barge and Monkey Boat," 89.

65. Board of Education, *Report of the Board of Education . . . 1919–1920* (London: HMSO, 1921), 18. As established in 1919, the Ministry of Health absorbed a myriad of different agencies that had been associated with those living on the canals. The LGB was abolished, and most of its authority transferred to the Ministry of Health, while the medical responsibilities associated with the Board of Education were as well. The Ministry also assumed the responsibilities that had been assigned to the Home Secretary as a part of the Children Act of 1908.

66. Committee on Living-in on Canal Boats (Ministry of Health), *Report of the Departmental Committee Appointed to inquire into the practice of Living in on Canal Boats in England and Wales and to Report Whether Any Alteration in the Practice is Desirable* (London: HMSO, 1921).

67. Committee on Living-in on Canal Boats, *Report of the Departmental Committee*, 3–4.

68. In addition to Owen J. Llewellyn, the committee contained a mixture of labor and progressive representation in Eleanor Barton and Thomas McKerrell and the more conservative perspective of the suffrage leader Louise Gilbert Samuel. Committee on Living-in on Canal Boats, *Report of the Departmental Committee*, 12.

69. Committee on Living-in on Canal Boats, *Report of the Departmental Committee* [Minutes of Evidence], 63.

70. Committee on Living-in on Canal Boats, *Report of the Departmental Committee* [Minutes of Evidence], 63.

71. Committee on Living-in on Canal Boats, *Report of the Departmental Committee* [Minutes of Evidence], 63.

72. Committee on Living-in on Canal Boats, *Report of the Departmental Committee* [Minutes of Evidence], 64.

73. Committee on Living-in on Canal Boats, *Report of the Departmental Committee* [Minutes of Evidence], 72.

74. Committee on Living-in on Canal Boats, *Report of the Departmental Committee* [Minutes of Evidence], 72.

75. Committee on Living-in on Canal Boats, *Report of the Departmental Committee* [Minutes of Evidence], 73.

76. Committee on Living-in on Canal Boats, *Report of the Departmental Committee* [Minutes of Evidence], 74.

77. Committee on Living-in on Canal Boats, *Report of the Departmental Committee*, 4.

78. Committee on Living-in on Canal Boats, *Report of the Departmental Committee*, 5.

79. Committee on Living-in on Canal Boats, *Report of the Departmental Committee*, 7.

80. Committee on Living-in on Canal Boats, *Report of the Departmental Committee*, 9.

81. Committee on Living-in on Canal Boats, *Report of the Departmental Committee*, 11.

82. Register of Inspections [Hinckley Urban Sanitary Department], BW 117/2/2, Waterways Archive, Ellesmere Port, United Kingdom.

83. Register of Inspections [Hinckley Urban Sanitary Department], BW 117/2/2.

84. Charles Hadfield, *The Canals of the East Midlands*. 2nd. ed. (Newton Abbott: David & Charles, 1970), 212.

85. Register of Inspections [Hinckley Urban Sanitary Department], BW 117/2/2.

86. Register of Inspections [Hinckley Urban Sanitary Department], BW 117/2/2.

87. See National Society for the Prevention of Cruelty to Children, "Canal Boat Children," *Child's Guardian* (January 1930) in Canal Boats—Legislation Bill 1929, HLG 52/135, National Archives, United Kingdom.

88. Both "watermen" and "lightermen" were associated with the River Thames and London. Watermen traditionally transported passengers across the river, while lightermen navigated a particular type of flat-bottomed boat, known as a lighter (or, sometimes, a barge). The Amalgamated Society of Watermen, Lightermen, and Bargemen had formed in the late nineteenth century but by the early 1920s had become part of the Transport and General Workers' Union.

89. "Summary of Opinions Expressed by Prominent Gentlemen Connected with Canal Boats," January 1, 1930, Canal Boats—Legislation. Bill 1929, HLG 52/135, National Archives, United Kingdom.

90. "Summary of Opinions Expressed by Prominent Gentlemen Connected with Canal Boats," January 1, 1930.

91. "Summary of Opinions Expressed by Prominent Gentlemen Connected with Canal Boats," January 1, 1930. Italics underlined in the original.

92. "Canal Boats Bill. Statement on Behalf of the Canal Carriers Association and the Canal Association," 4, [Undated], Canal Boats—Legislation. Bill 1929, HLG 52/135, National Archives, United Kingdom.

93. "Memorandum by the Paddington Boatmens' [*sic*] Institute," 2, [Undated], HLG 52/135, National Archives, United Kingdom.

94. UK Parliament, Hansard, Commons, Orders of the Day, Volume 234 (Friday January 31, 1930), column 1433–1434. U.K. Parliament, accessed October 1, 2023, https://hansard.parliament.uk/commons/1930-01-31/debates/95b23363-d36a-4f9b-be5a-f96acbb2c21d/CommonsChamber.

95. UK Parliament, "A Bill [As Amended by Standing Committee B] to Amend the Canal Boats Acts, 1877 and 1884, and the Education Act, 1921, So Far as it Relates to Children on Canal Boats." Canal Boats—Legislation. Bill 1929, HLG 52/135, National Archives, United Kingdom.

96. George Newman to Owen J. Llewellyn, August 18, 1930, Ministry of War Transport. Canal Boats. Health and Education of Children, HLG 52/1094, National Archives, United Kingdom.

97. George Newman to R. C. Leaning (Chiswick), October 2, 1930.

98. Lilian Wilson, "Inquiry into Health of Canal Boat Children at Chiswick and Brentford," 1–2, [October 1930?]. Ministry of War Transport. Canal Boats. Health and Education of Children, HLG 52/1094, National Archives, United Kingdom.

99. Wilson, "Inquiry into Health of Canal Boat Children at Chiswick and Brentford," 2–3.

100. Wilson, "Inquiry into Health of Canal Boat Children at Chiswick and Brentford," 3.

101. Wilson, "Inquiry into Health of Canal Boat Children at Chiswick and Brentford," 3.

102. Wilson, "Inquiry into Health of Canal Boat Children at Chiswick and Brentford," 3–4.

103. Wilson, "Inquiry into Health of Canal Boat Children at Chiswick and Brentford," 4.

104. Wilson, "Inquiry into Health of Canal Boat Children at Chiswick and Brentford," 5.

105. Wilson, "Inquiry into Health of Canal Boat Children at Chiswick and Brentford," 4.

106. Wilfrid Henry Curtis, "Canals," *Journal of the Royal Society of Arts* 83, no. 4313 (July 19, 1935): 837.

107. Curtis, "Canals," 841.

108. Curtis, "Canals," 847.

109. Curtis, "Canals," 850.

110. Curtis, "Canals," 850.

111. Curtis, "Canals," 850.

112. Curtis, "Canals," 850.

113. Curtis, "Canals," 853.

114. "Drowned in the Canal at Widnes," *Runcorn Weekly News and District Reporter*, June 9, 1922, 1. British Newspaper Archive (BNA).

115. "Drowned in the Canal," *The Paddington, Kensington, and Bayswater Chronicle*, April 28, 1923, 1. BNA.

116. "Drowned in Canal," *The Dudley Chronicle*, April 9, 1925, 7. BNA.

117. "Drowned in the Canal," *The Derbyshire Advertiser*, August 14, 1926, 9. BNA.

118. "Drowned in Canal," *Liverpool Echo*, January 6, 1931, 7. BNA.

119. "Woman Who Suffered From Dizziness," *Liverpool Echo*, April 9, 1931, 4. BNA.

120. "Drowned in Canal," *The Aetherstone News*, June 21, 1935, 7. BNA.

121. "Drowned in Canal," *West London Observer*, November 4, 1938, 1. BNA.

122. C. I. Savage, *Inland Transport* (London: HMSO, 1957), 171.

123. Savage, *Inland Transport*, 173.

124. Bagwell, *The Transport Revolution*, 291.

125. Wendy Freer, *Women & Children of the Cut* (Gwernymynydd, Mold: Railway & Canal Historical Society, 1995), 75.

126. "Women Want Canal Work," *Hartlepool Northern Daily Mail*, April 2, 1941, 3. BNA.

127. "Women Canal Workers," *The Birmingham Mail, October* 7, 1942, 3. BNA.

128. "Canals save petrol and rubber," *Aberdeen Evening Express*, November 4, 1942, 2. BNA.

129. "Women 'Bargees' Find Canal Life Tough but Jolly," *Daily Herald*, March 31, 1944, 3. See also, "Need for Women to Man Canal Boats," *The Scotsman*, March 28, 1944, 6. BNA.

130. "Boat Crew Trainer," *Suffolk & Essex Free Press*, January 11, 1945, 3. BNA.

131. Bagwell, *The Transport Revolution*, 291. For statistics on overall tonnage during wartime, see also Savage, *Inland Transport*, 443. As a result of new women workers on the canals, some issues of living conditions appear to have been revisited. Mance was contacted by some of the women who had volunteered to work on the canals and had found the working and living conditions of families shocking. (Owen Llewellyn had retired from his position some years before and had not been replaced.) After some prompting by the Ministry of War Transport, and with some reluctance on their own part, the Ministry of Health took steps to see that better care was offered to expectant and new mothers on the canals. See Freer, *Women & Children of the Cut*, 75.

132. Frank Pick, "Report on Canals and Inland Waterways to the Minister of War Transport" (as issued June 1941), 1. MT 52/109, "Mr. Frank Pick's report to the Minister on Canals and Inland Waterways," 1941–1954. National Archives, United Kingdom.

133. Pick, "Report on Canals and Inland Waterways," 4.

134. Pick, "Report on Canals and Inland Waterways," 6–10.

135. Pick, "Report on Canals and Inland Waterways," 13–16.

136. Pick, "Report on Canals and Inland Waterways," 17.

137. Pick, "Report on Canals and Inland Waterways," 25.

138. Pick, "Report on Canals and Inland Waterways," 26.

139. Pick, "Report on Canals and Inland Waterways," 33.

140. Pick, "Report on Canals and Inland Waterways," 37–38.

141. Pick, "Report on Canals and Inland Waterways," 40–41.

142. Pick, "Report on Canals and Inland Waterways," 41.

143. Pick, "Report on Canals and Inland Waterways," 50.

144. Pick, "Report on Canals and Inland Waterways," 59.

145. Pick, "Report on Canals and Inland Waterways," 59.

146. Pick, "Report on Canals and Inland Waterways," 41.

147. See memo from S. W. Nelson, Ministry of War Transport to W. W. Ravenshear, Mines Department, July 9, 1941, MT 52/109, "Mr. Frank Pick's report to the Minister on Canals and Inland Waterways," 1941–1954. National Archives, United Kingdom.

148. Memo, February 18, 1942, "Mr. Frank Pick's Report to the Minister on Canals and Inland Waterways," 1941–1954. National Archives, United Kingdom. This policy continued into the postwar era, with a series of requests being made for publication during the time the Transport Bill of 1947 was under consideration.

149. Savage, *Inland Transport*, 447.

150. Savage, *Inland Transport*, 447.

151. Martin Pugh, *State and Society: A Social and Political History of Britain since 1870*, 3rd ed. (London: Hodder Education, 2008), 258–259.

152. Alun Howkins, *The Death of Rural England* (London: Routledge, 2003), 124. Sir Dudley Stamp offers somewhat different statistics of 8,877,712 acres in 1938 and 14,588,000 acres in 1944 under "steady till." The total in 1944 was near what he indicates was the maximum of 15,275,513 acres "under the plough" recorded in 1869. See Sir Dudley Stamp, *Nature Conservation in Britain* (London: Collins, 1969), 94–95.

153. Howkins, *The Death of Rural England*, 124.

154. Howkins, *The Death of Rural England*, 125.

155. Sir Dudley Stamp, *The Land of Britain: Its Use and Misuse* (London: Longmans, Green, and Co., 1948), 405.

156. Stamp, *The Land of Britain: Its Use and Misuse*, 405.

157. Howkins, *The Death of Rural England*, 137. For a further analysis of the report, otherwise known as the "Scott Report" after the committee's chair, Lord Justice Scott, see John Sheail, "Scott Revisited: Post-War Agriculture, Planning and the British Countryside," *Journal of Rural Studies* 13, no. 4 (1997): 387–398.

158. Sir Dudley Stamp, "The Scott Report," *The Geographical Journal* 101, no. 1 (January 1943): 16–17.

159. Howkins, *The Death of Rural England*, 137.

Chapter 4

An Independent Spirit Reclaimed?

The Cressy *and the Re-Imagining of Canals Before 1951*

In the early twentieth century, images of canals as sites of danger and industrial decay were blunted by a nascent understanding that they might be unique channels of access to the natural environment as well as amalgams of pre-modern engineering and work and life experience. These forces, whose roots lay in the nineteenth century, at first amounted to only a small counter-current against a much larger tide of belief that contended if they could not be made to pay, canals ought to be shuttered. For some canal supporters, efforts to reclaim the waterways gained traction as new efforts to modernize rural life appeared to be shredding that life's fabric and undermining the lives of boatmen, whose closest cultural kin appeared to be those living in small towns and villages. Approaching their nadir by mid-century, many underutilized canals saw their first signs of hope amidst the popularity of Lionel Thomas Caswall (L.T.C.) Rolt's 1944 narrative *Narrow Boat*, an impassioned call to save working canals and the traditions they represented. Rolt's work inspired a small but determined band of followers, who in the creation of the Inland Waterways Association (IWA) challenged canals' political and financial administration and forwarded arguments regarding their cultural and economic value in a world they believed was being flattened by the rush to a planned state. In 1947, many canal routes were taken over by the Ministry of Transport when their role in transporting the nation's raw materials and wares had come to be seen as negligible. Yet, in this same year, the IWA found its footing and began to call for resistance to impending closures. By 1951, the organization had shaped a message that, while not yet re-directing national policy, had found a home amidst conservation groups ill at ease in the postwar era.

It was in the early twentieth century that canals began to find a niche within understandings of the countryside. The "countryside" as a space separated from the urban world had first become explicit in the eighteenth century at the dawn of the industrial era.[1] It was seen primarily as a setting for the recreational and aesthetic pursuits of the landed classes, and its ideals magnified through both pastoral literature and landscape painting. In small but demonstrative ways, canals made their way into the observations of writers focused on the countryside. A letter writer remarking on the opening of England's northern Lancaster Canal in 1819 highlighted the ceremony accompanying the official opening as well as the experience of the tunnel and countryside.

> Sedgwick [Aqueduct]'s retired beauties form a fine contrast to the old Baronial towers of Sizergh [Castle], rearing their lofty heads amongst the surrounding woods and crags: the scenery of the valley appeared bolder and to more advantage than on our passage this morning. At a turn in the Canal we got a sudden and a first view of Kendal: the Castle appeared to stand in the midst of a vast amphitheatre of hills and mountains of which it formed the centre.[2]

That same year, John Hassell, a noted watercolorist, published a narrative of travel on the Grand Junction that reflected a number of other accounts he had published since the early 1790s, recounting travels in Bath, the Isle of Wight, and London, among other locations. For him, the canal represented an experience that had, in the tradition of the Romantics, nature at its center. "Deviating from the tedious monotony of the turnpike road, the course of the stream destined for inland navigation must necessarily be directed through a succession of the richest scenery—whether stealing through the glades and glooms of rural retirement, winding round the brows of hills, or gliding through the vallies by which they are surrounded, alternately visiting the recesses of pictorial abode, or the populous town, and the busy 'hum of men'."[3] Upon leaving London on a journey, he noted a "beautiful burst of scenery" as views of the Hampstead hills emerged.[4] Passing Aldenham Common, he highlighted changes in the landscape brought by canals but considered them in accord with "improvement." He wrote,

> This noble sheet of water occupies a space of considerable extent . . . which thirty years ago was a barren waste; here the improvements in agriculture are indeed conspicuous, for at this place a poor, sandy, meagre, wretched soil has now by good husbandry been converted into rich pasturage.[5]

Although much of Hassell's narrative described the homes and geography through which the canal passed rather than the canal itself, it appeared that an amalgamation was present.

> Aylesbury is distinctly seen, though at seven miles distance, to which a branch of the Grand Junction bears its way through the valley before us. This stream, and the navigable feeder to Wendover, are alternately seen meandering through

> the rich garden beneath. Hedge-rows traversing the different pastures, at length fall into each other's focus, and form a concatenation of sylvan scenes to the extremity of the horizon.[6]

Approaching Stoke Bruerne, where a series of locks lifted boats to the Blisworth Tunnel, he remarked:

> It was harvest time, and the avocations of the farmer appeared on every side. The effect on the landscape was good, and the incidents natural. These combinations charm the spectator in a common scene, but where a number of local beauties are joined, they give a consequence to the landscape. . . . Every object that encountered our sight, as we passed to the village, appeared of the picturesque; the team retiring from the glebe, had its rustic mate and tired plough boy, with all their paraphanalia [*sic*], seated on the leading horses; the navigation attendants and their cattle were bustling to pass the lock; the freighted boats presented a motly [*sic*] group of passengers on their decks; while a number of female attendants on the canal horses, reminded me of those hardy Cambrian lasses I have so often seen following their laborious avocations.[7]

By the mid-nineteenth century, industrial transformation brought the rise of a middle class which sought to separate itself from a congested and unhealthy urban environment. The idea of a natural environment that existed apart from, and often in opposition to, the power wielded by science and technology had taken root. Many among the middle classes sought to seclude themselves in the countryside, where nostalgic visions of a rural past were idealized in contrast to the nation's modern and corrupt industrial capitalism. In the eyes of some, both the engineered waterways and the life they engendered could be integrated into the landscapes through which they passed. Such was emphasized by the Scottish-born writer William Black, whose plethora of novels and stories in the mid to late nineteenth century emphasized outdoor experiences and characters' interactions with nature. In *The Strange Adventures of a House-boat*, published in 1888, a group of characters travels on the River Thames west of London, then joins the Oxford Canal and Grand Junction canals, and, after a detour along the Stratford-upon-Avon Canal, joins the Worcester and Birmingham to head south. Later, traveling on the River Severn to Bristol, they head back eastward along the River Avon and Kennet and Avon Canal.

The skipper hired for the journey (although his character remains undeveloped in depth) is an experienced boatman nicknamed "Columbus" by the passengers for his role in guiding them into unknown lands.[8] And, indeed, as the trip commences, his knowledge of the local environs of the canals is essential for navigating as well as finding intermittent mooring and stabling for the horse pulling them. The route traced by the "Nameless Barge" (a name given to it by the passengers for most of the voyage) is one that traverses little of the industrial Midlands, leaving its passengers able to enjoy the landscape and ruminate on the historical backdrop of the lands through which they travel. Black, who had early training as a landscape painter, placed descriptions of nature at the center of the work. To the west of London on the Thames, the novel's narrator notes,

> There was not a breath of wind to ruffle the smooth-flowing river; and the perfect reflections of the trees and bushes—in warm hues of yellow-green and olive—were only disturbed when the towing-line dipped and hit the surface into a shimmering silver-white. It was a peaceful landscape; very English-looking; in the distance there was a low line of wooded hill, with here and there a church-spire appearing among the trees.[9]

Later, he notes that "ancient" highways, including rivers and roads, had had centuries in which to draw populations to them and that, in contrast, railways had generally been constructed deliberately to link existing towns together.[10] "But the peculiar difficulties in the construction of canals have, in the majority of instances, prevented their projectors from doing much beyond aiming at the chief objective points," he adds.[11]

One of the very few non-fictional nineteenth-century accounts of canal travel for pleasure was published by Reginald Blunt in 1893, detailing the journey he and his brother took on a circular route in southern England, starting from Reading, continuing west on the Kennet and Avon, and then proceeding northward on the Wiltshire and Berkshire Canal where they met the Thames again and proceeded back to Reading. The journey, encompassing a region of southern England removed from the rugged and industrial world of the Black Country, was primarily rural and undertaken by both men (partly inspired by Black's *The Strange Adventures of a House-boat*) as a deliberate attempt to choose a unique holiday. With the help of a Thames waterman who accompanied them, they set out on a nearly deserted Wiltshire and Berkshire Canal (Wilts and Berks Canal) where traffic had already significantly diminished, and they were warned that the wooden bridges were too low and "out of order."[12] The brothers had altered their "monkey boat" (narrowboat), to house the crew of three that accompanied them in the cabin and themselves in a "tented cabin" in the boat's open hold in the middle of the vessel. They had set off pulled by "Methuselah," who had come from a hansom cab, and were joined by "Bob" the collie.[13] Few of the travails of life on the industrial canals were apparent in Blunt's account. He took time to recount the domestic routines of the boat as well as sight-seeing, which included,

> our explorations of sleepy little hamlets, and quaint old cottages and inns; our gossips with the natives, and yarn-spinnings with the lockmen. . . . We generally

> took our turn for an hour or two at the tiller, and in working the locks and bridges, some time or other during the day.[14]

As written, the account might have anticipated any number of late twentieth-century testimonials to the peace to be found on canals of the post-industrial era, with Blunt writing,

> the charm of our progress lay in the fact that no one day was like another; and its rule was the absence of all hard-and-fast rule or plan. If we liked a place, we stopped there, and stopped as long as we liked; . . . At Bradford-on-Avon, to which, in our ignorance of its unknown delights, we had allotted only a luncheon hour, we stayed three days; other places, where we might well have been tempted to linger—such as Bowood, or Dauntsey (near Malmesbury), we were content to glide idly by.[15]

Blunt concluded his account:

> For those, then, who are glad to get away from the madding crowd, the beaten track; to escape for once, from noisy trams and crowded coaches and dusty roads; from unknown hotels, with their strange beds, their constant packing and unpacking, and paying of bills; and from the boredom of fellow-travellers and crowded *tables-d'hôte*; from the perpetual planning of journeys, catching of connections, and all the small worries and turmoils, in short, of ordinary travel;—for those who want to slow down the pulse of life a little, and are content with a leisurely pilgrimage at an average rate of less than three miles an hour; who are true lovers of our quiet English country and its slow, strange life; for these, if such, besides ourselves, there be—and I think there are—some such a peaceful progress as I have tried to describe in outline may safely be commended, and might well prove a happy variation of the conventional holiday.[16]

The late nineteenth century also spawned many of the first organizations dedicated to the preservation of the nation's cultural and environmental resources, often bound together in the idea of the "countryside." Among these, the Royal Society for the Protection of Birds established in 1889, and the National Trust established in 1895, came to be among the most influential.[17] Small movements sought to found idealistic and communal settlements at a cultural level, while other movements emphasized the opportunities for recreation and escape provided by extra-urban travels to natural environments. Early organizations included the Commons Preservation Society (1865) and the Footpaths Preservation Society (1884).[18] After 1900, the urban working class increasingly found its way to the countryside seeking relief from industrial life. The railroad acted as a catalyst, providing cheap transportation. Legislative change and the trade union movement provided more time off for many.[19] For working-class visitors, rambling (hiking) was among the most popular activities.[20]

By the early twentieth century, the vision of a serene, rural England was promulgated by those who were convinced that the nation's identity might be found in its pre-industrial past. It was in this context that the prolific writer and playwright E. Temple Thurston set out on a canal journey in 1911. His account was one of the first of the twentieth century to record a personal journey through the nation's canals as a tourist. He recounted in *The "Flower of Gloster"* both the landscape and the customs of the boatmen, demonstrating the degree to which they had evolved in the nineteenth-century imagination to be foreign to the modern and industrial present.[21] Thurston recounted an exchange he had with the owner of his rented craft regarding boat people. "'They're not like ordinary people anyhow,' said he—'I've studied them here and there, and I've noticed that. They're all dark—a brown-eyed lot, I call 'em. Yer see, they never 'malgamate. They fruitify amongst themselves.'"[22] For his part, Thurston concluded that he found some of his beliefs about the boating population confirmed, writing,

> Undoubtedly they are of Southern extraction. Their dark, black hair, their olive skin, that soft expression of lethargy in the eyes, all point to the blood of some race other than the fair haired Saxon. Perhaps vagabondage is in the blood and these are the outward and visible signs of a grace that I for one would not be sorry to count amongst my virtues.[23]

Though Thurston's journey along Britain's canals brought him face to face with those whose lives were founded upon one of the earliest engineering works of Britain's industrial empire, the solitudes he found were closely linked with nature. He remarked upon canals' changed place in the nation's identity and reflected on what Anna Laetitia [Aiken] Barbauld had written over a century earlier (see chapter 1):

> I would not for a kingdom have missed those few weeks in the heart of England, far distant from any of those main thoroughfares where the dust of motors powders the face of Nature till she is worse than some painted thing. . . . They [the canals of England] will lead you through the quietest villages, the most remote of hamlets, with which, to know all the thousand little clusterings of people, a man might well spend his lifetime and learn but half of the country in which he lived. . . . And through these countless places, hidden in the nooks and crannies of England, the *Flower of Gloster* bore me gently, at her peaceful, sleepy pace.
>
> So long have the canals been made by now, that but little if any of the signs of their making remain. Only for their tideless waters, many of them might well be rivers—threads of silver on which are strung the brilliant emeralds of many meadows.

> Were Anna Laetitia to stand at the topmost lock of the Golden Valley, where the Sapperton Tunnel pierces its dark way into the very heart of the hills, were she to stand there to-day and look down that wonderful valley of glorious gold where in the month of May the deep woods have laid down their royal carpet of blue-bells, she must indeed retract all of that little miscellaneous piece of prose she wrote.[24]

Among those who joined Thurston in writing on canals as World War I approached was Austin E. Neal, who made two trips aboard the petroleum-powered boats "Hectic" and "Hectic Too," just before 1914 and during the summer of 1914, respectively. The first journey took him on the Shropshire Union and Llangollen (Ellesmere) canals, while the second took him on the Trent and Mersey and Ashby canals. The details of his work reflected on the processes of daily living and the specifications of the boat, as well as his interaction with the landscape, the engineering systems of the canal, and the men running the locks. "The word 'canal' to the uninitiated conjures up a picture of an evil-smelling dirty stretch of water edged in by the slums of any town, the only too-obvious repository of the town's surplus bottles, tins, and canine and feline quadrupeds," he wrote, adding,

> Poor souls! they have never seen that wonderful two miles of fairyland on the Shropshire Union Canal from Llangollen to the Horseshoe Falls, where every pebble in the canal bed shines through the pellucid water, and where moss and ferns and flowers crowd on either bank, and the silver birch entwine.[25]

Neal emphasized the engineering achievements of the lock systems which now were used with nary a thought for the genius behind "their wonderful simplicity and efficiency."[26] Yet, as was to be emphasized by other writers, this engineering was now distinguished in its age and design among other significant elements of the canal system, "A lock must be treated with the respect due to its time-honoured right to be considered one of the Old Things of England. It must not be hurried or brutally driven, or even cajoled."[27] Likely re-writing his account just before publication in 1921, Neal emphasized something of the passing of an old world on August 3, 1914, just before Britain entered into World War I. Traveling several miles up a weedy Ashby Canal from Hinckley (where Owen J. Llewellyn had had such difficulty maintaining consistent inspection of boats) he wrote of his encounter with a now derelict factory,

> A pathetic sight, truly, these places of a past generation – it was no doubt a flourishing and busy concern when the canal was utilised as a means of transport, but the growth of weed is a silent witness of how little traffic there now is

> on this waterway, and so these works—whatever they were—found themselves declining—and then it was not long ere the fall.[28]

Nonetheless, canals had a role to play as a counter to the new post-World War I era. Neal concluded:

> If you suffer from war-strain, shell-shock, too much married life, too much single boredom, too much income-tax, too many excess profits, or no profits at all . . . look up the honoured old waterways of England. They will be charmed to see you; you will be charmed to see them; and they will repay you for your trouble in increased health, contentment, and a storeful of joyful memories to carry with you until the time comes for you to take that last cruise with the oldest ferryman of them all—and may He then pilot you safely to the "Land where it is always afternoon."[29]

Neal's final thoughts spoke to treasured memories of escape on canals, perhaps personally very important to him in the postwar era, and the sense that although heavenly, the age of the canal boatman might really be coming to a close. Achievements in technology and transportation bound the nation together as never before after 1918. Radios and automobiles became steadily more common among the middle classes and so too did aspirations for home ownership outside the narrow and congested streets of the nation's industrial cities. In the interwar years as a whole, the number of automobiles rose from 110,000 to 2.3 million.[30] With this, as well, came a belief, skeptical of contemporary urban life, that the social harmony to be found in rural England was being fundamentally disrupted by new technology.[31]

A cascade of writings bearing witness to a world that appeared to be passing away—not only through the disappearance of the agricultural village, but through the loss of its attitudes, values, and way of life—reached a peak in the 1930s and has been broadly defined as "ruralist."[32] Writers in this tradition bemoaned the increasingly porous boundaries between urban and rural and the intrusion of urban amenities and values into the countryside. They championed the cause of a countryside that struggled to maintain its traditions amidst the demands of capitalism and the state in the twentieth century.[33] In 1923, George Sturt's *The Wheelwright's Shop* had depicted the reverence he held for his own family business, which he had fitfully tried to run at the close of the nineteenth century. He wrote that the craftsman had once been as "a colonist pioneer with the materials of his own neighbourhood—the timber, the clay, the wool—other materials to supersede the old ones were now arriving from multitudinous wage-earners in touch with no neighbourhood at all, but in the pay of capitalists."[34] He added, "Village life was dying out; intelligent interest in the country-side was being lost; the class-war was disturbing erstwhile quiet communities; yet nobody saw what was happening."[35] As Prime Minister Stanley Baldwin proclaimed early in 1924:

> The sounds of England, the tinkle of the hammer on the anvil in the country smithy, the corncrake on a dewy morning, the sound of the scythe against the whetstone, and the sight of a plough team coming over the brow of a hill, . . . These things strike down into the very depth of our nature, and touch chords that go back to the beginning of time and the human race, but they are chords that with every year of our life sound a deeper note in our innermost being.[36]

In 1926, the Council for the Preservation of Rural England (CPRE) was founded as an advocacy organization to protect against the increasing intrusion of the city into the countryside, a widening social and economic process that, in the eyes of some observers, disrupted the organic relationship between local communities and the lands sustaining them. In 1928, Clough Williams-Ellis, an architect best known for his design of an Italianate village in northern Wales, published *England and the Octopus*, a now famous work that decried the intrusion of the city into the country, particularly along highways in a process more widely known as "ribbon development."[37] The desire to protect both the landscape and traditions of the countryside emerged among a group of ruralist voices led by H. J. Massingham, who first published his ideas in *Wold Without End* in 1932.[38] Increasingly, understandings of the rural past emphasized the experience of idealized settings that, divorced from complex narratives of economic development, saw industrialization as a foreign agent. Critical to the ruralist understanding of Britain was the significance of a life in balance with nature—what many supported as an "organic" relationship in which human health and well-being were connected to the health of the agricultural and natural worlds.[39] Philip Conford has described the interwar "organic" movement as inheriting traditions from Romantic poets, as well as the ideas of John Ruskin and William Morris, adding that the "organic school" saw the countryside as a "work of art and as a source of social stability, wisdom and spiritual renewal."[40] Diverse voices contributed to the emergence of a belief in an organic England, particularly at the village level, that was either under threat or had already been destroyed by an onslaught of the forces of modernity, including industrialization, technology, capitalism, commercialism, and statism.

Guidebooks emphasized the experience of the rural as the experience of the true nation, but by the early 1930s, often embraced experiences facilitated by the use of motorized road transport.[41] In particular, automobiles, though themselves a product of a new chapter in the era of industrialization, suggested an opportunity to break away from the controlled environments of urban and industrial life (as well as mass transportation) and experience the countryside directly.[42] Nonetheless, when significant numbers sought the same destination, the automobile seemed to become the enemy of the countryside. One of the most significant accounts of travel in England was by Henry Vollam (H.V.) Morton, whose *In Search of England* was published in several editions between 1927 and 1935. Morton relished the new era of discovery now available and, though warning of the danger of the "vulgarization" of the countryside, argued that knowing the English village (which was now itself integrated into the countryside) was necessary for an understanding of the nation.[43] He chronicled his experiences of visits to the most significant symbols of the English past, ranging from Cornwall to Stonehenge to Durham Cathedral, including, as he wrote, his ruminations on their significance in the nation's deeply rooted heritage. Those whom he met who showed him over sights or represented a piece of the fabric of a local community were essential to the story. In Stratford-upon-Avon, the significance of the past and the dilemmas of the present collided for him. In the interwar period, remembering a friendship with Frank Benson, an actor who became central to the wider development of Shakespearean theater in the early twentieth century, he wrote,

> I remember him [before the First World War] waving his arms at me in a storeroom hung with hams which he used as an office, and telling me . . . that only through Stratford, the common meeting-place of the English-speaking world, could we heal the pains of Industrialism and make England happy again. . . . Then, I seem to remember, shortly after that war was declared.[44]

In the aftermath of World War I, the optimism had vanished for Morton. He wrote,

> I drove through a perfect death-rattle of motor traffic into Stratford between those well-loved Warwick hedges. My quiet old Stratford was suffering from a rash of trippers. Charabancs from everywhere were piled up in the square; half the motor-cars of the Midlands were either coming or going; and the hotel was full of long-legged girls from America, and sallow fathers and spectacled mothers.[45]

The countryside was beleaguered, Morton noted, with

> mortgages on farms; no fluid capital; the breaking up of famous estates when owners die; the impossibility of growing corn because of the expense of labour and the danger of foreign competition; the folly of keeping cattle when the Roast Beef of Old England comes so cheaply from the Argentine.[46]

The decline of the countryside was equated with the decline of England. He wrote,

> The 'Back to the Land' cry is a perfectly sound instinct of racial survival. . . . We may not revive the English village of the old days, with its industry and its arts. . . . But the village is still the unit of development from which we have advanced first to the position of a great European nation and then to that of the greatest world power since Rome.[47]

Canals, firmly entrenched in the minds of most in England as a resource of industrial transport, were mentioned only intermittently by ruralists seeking to shake off the creeping grasp of the city. Common conversations regarding canals were much more likely to have been about incidents of drowning (see chapter 3). For some of those who traveled on canals, there could be uncertainty regarding their multi-faceted identities. They could be both products of the industrial world and representations of the natural world populated by a pre-modern workforce, at the same time.[48] For these writers and many others, one end of the Stratford-upon-Avon Canal may have been in Stratford-upon-Avon, but the other end was in the industrial West Midlands. In 1928, T. W. Cubbon navigated a small petroleum-powered boat up the River Dee, onto the Shropshire Union Canal, and then the Staffordshire and Worcestershire Canal, with the goal of reaching the River Severn. The journey was one that had the solitudes of nature as its goal but skirted the fringes of the Black Country. As he traveled to the west of the conurbation of Birmingham, he remarked,

> The whole of that once lovely—but now hideous—region is interlaced with a network of canals, whose shining threads act as a spider's web, to entice labourers. . . . The sweet spring water from the surrounding hills is caught and impressed in dozens of reservoirs to feed these water arteries, which soon lose their crystal clearness by contamination with coal dust and the exhaust of steam engines, until they become evil-smelling and cankerous, as they reach the heart of the great Black City.[49]

Cubbon was thankful that his travels simply skirted the Black Country and, like Morton, spent a good deal more time idolizing the upper Avon [Shakespeare's world], writing, "These autumn evenings on the Avon were a revelation of beauty; . . . No roar of town, or shriek of railway-whistle to disturb the serenity of the night!"[50] And yet, likely to a much greater degree than what Morton had encountered, Cubbon's journey was one of personal persistence as failing locks and weirs and the difficulties of getting permits limited the journey.[51]

Later to be a post-World War II advocate of canals, Alan Patrick (A.P.) Herbert published his work *Water Gipsies* in 1930. A humorist, playwright, and eventual independent MP, he had served as a naval officer in World War I and would go on to serve aboard a boat in the Thames in World War II. (In the years after 1945, he would serve as president of the IWA, thus playing at least a symbolic role in shaping the movement to restore Britain's canals.) The novel focuses on two sisters, Jane Bell and her sister Lily, who operate a barge on the nation's canals and rivers, and addresses their personal and romantic adventures. As they begin their travels on the newly designated Grand Union in London (old Grand Junction Canal), they also begin an exploration of both the canal system and the lives of those who work there, passing "into another world, peopled by another race."[52] Treated respectfully by the boatmen, one of the sisters remarks that it was the women onboard the barges who "frightened her."[53] In these "little kingdoms," the women were "dark-eyed, gipsy-looking, weatherworn creatures, with shawls over their heads and children crawling about at their feet, . . . "[54] The decorations of the boats were described as part of an age-old tradition,

> They were painted, every one, in the ancient traditional style of the canal, handed down by generations of boatmen. "Castles," "hearts," and "roses" were the three indispensable themes of the decoration—roses for beauty, hearts for romance, and castles, maybe, for worldly honour.[55]

The emphasis on tradition and ritual was pursued as the "mysteries" of the handling of the locks were demonstrated for one of the characters, "a precise and delicate ritual in which every member of the family had his proper part. . . . The locks are the landmarks and milestones and strenuous obstacles of the water highway."[56] In a romanticized vein, one of the boatmen in the book reflects on his surroundings:

> Mr. Green looked proudly round his tiny home and said, "Well, I was born in this boat and my father was born in this boat, and his father before him, and I don't ask for nothing better. Water gipsies, that's what we are. I don't want no pictures—there's all the picture I want," and with his bread and cheese he pointed at the fairy castle on the cupboard door. "Castles and hearts and roses, that's the old style, Miss Jane, and good enough for me. And, come to that, we get pictures all day, don't we, old lady, moving along?"[57]

This interlinking of water and craft tradition, in a culture that was both dependent on the waterways and an expression of it, reflected the heart of interwar ruralism and organicism. Through most of their history, canals had been defined by their significance as engineering works. Now, the perceived simplicity of their operations and the gradual departure of their workforce had marked them as one of Britain's disappearing traditions. In this vein, William Bliss, a solicitor who began to traverse English waterways using a simple canoe rather than a canal boat, published an account of his experiences in *The Heart of England by Waterway* in 1933. Bliss saw canals as woven into the experience of nature through waterway travel and found little to mark them as different from the nation's system of natural waterways except his understanding that many were becoming derelict and that the opportunities for navigating them were diminishing. Noting the recent abandonment of the Thames and Severn, as well as other canals, he wrote,

> They are going one by one, and I am writing thus about them before they go, so that anybody who reads and who has never yet experienced the peaceful beauty, the sleepy contentment that is peculiar to these English waterways, may do so before it is too late.[58]

Bliss wrote that he felt one day a Government would wake up to the role canals could play and thus intervene to improve their capacity, along French lines. Though this might mean some compromise in their sleepy nature, he added, "They will always still be beautiful in spite of all this, and for the plain reason that they must still pass through the most beautiful and peaceful and unspoiled—yes and unspoilable—secret places of England."[59] Bliss's account emphasized an organic connection between rivers and the engineered waters of canals which they supported:

> And then, when at last the stream whose waters it has drunk can give to it no more, and this intimate transfusion of blood must cease and the canal must find its way across the grassy uplands of an English watershed, seeking (like a beautiful vampire) another stream to drain and to cajole, you see how carefully it must go, and with what delicately adjusted meanderings among the quiet hills. . . . The summit-level of a canal is therefore always, because it simply cannot help it, a place where time stands still and distance does not seem to matter; a place, too, of absolute quiet and stillness and untrodden ways.[60]

For Bliss, the human presence on canals was entwined with their natural setting. He wrote,

> I shall be sorry when I can never see again the slow barge coming towards me, pushing the quiet water away in front of its dutch bows. . . . I shall be sorry when I can see no more the bargee or the bargesse . . . leaning against the tiller and looking out across the quiet fields with that bemused stare of an inward and sort of mystic contentment that only bargees (and cows when they are standing to their knees in water) have ever yet, of all God's creatures, perfectly achieved.[61]

His nostalgia for a passing rural world reflected that of his contemporary Massingham. He wrote that having defeated "Prussianism," England seemed to have become "prussianized" with a government now more intent upon an intervention and regulation that had already undermined the life of the rural worker. He wrote that the rural laborer was still the "salt of England:"

> The ploughman, the reaper, the hedger-and-ditcher, the thatcher (what few of him are left); the man who throughout the revolving year can—and year in year out does, as his fathers before him—plough, harrow, drill, sow, weed and reap the harvest; who cleans and dungs the thankful ground; who can pleach and lay a hedge, scour a ditch, thatch a rick or barn, mend a gate, mud out a brook, and do the hundred other odd things that come to his hand to be done—he is still England, and to be considered above all the factory hands of the Manchester school's plague of industry, who can each do but one thing, and far before all the little middlemen who but sell at a profit what he produces.[62]

In 1937, Williams-Ellis edited a collective work entitled *Britain and the Beast*, which offered a thorough and multi-pronged attack on a rapidly evolving nation. He wrote that "indignity and discomfort" in a nation where "disorder, ugliness, and inefficiency" were generally tolerated was the common subject of the twenty-six contributors.[63] Collectively, the writers embodied themes of the protection of the countryside, restraint of urban intrusion into the countryside, and protection of nature. Among the essays, the economist J. M. Keynes argued against the unwillingness of the government to invest in the preservation of monuments, Massingham reiterated his plea for the preservation of the English village, the philosopher Cyril Edwin Mitchinson (C.E.M.) Joad called for the cultivation of taste in leisure, the architect Patrick Abercrombie offered a defense and endorsement of planning strategies, the writer and broadcaster Stuart Petre Brodie (S.P.B.) Mais decried creeping suburbanization into rural areas, and the ornithologist Kenneth Spence renewed the call for protection of the Lake District. The book suggested the need for an active agenda blocking the destruction of the countryside, containing at its close essays on the aims and activities of the National Trust and the CPRE. The only appearance of canals, however, was as a reference to industrial decay. In his introduction, Williams-Ellis quoted Evelyn Waugh's *Vile Bodies*.

> Nina looked down and saw included at an odd angle an horizon of straggling red suburb, arterial roads dotted with little cars, factories, some of them working, others empty and decaying, a disused canal, some distant hills sown with bungalows, wireless masts, and overhead power cables.
>
> "I think I am going to be sick," said Nina.[64]

Tension between the claims of local village life and the power of a centralized state was clear as Williams-Ellis suggested that all the contributors to

Britain and the Beast had at least given implicit acknowledgment that what might be needed was a "Ministry of Amenities."[65] More directly, in the penultimate personal contribution to the book, Lord Howard of Penrith outlined what he saw as the definitive steps taken in the protection of nature with the German Reich Conservation Act of 1935, noting, "The instinct to preserve these precious possessions seems to be, so to speak, in the blood."[66] For some organicists in Britain, the intersection of perceived national decline, the debilitating effects of capitalism, and a belief that the qualities of the English "race" were being undermined meant a turn to far-right politics by the late 1930s. Massingham, though he seems to have steered clear of far-right politics himself, joined organicist figures including Jorian Jenks, a significant presence in the British Union of Fascists, and Nazi sympathizers Rolf Gardiner and Gerard Vernon Wallop (Viscount Lymington), in founding an organization known as the Kinship in Husbandry in 1941.[67] Seeking to promote the revival of agriculture, the Kinship in Husbandry argued against the increasing mechanization of agriculture, the use of chemicals and artificial fertilizers to achieve greater yields, and the loss of rural tradition predicated on the direct application of physical labor to the soil.[68] Fear of scientific and bureaucratic management percolated through their work. Lymington was a member of the nationalist English Mistery in the late 1920s and continued to be active in efforts to revive both rural agriculture and society; the health of the former, in particular, be believed essential to the strength of the nation. In the mid-1930s, he joined the right-wing English Array.[69] His work *Famine in England* in 1938 called for a revival of agriculture in Britain that would both nurture the English "race" and provide a bulwark of strength in case of war.[70] Conford writes,

> For Massingham and his friends in the Kinship in Husbandry, rural reconstruction was a matter of reversing a historical process which he described as "War upon the English tradition." A growth of regionalism was required, in order to oppose the unholy alliance of big business and government bureaucracy whose abstract plans were destroying rural life and "authentic national character."[71]

The introduction of canals into the heart of the "English tradition" soon came with L.T.C. Rolt, who made a record of a 1939–1940 journey he and his wife Angela had made on portions of the Oxford, Grand Union, Trent and Mersey, Shropshire Union, and Coventry canals, in *Narrow Boat*, published in 1944. Rolt, who had already developed an interest in automobiles and trained as an engineer, became fascinated by canals after a relative's purchase of the *Cressy* (a horse-drawn narrowboat), that relative's installation of an engine, and the subsequent conversion of it to a floating home. (The engine was a Model T Ford marine conversion engine.)[72] Later, acquiring the *Cressy* and planning a journey on what seemed an increasingly derelict system of waterways, Rolt felt time was of the essence. He realized that the route taken by Thurston on the Thames and Severn Canal was now impassable and that if he did not travel soon, similar routes might be lost to him.[73] Similar to other accounts of canal travel for pleasure or exploration, Rolt's journey was marked by landscape appreciation, reflections on the history of the localities through which the canals passed, engagement with canal technology in various states of repair, and descriptions of those who worked and lived on the waterways. As he remembered later in life, upon first being introduced to the *Cressy*,

> In memory it has become the still and silent landscape of a dream and through it *Cressy* glided smoothly and quietly along, her little compound engine ticking over with no more noise than a well-oiled machine. She did not intrude upon the landscape; she became a part of it like the canal itself.[74]

For Rolt, canals represented an amalgam of industry and nature that had become central to Britain's identity; yet it was an amalgam being overpowered as mechanized production made inroads into the countryside as never before. He wrote that canals had become the last bastion of a people whose way of life had survived the process of industrialization almost unchanged, adding that canals had become such an integral part of rural life that it was nearly impossible to separate them from the towns they had nurtured.[75] In a much more conscious fashion than previous writers, Rolt delineated the experience of the canals as a counterpoise to the worlds of suburbanization, increasing technical complexity, and large-scale business and marketing that marked interwar Britain. Though he had, and would in the future, explore interests in automobiles and railways, canals were a valued world apart. As he wrote later in life:

> The craftsman-built cars in which I had delighted had spawned an ugly crop of mass-produced successors which had become one of the chief agents of destruction. Nor could my other love, the steam locomotive, supply a solution. Even if I would, I could hardly buy myself a steam locomotive and a saloon carriage and trundle round the railway byways of England, stopping where I would. But on the canals and rivers of England we could do just this. They seemed to me then so many secret ways leading into the heart of England.[76]

In *Narrow Boat*, Rolt described the world of the canals as not just one that had been bypassed but one that was on the verge of extinction. In Tooley's Boatyard, a well-known stop on the Oxford Canal at Banbury, he found a small family business based on skill struggling for existence in a world now focused on quantity rather than quality.[77] And, while waiting for his boat to be fitted out in Banbury and investigating the southern section of the Stratford-upon-Avon Canal, he met a nearly deaf old man who had once worked for the canal and inquired of him whether any boats still passed through. Rolt recounted, "He nodded in a most encouraging fashion, 'oh, ah', he affirmed, 'there comes one sometimes', but when I asked how long it was since he had seen the last he confessed that it was ''bout four years back'."[78] Rolt suggested that the countryside had experienced two forms of intrusion: one associated with the building of the canals, which had been absorbed, and one associated with modern mechanics, which had been essentially destructive. Remembering the first part of his canal journey on the Oxford to Braunston, he remarked it had been said an "English drunkard made the rolling English road," but added,

> There is not a wandering lane that can vie with the canals in this respect, and this meandering is a great part of their charm. Like the old roads, they were built before the age of hurry, and the way in which they follow the lie of the land is particularly characteristic of the earlier waterways built or surveyed by James Brindley.[79]

As Rolt traveled north on the Oxford and eventually joined the Leicester section of the Grand Union Canal (the "Leicester Cut"), Britain was on the cusp of war with Germany. The months following September 3, 1939, would bring intensive mobilization and momentous changes for the countryside. This was a mobilization Rolt would eventually join by 1940, taking an initial job in war production. Though the war seemed distant in most of his account (as eventually published in 1944), in describing a portion of his journey on the Leicester section of the Grand Union Canal before the war's outbreak, he seems to have considered the impact of a rutted and militarized landscape:

> A most monstrous machine the height of a house was dragging itself over the ridge and furrow on caterpillar tracks larger than any tank. The simile, indeed, was fitting, man having apparently declared war on Nature and brought up his heavy artillery. In the tail of the machine two sets of revolving steel tines, each as long and thick as a man's arm, were tearing the ground like vast talons. High up under a canopy in a haze of pungent diesel exhaust fumes the modern husbandman, a mechanic in blue overalls, sat before a host of levers, as far removed from the land as a passenger on a luxury liner is from the sea.[80]

The presence of the tractor represented the shocking entrance of a machine into a largely rural world. Where the technology of the canal had gently entwined itself with the landscape, the new tractor sought to sever and dominate.[81] Though in this and further writings Rolt would emphasize his commitment to a working landscape, not a rural idyll, the shock of a new layer of technology in which the workman had been empowered but further removed from the results of his endeavor was troubling.[82] The people of the canals represented both a culture and form of labor threatened by the machine-dominated present and were unprepared to grapple with it. Rolt argued they had thus far escaped the leveling influences of the urban world, choosing to maintain themselves by marrying within their own community.[83] Few could read or write, and although appearing "surly and taciturn to strangers from the towns," had an intelligence "whose great charm lies in the fact that it has not been acquired from Council schools and newspapers, but is in part traditional and in part evolved during many slow journeyings with only heron and plover for company."[84] The natural landscape of Britain, traditional arts and crafts, and the distinctive life of the canal boatman were all entwined and, as Rolt would write in the 1944 conclusion to the book, at risk in the postwar world. Decorative paintwork, including images of roses and castles (often associated with "gipsy" designs), was maintained with great care by many workers. He wrote, "Their inborn gipsy love of colour and polished metal finds expression in the gaily painted cabins of their boats and in the wealth of glittering brass ornaments which adorn them," later adding,

> Who first established this convention of tall stuccoed towers and wide-eaved red roofs? Perhaps it was some old wandering Romany who exchanged his caravan for a narrow boat when the canals were young and adorned the walls of his new home with memories of fairy-tale castles in the Carpathians.[85]

The theory of canal boatmen as the specific descendants of Romani ("Gypsies"/"gypsies") would be debunked by the late twentieth century, but for Rolt an early acknowledgment of such a myth no doubt re-affirmed boat workers' distance from the modern world he inhabited.[86]

Rolt also embraced the tradition that so-called "Number Ones" formed the bulk of those captaining boats on the canals in the nineteenth century. A "Number One" was the designation for a captain who owned his own boat and hired his labor out to carrying companies. Rolt's conclusions regarding the predominance of "Number Ones" among canal boat workers have faced significant challenge from historians.[87] Nonetheless, his view of their world was critical for understanding what he believed to be under threat in the post-World War II era. He wrote that in years past, independent boatmen had handled the bulk of canal traffic and that carrying companies had only developed recently.[88] Later, he added, "they cannot survive the cut-throat competition of the modern haulage business. Many have given up the unequal struggle by selling their boats and their labours to one or other of the companies."[89] In recounting a conversation with an innkeeper who referred to the illiteracy of boat people as a scandal, Rolt remembered that he argued education had only made the life of the unemployed townsman more intolerable. He wrote:

> I certainly consider that the simple forthright mother wit of the illiterate boatman is preferable to a "culture" born of the cinema and Sunday newspaper, and though I could not lose the company of books, I have often envied him his happy immunity from the howling bedlam of the hoarding and the popular Press.[90]

In the 1944 conclusion of his work, Rolt remarked upon the way of life he witnessed on his journey, one then passing away. Those he had met had revealed that way of life to be one "which was the antithesis of the stereotyped and rootless existence of twentieth-century 'economic man', and it is the spirit of which these men are the unthinking guardians which must not be permitted to perish from the earth."[91]

As Rolt clearly understood by 1944, the opening of World War II was a watershed moment in which the roles played by both rural and urban spaces in Britain, and methods of agricultural and industrial organization, were to be transformed. In October of 1939, Morton set off on another journey through England; but setting forth from London, he found all the talk was of war, and that there was a new martial presence in the countryside. In Surrey, he visited a plant making Wellington bombers. Then, traveling on to Salisbury and noting the lack of traffic (like being transported back to 1919, he said), he wrote of both the beautiful countryside and the darkness that descended with the blackout.[92] Here, he was able to ride in one of the new tanks being produced. After visiting Weymouth, he traveled on to Gloucester and ruminated *en route* about the gap between rural and urban now being exposed particularly with the arrival of evacuees from the cities. He noted that in some sense the war was far away from rural England, as little serious attention was paid to air raid sirens and gas masks among its residents. Nonetheless, his writing revealed changes that were soon to have a big effect (even if they still seemed parochial at the time, he wrote) such as the decisions of the War Agricultural Committee on what was to be plowed up.[93] Despite fears of wartime regimentation, many ruralist writers believed foundational images of England stood in marked contrast to a mechanized German society. Among some, the unkempt but idyllic English countryside was called into service as a foil for the perverse state-directed modernization that seemed to be emerging in Germany. In a 1943 Home Service broadcast, Peter Scott (both son of the Antarctic explorer Robert Falcon Scott and, later, a postwar champion of canals) spoke of his time serving aboard a destroyer in 1940,

> I remember as dawn broke. . . .thinking suddenly of England in quite a new way. . . . I thought of Devon . . . the wild moors and rugged tors . . . the narrow

> winding valleys. . . . That was the countryside we were so passionately determined to protect from the invader.[94]

Flora Thompson chronicled the changing world of rural Britain in three novels between the late 1930s and the mid-1940s entitled, *Lark Rise*, *Over to Candleford*, and *Candleford Green*. Each of the novels drew readers into a nostalgic view of the English past and yet spoke to contemporary anxieties as the city and modern world intruded into the countryside. Published collectively as *Lark Rise to Candleford* in 1945, the trilogy included an introduction by Massingham in which he praised her ability to describe a disappearing rural England. He wrote,

> What Flora Thompson depicts is the utter ruin of a closely knit organic society with a richly interwoven and traditional culture that had defied every change, every aggression, except the one that established the modern world. . . . In the shell of her concealed art we hear the thunder of an ocean of change, a change tragic indeed, since nothing has taken and nothing can take the place of what has gone.[95]

Planners raced to come to grips with changes to rural Britain with initiatives which were to continue into the postwar era and have sometimes been seen as a "second agricultural revolution."[96] The use of artificial fertilizers increased approximately four times between the late 1930s and the late 1960s, with mechanical innovations such as combine harvesters, tractors, and milking machines all serving to magnify production.[97] Though chemical pesticides and herbicides would not come into widespread use until the 1960s, the groundwork for their use had been laid in wartime.[98] They would also become a critical source of water pollution.[99] Upon their arrival, "ruralist" and "organicist" voices would often be raised against them, most prominently in the Soil Association. Established under the leadership of Lady Eve Balfour in 1945, its aims soon after its founding included "working for a fuller understanding of the vital relationship between soil, plant, animal and man."[100] Similarly, Massingham and other ruralists rejected the rise of industrial and scientific planning in Britain and devoted themselves to the enduring qualities they found in rural life, particularly an eschewal of modern methods of farming.[101] Remembering later in life the new swaths of land he saw being put under till during the war, Rolt wrote,

> After the agricultural depression and consequent rural dilapidation and neglect, any renewal of activity seemed welcome. On the other hand, the frantic speed with which this primeval landscape was being transformed into a huge mechanized ranch farm, ruthlessly exploiting its stored fertility, sickened and saddened me.[102]

Massingham's decision to pen the Foreword to *Narrow Boat*, offers one of the earliest signals of the unique place canals were to hold at the intersection of industrial heritage and the rural landscape in twentieth-century Britain. Massingham wrote that the culture of the narrowboat operators was now disappearing under the "brutal impact of modern industrialism as all our traditions are perishing or so threatened."[103] And added,

> These canals are falling so rapidly into disuse, partly by underhanded sabotage and partly as an extreme example of rural dereliction as a whole, that a merely geographical account of them alone would have filled up an important gap in our knowledge of our own country.[104]

He concluded, "To regard Mr. Rolt's book as nostalgic is, therefore, wholly to misinterpret it. He is pleading for something that is part of the soul of England."[105]

It was the relationship of people to technology and methods of production that was critical for Rolt. In his training and experience in mechanical work, he well understood the nature of technological innovation. Likely at about the time *Narrow Boat* was readied for publication, Rolt prepared an article for the June 1944 issue of *The Field* (a publication focusing on field sports as well as the rural landscape). He argued that canals carried with them the strength of industrial Britain but with the lingering craftsmanship reminiscent of the earliest days of the Industrial Revolution.[106] The construction of the canals had been planned by engineers who sought to control both the landscape and the water itself; nonetheless, they (the canals) had evolved to find a place in the social and geographic landscape of the countryside, he wrote. This organic connection to the countryside and rural tradition had not been replicated by the aggressive capitalist intent of the railways. A month after the article in *The Field*, *Country Life* drew attention to the cultural impact of traditional canal life as it published an extended story placing the colorful rose and castle paintings of canal boats within a folk art tradition that the journal argued had once made Britain much more colorful.[107] Of the canal engineers, Rolt wrote,

> They were craftsmen and, consequently, assessed the value of their work by those intrinsic qualitative standards common to all craftsmen; they could never have conceived that in the comparatively short space of a century many of the results of their monumental labors would already be neglected and forgotten.[108]

In the conclusion to *Narrow Boat*, he added:

> The future of the English canals, . . . depends no less than that of the countryside on the order we build after the war. . . . In a society framed to cherish our national heritage the canals can play their part not only as a means of transport

> and employment, but as part of an efficient system of land drainage and a source of beauty and pleasure. But if the canals are left to the mercies of economists and scientific planners, before many years are past the last of them will become a weedy, stagnant ditch, and the bright boats will rot at the wharves, to live on only in old men's memories. It is because I fear that this may happen that I have made a record of them.[109]

Indeed, Rolt's interest in canals as working infrastructure rather than as sites where the past would be re-created or preserved was reflected in the film *Painted Boats*, released by Ealing Studios in 1945.[110] Formatted as a docudrama, it offered an enduring visual documentation of life upon the canals. In the movie, the two main teenage characters, Mary Smith and Ted Stoner, have grown up in the world of canal boat families but find their world challenged by changing times and the experience of war. Continuing to work with his family, Ted dreams of starting a new life on land with greater job opportunities. Mary, also a part of the crew of her own family's boat, can hardly imagine leaving "the cut" with its enduring social connections and rhythms of work. Though World War II is present in the background of the film, it is rarely explicit in the day-to-day life of the canal. The simple romantic plot of the movie sees the young couple grappling with the pressures of a new world and questioning their own roles within it. The couple's final decision about whether they will stay together and seek either a home on the water or a new home on land is pre-empted by Ted's call-up for service and quick departure. The rest of his family finds itself leaving the cut and turning the working of their boat over to a female crew recruited to staff it in wartime. For her part, Mary is saddened but receives personal encouragement (as well as what seems an enduring reassurance regarding the canals) from Ted's younger brother who says, "Ah, the land's no place for a man. . . . He'll come back, Mary. He won't stick to land."[111] The movie, combining footage of work on the canal and descriptions of its technical workings provided by a narrator, is thin in plot but rich in images of the canals themselves. The independence and weathered character of those working the canals are represented as the core of a hardworking industry seeing Britain through the war. Rolt was an unacknowledged consultant on the film and apparently had some impact in suggesting shooting locations and boats to be used.[112]

In 1945, Rolt, on his own, could only have had a slim to negligible effect on the future of canals. However, that future was changed by the many letters of interest he received regarding *Narrow Boat*. Among them was one from Charles Hadfield, who himself had become a canal enthusiast in the pre-war era and in 1944 had written a book entitled *English Rivers and Canals* with Frank Eyre[113]. It was nearing publication at the time *Narrow Boat* was published. Recalling his enthusiasm for Rolt's work years later, Hadfield said, "It made me realize back in 1944, perhaps more than Tom himself appreciated, the possibilities that existed for using the smaller canals for pleasure cruising once the war was over. But, above all, I was delighted to have found another real live canal enthusiast."[114] Writing to Rolt to advocate for the formation of a canal association after the war, Hadfield found that someone else had just written to Rolt to suggest the same thing.[115]

The second correspondent was Robert Aickman, a writer and literary agent in London who had picked up *Narrow Boat* soon after its publication and quickly communicated with Rolt seeking a meeting. In Aickman, Rolt found another fellow canal devotee with whom he could work (at least initially) in answering the call to preserve the waterways. Aickman soon exerted the most influence upon Rolt in seeking the creation of an organization that would bring the cause of canals to the attention of both the government and the public. Indeed, more than Rolt, Aickman focused on defining the scope, administrative procedures, funding, and relationship of a waterways organization to other societies, which themselves sought to preserve public access and traditional country pursuits against a backdrop of large-scale state and commercial interests. Like Rolt, he believed that not only the future of canals, but that of society more generally, was at stake. As he would write later, "I thought that in the waterways system, vast in relation to the size of the country, beautiful, potentially useful, neglected and forgotten but still legally and technically alive, I might have found a field for a modest social initiative and adventure: . . . "[116] In his early work with Rolt, Aickman pressed for an integration of a nascent waterways movement with other sympathetic organizations, a position which would eventually lead to an early affiliation with the CPRE and ongoing communications with newspapers and magazines such as *Country Life* and *Time and Tide*.[117] Moreover, members of the waterways movement were encouraged to establish links with chambers of commerce, angling societies, preservation societies, and rowing clubs.[118] Having received a steady correspondence from admirers of *Narrow Boat*, Rolt must have found not only fertile ground for a waterways movement but a dedicated organizational partner in the form of Aickman.

Yet, intervention in the world of canals held risks for Rolt's view of the countryside. He had attempted to live within the rhythms of independent life and labor by which he believed the boat population had withstood the disruptions of modern society. This, despite the fact that some members of the modern world believed they might be able to alleviate conditions afloat by direct action. From Rolt's point of view, intervention to sustain the canals might simply turn life upon them into an artificially sustained display in which the rudiments of independent life and labor, heretofore symbiotic, would be replaced by efforts of outsiders to preserve the exquisite and diminishing world they saw afloat.[119] The movie *Painted Boats* prompted stories such as that of December 8, 1945, in *Illustrated*, chronicling the tradition of boat painting (drawing connections to a supposed Romani ("Gipsy"/"gipsy")) history and emphasizing a set of fading craft traditions. "The men and women who spend strange, nomadic lives wandering over the thousands of miles of Britain's canals have an innate love of color. Perhaps it is because many of the bargees are of gypsy origin; certain it is that their barges, the long narrow boats which have played such a part in inland transport since the days of the Industrial Revolution, are quite the most colorful craft afloat."[120]

Yet, seeing within Aickman's proposal an effort that might save the boatmen's very existence, Rolt followed Aickman in developing plans for a new organization. As he was to write later, "Anything which might help the canal boaters could not, I thought, fail to be a good thing so I welcomed the idea enthusiastically."[121] Rolt and Aickman would eventually meet and undertake a journey on the waterways together, with Aickman writing later, "I truly believe it is the best time I have ever spent on the waterways. . . . It is often the first step that is the best."[122] Though enthusiastic in his early response to Aickman, Rolt remained cautious, if not saddened, by the eventual turn of the organization the two soon helped to found—the IWA—toward preservation, years later. Moreover, he would grow increasingly anxious (as others would as well) regarding Aickman's steadfast and unyielding approach to the management and organization of the IWA. As Aickman was to write some years later in his autobiography, "If one really means business with one's cause, one must decide upon a precise aim or set of aims; . . . The inconsistency and mutability which are essential to the proper conduct of life as a whole, . . . should have no place in one's work for the cause."[123]

In the immediate term, however, the most important overall development for Rolt and Aickman came on July 26, 1945. The Labour Party, having garnered support with its collectivist plans for the postwar era, won the general election. The ascension of Labour ushered in a period in which the power exercised by governing institutions over local and regional social, cultural, and economic practices became a profound source of concern for canal advocates. The fear of a triumphant "civilization" undermining and obliterating culturally distinctive histories on the canals drew upon the themes presented in *Narrow Boat*. Although Rolt and others had considered modern capitalism a threat which had undermined life and culture on the canals, after July 1945 Labour's platform of nationalization suggested the future was now to be determined by collectivism exercised through Parliament. It now seemed imperative that an organization through which to advocate for canals' future be initiated. Continuing correspondence between Rolt and Hadfield and Rolt and Aickman led to the decision for a press release which would issue a wider call to action:

> The Inland Waterways of England, Scotland and Wales, combining as they do both use and beauty, are a priceless part of our National heritage. Owing to the influence of centralised trade and transport monopolies, canals and navigable rivers have been allowed to fall into a state of neglect and decay. Again, many of our rivers and streams, navigable or otherwise have suffered grievous harm at the hands of sectional interests. Indiscriminate building on their banks has destroyed their amenities, excessive pumping and ill advised dredging has been detrimental to local agriculture, while pollution by industrial wastes has destroyed fish and proved harmful to the wildlife of the riverside.
>
> It is therefore suggested than an Inland Waterways Society should be formed with the object, not only of protecting our rivers and canals from such abuses, but of encouraging water-borne trade and stimulating interest in waterways generally.[124]

The release indicated anyone interested in joining was welcome to contact the organization at a specific address (which was to be Aickman's home in London).[125] The release was sent out to a select group of magazines including *The Field*, *Country Life*, and *The Motor Boat*, which were to carry significant coverage of the emerging organization in the ensuing years. The small group that convened at Aickman's home in early 1946 founded the IWA, selecting Aickman as chair (handling most day-to-day activities of the organization) to be assisted by Hadfield as vice-chairman and Rolt as Honorary Secretary.[126] In November 1946, the first issue of the IWA *Bulletin* was published, which, much of it reflecting the work of Aickman, combined the themes of reclaimable industrial potential, aesthetic and cultural significance, and leisure potential on the waterways. The group had included rivers in their initial statement, but remained focused on canals, which shared many common difficulties. They promoted a number of canals on behalf of which they indicated they were working. The Kennet and Avon in the south of England, "not only flows through delightful countryside and is a work of great engineering importance, but provides also a wide boat link between the Bristol Channel and the River Thames, obviating the dangerous passage around the south coast."[127] The Stratford-upon-Avon Canal, "connects the riverless multitude of Birmingham with their leading pleasure resort [Stratford-upon-Avon]," and "The River Stour has unique association with the painters Gainsborough and Constable; . . . And is as well one of the most attractive rivers in England for pleasure boating."[128] Although some of the IWA's most passionate arguments for the preservation of canals were to be made increasingly in terms of their connections with the natural world, the organization still seemed ready to give precedence to their function in industry. This may have reflected not only Rolt and Aickman's desire to see canals put to their original use, but concern that the canal industry still had not welcomed their organization's presence. As the IWA would later include in the second issue of the *Bulletin*, "As we use the word 'amenities' and misunderstanding seems to exist in various quarters, notably among our friends in the canal carrying trade, upon our views and purposes in this matter, we emphasize our opinion that the inland system is unlikely fully to prosper and to play the part it should in the national life, unless it serves primarily the purpose for which it was constructed: namely, the carrying of waterborne trade."[129]

Under Aickman's leadership, the IWA consistently emphasized a crisis on the waterways and a need to act in the face of harmful government action/inaction. A sense that the loss of the traditional world of the boatman was imminent deepened a cultural message beyond just the canals' daily maintenance and use and helped to draw wider support from those linked with the arts and humanities. Narrowboat artwork—distinguished by the presence of roses and castles—the intricate designs of which were painted by craftsmen in boatyards, was critical. Rumination and debate about its origins, and thus the potential origins of boatpeople, whether as Romani ("Gypsies"/"gypsies") or otherwise, swirled. In the December 1946 issue of *The Architectural Review*, a publication that would continue to argue on behalf of the cultural world of canals, the artistic world of canal boats and the men and women on them was profiled by Barbara Jones.[130] Jones, whose murals and paintings appeared in a number of industrial design exhibits as well as on book covers in the postwar period, drew inspiration from the popular arts in Britain and investigated and wrote about topics as diverse as taxidermy, carousels, seaside architecture, and the art associated with funeral rituals.[131] She described a visit to the waterways where she not only charted the external paintwork of the boats but assessed the richly ornamented (and cramped) domestic spaces of the boat cabins. She wrote, "The table cupboard is the most noticeable thing: in the centre is a recessed panel enclosed in yellow and scarlet moulding, on which is painted a castle, at once English and exotic."[132] Though she believed individual versions of castles suggested diverse origins in Central and Eastern Europe, she was forced to conclude, "The origin of one castle may be perfectly obvious, and then another boat will come in which was painted at a different wharf or yard and an entirely new castle will start a fresh wave of thought. No one really has the slightest idea where they come from."[133] The drawings accompanying her article, delineating the rich decoration of the boats and the clothing and objects used by their occupants, drew the reader into a culture she recognized as very much embedded in the English countryside and identity. As she noted, "Certain sections have fallen quite into decay, and the whole system is said not to pay but to be so deeply imbedded in the life, agriculture and drainage of England that abandonment would be impossible."[134]

Jones' focus upon the folk arts and traditions she believed evident in life on the narrowboats enhanced the arguments of those who saw within the boatmen's diminishing numbers the loss of an organic society in the face of a powerful and homogenizing centralized state. In 1947, Rolt made a contribution to Massingham's work *The Small Farmer*, and also published his own much more wide-ranging historical and philosophical work, *High Horse Riderless*.[135] In the latter, he argued that the loss of belief in a medieval order had paved the way for an acquisitive society and the brutality and thirst for power in the twentieth century. He spelled out his concerns about centralized planning and the emerging tyranny of bureaucracy. These ideas continued to make their way into his concerns both for the preservation of canals and the threat represented by impending nationalization. To restore stability meant a return to an agricultural society in which the natural order would again be respected, he argued. As Charles Langbridge Morgan, reviewing the book in *The Sunday Times*, noted, "He is persuaded that collective materialism is a source of the instability and disintegration of modern society, and says firmly that the Christian philosophy of moral law, which implies man's cooperation with Nature and not his attempt to exploit it in pursuit of power, "constitutes the only sound basis of human contentment and social stability."[136] Langbridge continued, "That this is true, that the separatism of the servile state with its ration of bread, of circuses and of culture, is not a substitute for the natural harmony of a genuinely creative individualism is now beginning to be understood, and Mr. Rolt's book may greatly assist the understanding of it."[137]

Indeed, Rolt's suspicion of the centralized state had already been in evidence in his writings directly focused on canals. In penning the early IWA pamphlet, *The Future of the Waterways*, he once again contended that families working canal boats once largely owned their own craft, "In the past, the majority of these canal families owned their own boats but, since the trend of the age is adverse to small ownership in any form, the majority are now the employees of canal Carrying Companies."[138] Such a contention was largely debunked by scholars in the late twentieth century not long after Rolt made his argument.[139] Nonetheless, in his eyes, independence such as that among canal boatmen was essential in helping to ensure that links to a more stable social and economic order were maintained. He continued, "Every well kept and well decorated boat is thus evidence that one trade at least has not yet lost its sense of dignity and self-respect. . . . Yet if canal transport falls into the hands of a private or a state monopoly possessing no real standards of value, disappear they certainly will."[140] The understanding that capitalist as well as socialist dominance could undermine the qualities of individual craftsmanship that had been so central to a previous era was important. Rolt did not suggest that previous ages had been static, though he infused his work with a sense of foreboding regarding change. The environment that had spawned canal construction, he argued, might very well have contained within it the seeds of destruction for the world of the craftsman. As he had written in 1944:

> The promoters of the canal schemes, unlike the engineers, were not concerned with the canals themselves, but with the potential monetary profits to be derived from them. . . . This obsession with illusory monetary values, . . . was the *primum mobile* of the Industrial Revolution and a force which still masters us. It has logically resulted in the most wasteful civilisation the world has ever known, and a conception of progress which consists, not in building upon the past, but in leaving it behind with ever-increasing speed, with the inevitable consequence that present values become more ephemeral and degenerate with every year that passes.[141]

By early 1947, the IWA had begun to build useful political connections. In February, Sir A.P. Herbert, now an Independent Member of Parliament (with conservative inclinations), was announced as the IWA's president, while Peter Scott was announced as the first vice-president.[142] During World War II, Scott had broadcast his fond reminiscences of Britain on the air; his most significant distinction would come later as the founder of what is now the World Wide Fund for Nature (WWF). In March, Rolt and Aickman were part of a delegation of five that, led by Herbert, met with Labour Minister of Transport Alfred Barnes to discuss the fate of canals in collectivist Britain. The group submitted a "Memorandum" spelling out the aims of their organization. At the outset, the group grappled with what they might have expected to be the view of a government anxious to modernize: "Our organization was recently described in the popular press as 'romantic' for advocating an 'archaic' method of transport. That neither of these adjectives is applicable we trust this memorandum will make clear."[143] It then went on to offer a dual vision for the canal system as a carrier of commerce and a source of recreation. Initially, it laid out arguments that canals could continue to offer industrial transport and highlighted the ways in which the IWA believed water transport could be more advantageous than rail. Embedded within this examination of the canals as an industrial resource was a discussion of the degree to which boating families had suffered as a result of government rationing that failed to adequately consider their conditions. Following this initial argument, however, was a key discussion of the recreational possibilities of canals and their place within the landscape of Britain. This was, indeed, a secondary component of the "Memorandum," but was written with a verve that indicated the IWA's commitment to canals as a unique ecological and cultural resource—an argument that was to be sustained in coming years, "Unlike our railway system, our inland waterways, part artificial 'cut' and part navigable river, are of their nature inextricably associated with the natural structure and ecology of the country. The canal or river is thus not merely a commercial highway; it is also at once a landscape amenity, the habitat of fish and other wildlife, and an important agent of water supply, conservation and drainage."[144] The Memorandum went on to suggest:

> The extraordinary development of pleasure boating on the Norfolk Broads during the last twenty years is proof of the demand that exists and can be reached by efficient management. There is no reason why a similar development should not take place in all parts of the country, with immense gain to the nation's health and happiness, and a huge alleviation of all kinds of holiday congestion; and it is precisely those waterways which, because they pass through predominantly rural areas are unlikely to carry a heavy commercial traffic, that would prove of the greatest value in meeting this growing demand. Many of these water-ways are not only of the greatest scenic beauty, but the quality of their engineering alone would justify their preservation as historical monuments quite apart from their practical value.[145]

The IWA had thus begun to mount a campaign founded on a desire to affirm the unique values the canals represented in the overlap between nature and technology. To this was now to be added the organic relationship between technology, labor, and human communities that the organization's leaders believed the canals represented. In the fall of 1947, they succeeded in setting up an exhibition of canal life at Heal's Gallery in London, which provided an overview of the canals and their history, as well as artifacts of their culture. The exhibition, which was likely in-step with Heal's original reflection of the Arts and Crafts movement, drew substantive press attention from a number of quarters, including from the *Birmingham Post*, which remarked that England's canals remained "a useful as well as a romantic heritage."[146] The physical artifacts on display drew the attention of *The Architect and Building News*, which, in an article entitled "English Primitives," remarked, "It consists, in addition to photos and drawings, [of] all sorts of decorated objects, tea-pots, poles, painted seat boards and assorted jerries of all kinds. . . . Refusing, like the gypsies, to quite die out, this English folk art now only survives on canals, fairgrounds and occasional pub-signs, the product of contented illiterate minds and brown horny hands, in moments of leisure after hard physical work."[147] Echoing *The Architect and Building News*' assessment, the *Evening Standard*, in an article entitled "Beauty and the Barge," remarked, "The flower patterns and primitive landscapes are scarcely English in style. They suggest European peasant art, though some trace them to the toy theatre so popular in the early nineteenth century, when the canals were being made. . . . Whatever their origin, they add gaiety to the landscape in the same way as the land gipsy's caravan."[148]

In addition to those who saw canal traditions as marked by pre-modern artistic craft traditions, the architectural choices of early engineers and the relationship of these choices to the landscape were of significant interest. Eric de Maré, an architectural photographer closely tied to the leadership of the IWA, wrote most of a 1949 special issue of *The Architectural Review,* a journal which, counting Nikolaus Pevsner on its editorial board, took a protective view of architectural forms it felt to be under threat in the postwar era. De Maré reflected the wide architectural and design interests of the publication by focusing on the form, function, and cultural heritage of canals. The aim, wrote de Maré, was to stimulate a reconsideration of canals from the standpoint of a "planner with aesthetic vision."[149] His work included assessments of the history, engineering, and built environments surrounding canals. Remarking on bridges amidst a myriad of assessments of built structures, he echoed Rolt's emphasis on early canal engineering that was woven into the landscape. "Canal bridges are perhaps the most beautiful and interesting of canal structures, . . . Purely functional, quite unself-conscious, perfectly proportioned and fitting snugly into the landscape, these bridges have a fine monumental quality which few engineering structures have since achieved."[150] In de Maré's eyes, the Romani ("Gypsy"/"gypsy") designs of the canal boats were an essential part of their identity, one now in direct contrast with a centralized state. He wrote that, "the roses and landscapes with their Carpathian castles" constituted the one remaining folk art alive in England, while speculating on its origins, citing theories it had been inspired by "gypsies" from the Carpathians, "Victorian oleographs of Windsor Castle" or even an artistic style brought back by the Crusaders.[151] He wrote, "The art is tending to die out, and is not, unfortunately, receiving encouragement from the new controlling bureaucracy. The boaters at one time also had their own mode of dress, notably the bonnets and wide skirts of the women and the brightly colored woven belts of the men; to-day these traditional garments are rarely seen."[152]

For the IWA, early discussions of canal craft and architectural artifacts, as well as cultural life, were conjoined with the daunting task of ensuring railway administrators did not overlook the legal responsibilities they had in ensuring remaining canals were kept navigable. In early 1947, the IWA took up the cause of the northern section of the Stratford-upon-Avon Canal, where the Lifford Bridge at Kings Norton, once a drawbridge that would allow boat traffic underneath, had been fastened down after a road accident had damaged it. Providing notice to the Great Western Railway of their interest in navigating the *Cressy* underneath, Rolt and Aickman set forth on the canal in May 1947. Though they were accompanied by a railway maintenance boat that repeatedly got stuck, they were successful in navigating underneath the bridge after a crew designated by the railway raised it. Aickman noted with pride several years later the warm response from the gathered crowd: "We made statements. We posed for photographs. We signed autographs. We accepted cups of tea. We proclaimed a great new future for the canals of Britain. The Association was only twelve months old but our cause had arrived."[153]

The "arrival" of the IWA's cause, however, was not due simply to the organization's initiatives, but linked with several other emerging discussions on the use of, and access to, the British countryside in the postwar era. Still present primarily as an element of industry, canals remained absent from substantive discussions of holidays during wartime. After the war, though much of its rhetoric sought to contradict the planning movement, the IWA found itself operating in conjunction with many aspects of it. Shaping this new context was John Dower, a planning leader who, during wartime, had outlined the significance that new opportunities for travel would soon have on the countryside. Self-trained as an architect, he was active in the movement for town planning in the late 1930s, but also closely connected with the movement to protect natural areas. In 1938, he penned a widely read pamphlet entitled, *The Case for National Parks in Great Britain.*[154] In 1942, he was appointed to a section of the Ministry of Works and Buildings, which was focused on post-war reconstruction. Here, he was given oversight of rural policy.[155] In this position, he continued to refine his case for national parks and argued that they were integral to a postwar Britain that would be increasingly urbanized. In an address to the Royal Institute of British Architects in 1943, he argued that the wise use of the countryside and coastline could be one of the most important components of postwar planning. The demands on both types of spaces were affected by three important changes now taking place, including an increase in leisure time, the widening appreciation of "natural landscape beauty," and the "rapid advance in mechanical transport."[156] Reflecting his conflicted feelings regarding the emerging interest in the countryside, he wrote of the automobile, "The ubiquitous motor car is obviously in many ways, one of the greatest blessings of the age, . . . But in other ways it is quite certainly one of the greatest curses of the age. . . . Of the many evil consequences I need mention only – the holiday season condition over wide areas which makes it impossible to enjoy any country peace and quiet and clean fresh air in or near any place where motors can go, or to walk the highway with any safety and comfort."[157]

In examining the idea of keeping the countryside "intact," Dower critiqued the efforts of the "Councils for the Preservation of Rural England and Wales," which he said had been unsuccessful in the interwar period.[158] Rather, with the aim of planning in mind, he believed the attempt should not be to dam the new flow of travelers to the countryside, but rather to direct it. Room must be made for both new towns and the maintenance of agriculture.[159] In his address, almost no mention was made of access to inland waters, and no mention was made at all of canals. Indeed, Dower was clear on the need for the planning of Britain's countryside but (perhaps to the relief of Rolt and Aickman who might have taken notice) argued that this planning ought not to extend too far.

> Control, direction, encouragement, facilities—to sum up, planning—yes; but do not let it be carried a stage further and become regimentation. I suppose we are in for a pretty thoroughly "planned" future, . . . Let us at least see that in our holidays and our leisure we retain, behind whatever planning we adopt as a matter of convenience and common-sense, an ultimate and basic freedom of action for the individual, the family and the voluntary groups.[160]

Within just a few years, despite whatever reticence there might have been, planning quickly became central to postwar Britain. In 1943, Sir Leslie Patrick Abercrombie produced his County of London Plan, to be followed by the Greater London Plan the following year. Woven into the plans was an attempt to relieve traffic congestion and to provide for housing and open space, all while avoiding urban sprawl. Whereas Winston Churchill's wartime government had kept commitments and planning to a minimum, by 1946, the Labour government was making commitments to these very same ideas, carrying through legislation that was to substantively affect the geography of urban and rural areas. Among important pieces of legislation was the New Towns Act of 1946, which provided for the administration and construction of towns designed to accommodate populations of several thousand.[161] Furthermore, profound reconstruction of the social and economic priorities of the British state came with the National Health Service Act of 1946 and the Coal Industry Nationalisation Act of 1946, which ushered in the economic planning and social support considered integral to shaping a new postwar world. Building upon this, the call to expand the ability of the common people to experience the nation's outdoor amenities gained new traction. Dower was successful in initiating a national parks committee as the war came to a close and assisting its efforts until his death in 1947. Two years later, the National Parks and Access to the Countryside Act of 1949 designated National Parks and Areas of Outstanding Natural Beauty, reaffirmed access to public rights of way, and indicated a common national commitment to aesthetic amenities.[162] In this, there was an emphasis on both preserving the rural economy and carefully considering restrictions on forestry and farming.[163]

Two pieces of parliamentary legislation, due to take effect in 1948, were critical for the IWA. The first was the Town and Country Planning Act of 1947.[164] By its terms, local planning authorities were consolidated, required to develop planning documents, and given significant powers to oversee development in their local areas. Though canals were not directly affected by the legislation, the legislation's terms signaled the arrival of a postwar era in which governing bodies took much more active steps in planning for commercial and recreational infrastructure within their boundaries. The second, and much more immediately important piece of legislation, was the Transport Act of 1947, which nationalized railways and long-distance commercial road transport.[165] This legislation effectively made the Ministry of Transport, through the newly established Transport Commission, a prime mover in canals' future as railway companies were nationalized. The Labour Party partly achieved, as a by-product of the nationalization of inland transportation, much of what the Royal Commission on Canals and Inland Waterways had called for four decades earlier: a nationalized administration of the canal system. Under the Act, nearly all railways and railway canals, as well as several important canal-based entities including the Aire and Calder Navigation, the Trent Navigation, the BCN and the Grand Union, were brought under the authority of the BTC.[166] Most canals soon came under the control of the Docks and Inland Waterways Executive operating under the BTC. Yet, this new and powerful canal administration was not comprehensive. Unlike most railway lines where a single company had controlled both the track and the services operating on them, only a small portion of canal carrying services had been attached to canal companies or railways. Thus, some significant canal transport companies (at least for the time being) continued to operate independently.

In years to come, canal proponents were both to argue for government action to secure the system's future and yet remain profoundly suspicious of governmental actions. For instance, frustration regarding traditions of canal boat decoration became prominent at the close of 1948 when the IWA involved itself in the emotional issue of the re-painting, in a common livery, of boats the then just created Docks and Inland Waterways Executive had inherited as a result of nationalization. Although many boats on nationalized waterways were still in the hands of private carriers (and thus not subject to the re-painting scheme), the issue triggered the suspicions of Aickman, Rolt, and others regarding apparent efforts of the Ministry of Transport to erase cultural traditions on the waterways. The IWA *Bulletin* contended in February of 1949, "The art of the boat painter is a true tradition: his products all differ one from another, but differ within settled conventions. They are neither uniform nor extravagant. The boat painter's tradition provides the only surviving popular or folk art left to us."[167] Aickman, in an article for the literary and political magazine *Time and Tide*, also endorsed the artistic value of boat painting as Jones had done. He noted that canal boat painting was "a uniquely gay and beautiful thing in a dreary and hideous world."[168] The IWA and others eventually lost their battle against a common livery for the

Docks and Inland Waterways carrying vessels. Nonetheless, Aickman's final conclusion, if overwrought, did draw the attention of artist Enid Marx. In a response written to *Time and Tide*, she championed the canals' painting tradition, arguing that Britain might not be following the examples of its European competitors in ensuring the survival of craft traditions.[169]

As the 1940s closed, the IWA increasingly attempted to meld the protection of folk tradition with new opportunities arising on the waterways. Conversations between the IWA and representatives of the British Tourist and Holidays Board evolved in 1948, with the IWA *Bulletin* noting in January that it had emphasized to one representative the need for mooring facilities and a friendlier reception for boaters (particularly from overseas).[170] Later, the IWA trumpeted a March 4 article in *The Times* indicating the Home Holidays Division was ready to discuss the use of Britain's waterways for holidays. The article stated, "The idea of barge holidays on canals is one that has attractions both for those who maintain that the countryside can never be properly appreciated through the windows of a moving motor-car and those in whom the instinct for change and mild adventure is not wholly dead."[171] It added, "The canals speak at once of leisure and the past, and those who decide to spend a fortnight on their solitudes in preference to the crowded beaches of the seaside will be rewarded by the feeling that they have traveled into a country they did not know existed although it has all their lives been lying under their hand."[172] In the spring of 1948, the *Bulletin* noted the publication of the first annual report of the Home Holidays Division, which it said, "contains a very fair account of our efforts to advance the case for more widely and readily obtainable canal holidays."[173] Though noting overall progress in the cause of holidays on canals had likely not been as satisfactory as some might have expected, it did state that the Holidays Division had proposed a memorandum to the BTC regarding holidays on canals. "It is reasonable to say that this document must be based largely on evidence supplied by us; and also that, but for us, the case for this most attractive and healthy form of recreation would have gone completely by default."[174]

In his 1949 *Architectural Review* work, de Maré linked appreciation of the canal system's craft and architectural features with what he believed could be their utilitarian future as part of a "nation-wide system of water supply, conservation and drainage."[175] Canals' waters could provide pleasure boating and their towpaths a "paradise" for walkers.[176] "This use of waterways for pleasure is no minor matter in view of the coming of the holidays-with-pay scheme and the expectation of increasing leisure as mechanization of industry grows," he added.[177] Indeed, this latter argument for holidays on canals had muscle. One of the sources of canals' support came from the increasing number able to travel on scheduled holidays. In 1937, there were approximately 15 million Britons taking holidays. In 1938, the Holidays-with-Pay

Act gave power to Trade Boards in some industries to establish holidays with pay; though not applicable to the entire workforce, the legislation began a significant expansion of tourist numbers.[178] Planning for use of, and access to, the countryside during holidays had continued during the war, with the full expectation that the war itself would be followed by a surge in numbers.[179] In a survey undertaken by the newly created British Tourist and Holiday Board's Home Holidays Division in 1948, approximately 15.3 million people (34 percent of the population) were taking holidays at the seaside by the late 1940s, while 5.6 million (13 percent of the population) visited the country.[180] Noting that the current demand for accommodation was for about 21 million, the survey suggested that demand could rise to about 30 million once all those who were able to take holidays did so. Despite the opening of new accommodations and holiday camps, as well as the Home Holidays Board's efforts to encourage holiday-taking through a wider range of calendar dates, congestion was inevitable. As the Fabian Society, describing the expansion of working-class holidays in the nineteenth century and the current crescendo in holidays, noted, "In the end all classes are beginning to push their claim to the undeveloped, 'unspoiled' areas on the coast and inland."[181] Nonetheless, though making the case for continued state involvement in the provision of holidays, it indicated holidays, "must be free from regimentation and without the stamp of state uniformity."[182]

The struggle against uniformity—defined broadly within experiences of the landscape, modes of travel within it, and types of destinations visited—was one that girded the IWA. The IWA's relationship with planners was particularly fraught where issues of the integration of a renewed canal system into the postwar landscape appeared to be ignored. Reporting on a letter Aickman recently wrote to a group of eminent town planners, of whom Abercrombie was seen as the most significant, the IWA *Bulletin* remarked that only G. A. Jellicoe [who had been president of the Landscape Institute from 1939 to 1949 and in 1948 had become president of the International Federation of Landscape Architects] had had the courtesy to reply.[183] Though the English canal system was not a focus of his work, Jellicoe did go on to emphasize contact with water and garden canal features in his later efforts.[184] The *Bulletin* also indicated that Thomas Sharp was one of the few town planners who had expressed interest in canals. For his part, Sharp had been president of the Town Planning Institute between 1945 and 1946 and in 1946 had published *The Anatomy of the Village*, a signature work assessing the evolution of distinctive features of English villages and their role in shaping planning for the future.[185] The book, however, gave little direct attention to how rivers or canals might be integrated into village design. The *Bulletin* indicated Clough Williams-Ellis, the author of the 1928 work *England and the Octopus* which had decried indiscriminate urban expansion, "among a number of other architects," was a member of the IWA.[186] It went on to say very generally that he and these others "do much good work for our cause," but offered little more.[187]

The IWA pressed on all fronts open to it, ultimately seeking to prevent closures and save an interconnected canal system, but veering between engagement with state authorities when a new use for the waterways might be in the offing and renewing anti-planning and anti-statist rhetoric when officials seemed to doubt the potential of the waterways. Though it indicated compromise with the organic and ruralist outlook of Rolt's *Narrow Boat*, for the IWA, an emphasis on tourism meant a potential for growth in waterways' use, this time driven by individual choice and exploration of the countryside.

This seemed essential as across much of the canal network, industrial use was continuing to falter. In the summer of 1948, Rolt and Aickman undertook a 230-mile journey surveying England's northern canals—a journey, the record of which they indicated was sought by Sir Reginald Hill, chairman of the Docks and Inland Waterways Executive.[188] Despite the apparent interest and enthusiasm for canals' new leisure potential, the survey revealed the dire conditions for those still working the waterways. The *Bulletin* (in an article likely written by Aickman) indicated, "Our Honorary Secretary [Rolt], in his travels about the waterways, encounters only many repetitions of the same happenings; families leaving the cut after generations, owing to uncertainties and diminishing traffic; traders going elsewhere owing to the same uncertainties and increased tolls. We come across these depressing events several times on our Northern tour."[189] The northern voyage also highlighted the damage that could be caused by decaying infrastructure, as Rolt and Aickman's boat began to leak after having been dragged through a pound on the Huddersfield Narrow Canal. Forced to stay in Ashton for a day, Rolt, Aickman, and those who accompanied them contended with cotton workers angry that the necessary supply of water to their factory had been cut off (as a result of the emptying of a pound to address the boat leak), and the ongoing rain. Despite his reservations about the navigation help provided by the Docks and Inland Waterways Executive overall, Aickman credited a maintenance crew sent to assist that allowed them to proceed the rest of the way through the canal. Nonetheless, in the canal's Standedge Tunnel, "Once more according to precedent, the narrow boat stuck in the very small bore about halfway through, so that we had a considerable wait in the noisy, smoky darkness before it was successfully eased and we could proceed."[190]

Despite their shared experiences on the canal survey, Aickman and Rolt responded differently to the canals' dilemmas. For Aickman, a collapsing canal infrastructure crystallized the need to resist relentlessly an uncaring and inert bureaucracy. Only when bureaucratic force was met with immediate and full-scale resistance would government bureaucrats' implicit goal of ending support for the canal system stall, and the way be opened to full physical revitalization. Keeping immediate resistance always to the fore, he understood the impact a determined stand for all canals had upon local groups seeking a dependable national movement to support them.[191] Only a network of canals that maintained their multiple links could herald an era of new use and enjoyment by a population anxious to travel among them and witness their interconnected operations and history. Aickman appreciated the world of the canals as reflected in both their natural settings and working culture, but he had not the organic outlook Rolt had; better than the latter, he could more peacefully live with changed experiences on the waterways for the sake of holding the system together. For Rolt, on the other hand, canals encompassed a unique engineering and architectural world, a "golden age" of human interaction with technology in which mechanisms had reflected human ingenuity while not preventing their makers from deriving satisfaction from, and aesthetic appreciation of, their operations. This technological accord with humanity was evident in the ways in which canals blended into the landscape – failing to alienate individuals from those things that brought them aesthetic fulfillment. The life of the canals was also in the routines of work and living of the boatmen, who were themselves an organic part of the waterways. As the boatmen disappeared, so too did an essential part of the canal world. Although continuing to argue for salvaging as much of the system as possible, Rolt was willing to sacrifice sections of canal where working and cultural life had departed, for the sake of bringing pressure to bear on saving active routes.

In his 1950 book *The Inland Waterways of England*, Rolt suggested that while networks of canals might resemble those of the railways, the boundaries of their experience were much larger, encompassing the experiences of the ducks floating upon them, those fishing on them, and the inns alongside them.[192] The beauty, grace, and functionality of canal architecture were little surpassed by succeeding railway lines.[193] Moreover, he inferred that humans' interaction with the landscape had changed from one of cooperation to one of conquest, an effect railway engineers found difficult to disguise and that planners in 1950 seeking to make a "ferro-concrete factory, the dual track highway, the "pre-fab' housing estate, [or] the pylon and the power station" conform to the "English scene" found nearly impossible:

> The canal builders, on the other hand, had an easy task. The scars of construction soon healed; the towpath hedge grew up; the wild flowering plants of marsh and stream soon graced the margins; trees extended their boughs over the water as though to gaze at their newfound reflections; the birds came: the stately swan; coot and moorhen; wild-duck; the bright enamelled kingfisher and the heron standing sentinel or flying over on slow-beating wings. Thus the natural world

> quickly adopted the new waterway and so, far from detracting from the landscape, it made a positive contribution to it. To make the canal conform to the whims of the eighteenth-century landscape gardener was therefore not a matter of concealment but merely one of adaptation or embellishment.[194]

Rolt found himself afraid that the traditions at the heart of canal operation, beyond their continued visual integration into the landscape, were at risk. Varied styles and materials were no longer being used, with oak and the skills of working with it disappearing.[195] The narrowboat remained the most enduring of varied types of craft that had once populated the waters, striking because of the images of castles which continued to be painted on various portions of it. He wrote, "In this drab, machine-dominated England of ours the narrow canal boat is as unique and striking as some vivid tropical bird seen amid a flock of sparrows, but, like the bright bird, it has strayed into an alien world and may soon perish."[196] He continued to associate canal boat workers with Romani ("Gypsies"/ "gypsies"), intensifying the experience of mystery and other-worldliness for many observers. He suggested that when building the Bridgewater Canal, Brindley had sought nearby communities of Romani as laborers, their work then evolving from construction to navigation. He identified various elements of Romani life, ranging from the layout of cabins to matriarchal social structures, as representative of the "gypsy" influence.[197] "To sum up, I believe that the gypsy was the first master boatman, and that he followed the network of still waterways as they spread eastwards and southwards across the Midlands. . . . I believe that the 'Number ones' of later years, whether they were gypsies or English countrymen, perpetuated those traditions which are now passing away."[198] This identification of a unique cultural world of canals corresponded to the unique gaze of both the countryside and city offered to the waterways traveler. Even in the midst of a major urban area such as Birmingham, canal travelers could still hear trains, buses, trams, and cars overhead and find themselves withdrawn from the world of steam and petrol and, in fact, find themselves somewhat disoriented. Rolt contended, "For years they may regularly have used some busy thoroughfare without realizing that beneath them, through some dark invisible culvert, there ran a waterway linking coast to coast."[199]

Themes of canals' folk traditions percolated through the IWA's summer 1950 rally of boats at Market Harborough on the Leicester section of the Grand Union Canal. The number of canal boat workers and their families was continuing to diminish; in place of advocacy for them, themes of cultural recreation upon the water began to emerge. The IWA had drawn upon ruralist and organicist voices but now increasingly sought to ensure canals' preservation within a society seeking leisure and escape. The rally program spoke to this new understanding.[200] Full-page ads for the magazine *Yachting World* and a company providing holiday travel in the Norfolk Broads accompanied similar-sized ads for books on canal travel and history by Rolt, Hadfield, and de Maré. The festival's program advertised the participation of Frank Jones, a boatyard painter whose work had become well known in the early twentieth century, yet contained little else that would have spoken to either the industrial past or present of canals. Activities for the week did include outings on canal boats, but activities encompassing plays by the "Inland Waterways Festival Theatre Committee," dances, and an exhibition of paintings by Peter Scott were included. Outside of Jones' presence, there was little focusing specifically on the lives of canal boat people, beyond the images made available on a women's scarf sold as, "Depicting the gay and colourful scenes of quaintly painted barges."[201] Though appearing poised between reaffirming the traditions of the boat people and the era of leisure, the IWA was adamant a national commitment to the waterways was needed. "There is still no sign of a national waterways policy. It is hoped that the Festival and Rally will contribute to a vigorous public demand for a policy of proper maintenance throughout the two thousand miles of waterways which survive; of efficient management; and of full use."[202] There was also little the IWA could do to stem the decline in boating families. As commerce on canals declined, carrying companies contracted or disappeared altogether, and the children of canal boat families opted for a life ashore.

As the IWA campaign evolved after 1950, it was without Rolt, who was prompted to leave the organization as he and a number of others became increasingly dissatisfied with Aickman's intolerance of dissent in policy matters. Most significant was Aickman's commitment to forcing the government to maintain the whole canal system. Something else may have been at work, as well, however. Rolt had sought the preservation of a living tradition in his effort to promote the cause of canals, a desire at the heart of the ruralist outlook. Yet, the ruralist voice was now on the very edge of being overwhelmed by a message of leisure and recreational use of canals that, while paying homage to an industrial past, began to leave it behind. After leaving the IWA, Rolt went on to work with the Talyllyn narrow-gauge railway in Wales, where his engineering skills and commitment helped to revive and sustain its working operations. Here, he attempted to ensure the needs of the local Welsh community were considered at least on a par with those of the tourists who visited for a day. Yet, perhaps seeking a return to a "golden age" of harmonious communities interacting with their technological surroundings, Rolt found himself confronted with a host of contradictions ranging from the presence of non-Welsh volunteers to a reliance upon tourists for income.[203] In such a world, Rolt struggled with what had first connected him with canals and the Talyllyn railway. As he wrote in his autobiography:

> What chiefly appealed to me about the canal system was its indigenous working life. On the narrow canals this meant the working narrow boats and their crews which were so essential a part of them. These working boaters, so many of whom I knew and admired, unconsciously supplied that subtle traditional patina of constant use . . . it was an essential part of that blend of utility and beauty which used to compound the particular magic of canals. This was something which some members in the I.W.A. could never fully appreciate. They could value canals aesthetically as an important contribution to landscape beauty, but they could not assess to how great an extent utility was responsible for this.[204]

By the 1950s, both canals and the canal boat population were entwined in understandings of Britain's landscape, though the ideals of industrial use Rolt continued to guard were rapidly diminishing. An understanding of this transition begins with the nineteenth century, where the "countryside" as a space of escape from contemporary urban and commercial pressures first emerged. For most Victorians, canals were a part of a dark world generated by industrialization and in need of the "civilizing" forces of education and moral reform. Yet, by late-century, many organizations dedicated to accessing and experiencing both nature and traditional recreational pursuits were being established; this at the very point heavy industrial traffic was beginning its decline on canals. Thus, a door was opening to reinterpretations of canals that, rejecting the idea of the threats and dangers to be had on them, instead emphasized their unique and colorful experiences. Thurston, on his journey recorded in *The "Flower of Gloster,"* set the stage for the treatment of canals as an aesthetic and cultural experience, by not only remarking upon the experience of nature to be found, but the eccentricities of the population afloat. The latter he embraced for the "vagabondage" that seemed to be in their blood. Neal, writing of a journey just before 1914 in which he had witnessed both the traditions of canals' simple engineering and a return of nature to their increasingly quiet banks, emphasized the antidote they offered to life in the post-World War I era in which numbers of radios and automobiles were steadily increasing.

By the late 1920s, writings on canals were situated within a crescendo of "ruralist" writings emphasizing traditional pursuits within agricultural society. Proponents of an "organic" society also emphasized the significance of human harmony with nature in achieving well-being. In his travels by land through the countryside, Morton found himself unnerved by the number of tourists at Stratford-upon-Avon and disruptions in the countryside that had made its population dependent upon far-flung commercial contacts. On the water, Cubbon (though traveling in a petrol-powered boat), Herbert, and Bliss pondered quiet evenings on the waters, the aesthetic qualities of the natural world which surrounded them, and interactions with boatmen whose lives seemed both foreign and anachronistic. Though canal journeys thus far remained only at the fringes of ruralist conversations (symbolized by the off-hand reference to them as a dilapidated reminder of the industrial world in *Britain and the Beast*) a confluence of economic change and physical decline, as well as the increasing presence of nature, ensured they continued to straddle the boundary between the industrial and pre-industrial worlds at the end of the 1930s.

Rolt undertook a journey on Britain's canal system in 1939 and witnessed both the peace of the waterways as well as the unnerving intrusion of industrial technologies he felt were steadily driving a wedge between the present-day population and the craft traditions of old. Moreover, his emphasis on the significance of "Number Ones" spoke to an implicit anxiety regarding expanding administrative centralization in both industrial and governmental affairs. Against the odds, Rolt would continue to argue for the continued industrial workings of the canals in the postwar world. He contended the cultural value of the canals was bound up with their continuing industrial use and that the nation's financial managers were too focused on present-day balance sheets to grasp the benefits new investment in the canals could bring. In July 1945, the Labour victory at the polls meant that a new era had dawned. Canals were at first to be of only marginal interest to those grappling with industrial transport. Rolt was hardly known in governmental circles; many both within government and without likely expected that more of the network would have to finally be closed.

It was in this context that a small group of canal enthusiasts, motivated by Rolt's *Narrow Boat*, organized the IWA, a group focused on the future of the canals. Struck by the engineering history of the system, intensely interested in and sympathetic to its working population, disturbed by a lack of governmental interest, and intrigued by the alternative the canals offered to an increasingly technological and homogenized Britain, they took up advocacy for the system with little acknowledgment from its industrial directors. They were soon prepared to engage both a wider public and a new government at the close of the war. The rural areas through which canals passed were increasingly affected by mechanization, large-scale farming efforts, and, as was increasingly the case with towns as well, a future directed by centralized planning. Yet, even if it was only a nod to a relatively small and diminishing component of the transport population contributing to the war effort, the movie *Painted Boats* suggested there were qualities inherent in canal boat workers and their families that were essential to the nation. The suggestion at the end of the movie that there was a future for the canals in spirit (even if their upkeep was uncertain) was likely important for observers. As he completed *Narrow Boat*, saw to its publication, and then witnessed the book's popularity at the close of World War II, Rolt likely realized, as many others did, that canals might very well be an antithesis to plans for a transformed Britain. His voice, as well as those of others who argued for the unique place of canals in modern Britain, would be significant in the postwar era.

The IWA, led by Rolt and Aickman, argued for an overall effort that would restore a valuable transportation alternative and reaffirm the dignity and commitment of the state toward canal workers, many of whom advocates imagined reflected the values of a pre/early industrial craft society. The preservation of waterways culture represented one plank of the platform from which their struggle was waged. In the "gypsy" designs on boats, debates regarding "officialdom" and "statism" became less abstract and more nuanced and powerful. The IWA's effort to prevent the BTC from instituting a common livery on canal boats trading under its umbrella harkened back to the romanticized image of canal boatmen and their families. Moreover, Barbara Jones, Eric de Maré, and Enid Marx reaffirmed the cultural significance of boatmen, as did Rolt in linking canals to his wider interest in early engineering. All had witnessed the submerging of some independent craft traditions in the turn to larger commercial enterprises, the arrival of new forms of engineering, and a widening presence of mass-produced consumer goods. Though beliefs such as that of the predominance of "Number Ones" and boatmen as descendants of Romani ("Gypsies"/"gypsies") would be overturned in the late twentieth century, the beliefs' impact upon a historical understanding of the space of canals was significant.

The IWA also argued that the independence to be found on the waterways stood in stark contrast to the constraints of an increasingly managed society. In the post-World War II era, many planners responded to the demands of re-building with a commitment to facilitating an expanding population within spaces that responded to the local landscape and existing architecture. Indeed, Abercrombie had during the war produced a County of London Plan, and John Dower had nurtured his call for the eventual development of national parks, emphasizing the arrival of new leisure time. Moreover, the New Towns Act of 1946 outlined processes for establishing new communities outside of major urban areas. Most canals had been devised in a far different era, designed with the constraints of the local landscape in mind, constructed with local materials, and characterized by unique engineering solutions responding to the dilemmas of water management. Such characteristics put them at odds with a new era of centralization and did make redevelopment difficult. Yet, their role as naturalized links in the landscape—offering routes between towns and inland ports which were reminders of a still largely rural eighteenth century—re-positioned them as a collective antithesis to the world of suburban and exurban development emerging after 1945. The natural systems at work in their design and engineering, as well as their reliance on pre-modern forms of locomotion, ensured that what was seen as anachronistic technology by many in Britain offered a foil to modern society.

Postwar society intensified and spread an expectation of holiday travel, which initially was focused on the traditional seaside resorts of Britain. Yet, sheer numbers suggested other options needed to be made available. The 1949 National Parks and Access to the Countryside Act set the stage for re-envisioning democratic accessibility to the countryside. Canals were not at first included, still much more closely bound to understandings of industrial use. However, significant sections of canals, such as those associated with the Kennet and Avon and the Stratford-upon-Avon, suggested tourist interest could be significant. The IWA was only one of a number of groups concerned with widening access to the countryside, but the sheer scope of the canal network suggested it had the potential to be one of the most significant. Rolt, Aickman, and others participated in both surveys of canals and direct actions, which drew attention to both failing infrastructure and rights in navigation that persisted on the waterways. Although they both argued canals could be returned to industrial use and championed the interests of boatmen and their communities, headlines regarding underutilized or disused canals often highlighted drownings and the presence of trash, piquing many local communities which then sought to have canals closed and/or eliminated.

By 1951, canals' identity had been significantly affected by the desire to preserve and revivify patterns of work and culture among boatmen and their families. Waterways had also become a powerful symbol of technology embedded in the landscape that offered a counterpoint to the rapid and powerful disruptions brought by the postwar period. Canals challenged trends of suburbanization and offered a counter-current to homogenization through distinctive experiences of local towns and villages. They emphasized a slow discovery of urban and rural spaces—their distinctive technical features reflecting the regions through which they traveled. Canals could, despite their industrial focus, engender engagement with nature as they crossed the countryside and, with their use of horse and human power, suggest a less destructive story of early industrial change. Each of these forces would become even more powerful in the 1950s and 1960s.

NOTES

1. See, in particular, M. F. Bunce, *The Countryside Ideal: Anglo-American Images of Landscape* (London: Routledge, 1994).
2. *Lancaster Gazette*, June 26, 1819, 3. British Newspaper Archive (BNA).
3. John Hassell, *Tour of the Grand Junction, Illustrated in a Series of Engravings* (London: Hassell, 1819), vi–vii.
4. Hassell, *Tour of the Grand Junction, Illustrated in a Series of Engravings*, 7.

5. Hassell, *Tour of the Grand Junction, illustrated in a Series of Engravings,* 11.

6. Hassell, *Tour of the Grand Junction, illustrated in a Series of Engravings*, 48.

7. Hassell, *Tour of the Grand Junction, illustrated in a Series of Engravings*, 83.

8. William Black, *The Strange Adventures of a House-boat*, Vol. I (London: Sampson Low, Marston, Searle, & Rivington, 1888), 23–24.

9. Black, *The Strange Adventures of a House-boat*, Vol. I, 47.

10. Black, *The Strange Adventures of a House-boat*, Vol. I, 190.

11. Black, *The Strange Adventures of a House-boat*, Vol. I, 190.

12. Reginald Blunt, "In Tow," *Pall Mall Magazine* 1, no. 4 (May–October 1893): 506. The canal was eventually abandoned in 1914. Blunt's account was reprinted for twentieth-century waterways supporters in the *Bulletin* of the Inland Waterways Association (IWA) 77 (October 1966). In error, however, the *Bulletin* indicated the account was published in the *Pall Mall Magazine* in 1888.

13. Blunt, "In Tow," 511–512.

14. Blunt, "In Tow," 513.

15. Blunt, "In Tow," 515.

16. Blunt, "In Tow, " 518.

17. For the Royal Society for the Protection of Birds and the National Trust, see John Blunden and Nigel Curry, eds., *A People's Charter*? (London: Countryside Commission, 1990), 14–15; and Merlin Waterson, *The National Trust: The First Hundred Years* (London: BBC Books, 1994). See also Sir Dudley Stamp, *Nature Conservation in Britain* (London: Collins, 1969), 9.

18. Alun Howkins, *Reshaping Rural England* (London: HarperCollins, 1991), 229–230.

19. For the interwar period, see John Sheail, *An Environmental History of Twentieth-Century Britain* (London: Palgrave Macmillan, 2002).

20. Bunce, *The Countryside Ideal*, 114–117. For analysis of Bunce, as well as a broader history of the experience of the environment in Britain, see Sheail, *An Environmental History of Twentieth-Century Britain*. Despite the strength of his work on the environment in Britain and his focus on outdoor recreation, Sheail gives little attention to canals.

21. E. Temple Thurston, *The "Flower of Gloster"* (New York: Dodd, Mead, and Co., 1912).

22. Thurston, *The "Flower of Gloster,"* 29–30.

23. Thurston, *The "Flower of Gloster,"* 30.

24. Thurston, *The "Flower of Gloster,"* 41–42, 45–46.

25. Austin E. Neal, *Canals, Cruises, and Contentment* (London: Heath Cranton, 1921), 10.

26. Neal, *Canals, Cruises, and Contentment*, 12.

27. Neal, *Canals, Cruises, and Contentment*, 13.

28. Neal, *Canals, Cruises, and Contentment*, 172–173.

29. Neal, *Canals, Cruises, and Contentment*, 189.

30. Stephan Kohl, "Rural England: An Invention of the Motor Industries?," in *Landscape and Englishness*, eds. Robert Burden and Stephan Kohl (New York: Rodopi, 2006), 194.

31. Kohl, "Rural England: An Invention of the Motor Industries?," 191–192.

32. For a short overview of "ruralism" in the interwar period, see Malcolm Chase, "This is No Claptrap: This is Our Heritage," in *The Imagined Past: History and Nostalgia*, eds. Christopher Shaw and Malcolm Chase (Manchester: Manchester University Press, 1989), 128–146.

33. One of the most significant discussions of the disruption of the "countryside" by the arrival of capitalism and the harkening back to a "Golden Age" of the organic community is offered in Raymond Williams' *The Country and the City* (New York: Oxford University Press, 1973).

34. George Sturt, *The Wheelwright's Shop* (Cambridge: Cambridge University Press, 1923), 153–154.

35. Sturt, *The Wheelwright's Shop*, 154. Nonetheless, proponents of progressive industrial science argued that the preservation of an "essential" England was bound up with investing in the power offered by the application of electricity to agricultural tools and methods. Still, resistance to the implements of the new age of electricity, particularly pylons, was intense. See Bill Luckin, *Questions of Power: Electricity and Environment in Inter-War Britain* (Manchester: Manchester University Press, 1990), 83, 102.

36. Stanley Baldwin, *On England and Other Addresses* (New York: Frederick A. Stokes Company, 1926), 7.

37. Clough Williams-Ellis, *England and the Octopus* (London: G. Bles, 1928). For a wider discussion of impacts along the River Thames, see Andrew Sargent, *The Story of the Thames* (Stroud: Amberley, 2015), 194.

38. Harold J. (H.J.) Massingham, *World Without End* (London: Cobden-Sanderson, 1932). Exhibiting a nostalgia for a set of values they believed had once been inherent in the rural village, Malcolm Chase has argued that ruralists' promotion of paternalistic landlordism, rural craftsmanship, and connection to soil reflected antipathy toward a collectivist society and the uncertainties of a Britain in decline. (Malcolm Chase, "This is no claptrap: this is our heritage.") David Matless has emphasized the significance of land, and pointed to the absence of water, in ruralist discussions linked closely with "organic" understandings of human societies and their surroundings.

> Earth is the key element of organic England. It is striking how little water and air figure in organicist accounts. While industrial if not agricultural pollution of water and air is a concern at the time, and while organicists seek an holistic, ecological understanding embracing all elements, earth takes priority. (David Matless, *Landscape and Englishness*, 2nd ed. [London: Reaktion Books Ltd., 2016 (1998)], 150)

39. Philip Conford, *The Origins of the Organic Movement* (Edinburgh: Floris Books, 2001), 21.

40. Conford, *The Origins of the Organic Movement*, 116.

41. Kohl, "Rural England: An Invention of the Motor Industries?," 200. See also a discussion of Shell guidebooks, Patrick Wright, *On Living in an Old Country: The National Past in Contemporary Britain* (London: Verso Books, 1985), 66–67.

42. Ben Knights, "In Search of England: Travelogue and Nation Between the Wars," in *Landscape and Englishness*, eds. Robert Burden and Stephan Kohl (New York: Rodopi, 2006), 170.

43. H. V. Morton, *In Search of England*, 22nd ed. (London Methuen & Co., 1935), viii.

44. Morton, *In Search of England*, 257.

45. Morton, *In Search of England*, 257–258.

46. Morton, *In Search of England*, ix.

47. Morton, *In Search of England*, x–xi.

48. The term "pre-modern" must be referenced with an acknowledgment of its constructed nature—a term broadly referring to an idealized understanding of feudal social and economic relations often assumed to precede modern, individualist, and acquisitive systems. See William Stafford, "This Once Happy Country:' Nostalgia for Pre-Modern Society," in *The Imagined Past*, eds. Christopher Shaw and Malcolm Chase (Manchester: Manchester University Press, 1989), 33–45.

49. T. W. Cubbon, *Only a Little Cockboat* (London: George Roberts, 1928), 64–65. A cockboat is a small and often (though not in this case) rowed craft.

50. Cubbon, *Only a Little Cockboat*, 112–113.

51. Cubbon, *Only a Little Cockboat*, 116.

52. A. P. Herbert, *Water Gipsies* (Garden City, NY: Doubleday, Doran & Co., Inc., 1930), 147.

53. Herbert, *Water Gipsies*, 147.

54. Herbert, *Water Gipsies*, 147.

55. Herbert, *Water Gipsies*, 150.

56. Herbert, *Water Gipsies*, 152–153.

57. Herbert, *Water Gipsies*, 155.

58. William Bliss, *The Heart of England by Waterway: A Canoeing Chronicle by River and Canal* (London: H.F. & G. Witherby, 1933), 9.

59. Bliss, *The Heart of England by Waterway*, 9–10.

60. Bliss, *The Heart of England by Waterway*, 10–11.

61. Bliss, *The Heart of England by Waterway*, 16.

62. Bliss, *The Heart of England by Waterway*, 143.

63. Clough Williams-Ellis, *Britain and the Beast* (London: J.M. Dent and Sons, 1937), xv.

64. Williams-Ellis, *Britain and the Beast*, xiv-–xv.

65. Williams-Ellis, *Britain and the Beast*, xvi.

66. Williams-Ellis, *Britain and the Beast*, 283.

67. See Conford, *The Origins of the Organic Movement*, 146–147, 224, 229. See also Frank Trentmann, "Gardiner, (Henry) Rolf (1902–1971), ecological campaigner and youth leader," in *Oxford Dictionary of National Biography*, September 23, 2004, accessed September 10, 2024, https://www.oxforddnb.com/view/10.1093/ref:odnb/9780198614128.001.0001/odnb-9780198614128-e-59314; Matthew Jefferies and Mike Tyldesley, eds., *Rolf Gardiner: Folk, Nature and Culture in Interwar Britain* (Farnham: Ashgate, 2011); Malcolm Chase, "Wallop, Gerard Vernon, Ninth Earl of Portsmouth (1898–1984), politician and environmentalist," in *Oxford Dictionary of National Biography*, September 23, 2004, accessed September 10, 2024,

https://www.oxforddnb.com/view/10.1093/ref:odnb/9780198614128.001.0001/odnb-9780198614128-e-59347. An earlier assessment supporting the overall discussion provided here can be found in Jules Gehrke, "Countryside, Recreation, and the Transformation of Canals in Britain in the Mid-Twentieth Century," *Journal of Tourism History* 11, no. 2 (2019): 167–186.

68. See H. J. Massingham, *Remembrance: An Autobiography* (London: Malvern Wells, 1941).

69. Conford, *The Origins of the Organic Movement*, 149–150.

70. See Viscount Lymington, *Famine in England* (London: The "Right" Book Club, 1938). Conford writes that in December 1938 Lymington's journal *New Pioneer* "enthusiastically" endorsed *Mein Kampf*. (See Conford, *The Origins of the Organic Movement*, 150.)

71. Conford, *The Origins of the Organic Movement*, 122.

72. Lionel Thomas Caswall (L.T.C.) Rolt, *Narrow Boat* (with a Foreword by H. J. Massingham), 2nd Rev. ed. (London: Methuen, 1972(1948)), 197.

73. Rolt, *Narrow Boat*, 15.

74. L.T.C. Rolt, *Landscape with Machines: An Autobiography* (London: Longman, 1971), 113.

75. Rolt, *Narrow Boat*, 119.

76. Rolt, *Landscape with Machines*, 226.

77. Rolt, *Narrow Boat*, 21.

78. Rolt, *Narrow Boat*, 51.

79. Rolt, *Narrow Boat*, 57.

80. Rolt, *Narrow Boat*, 72–73.

81. This shock of the new machine in a rural landscape has been explored in Leo Marx, *The Machine in the Garden: Technology and the Pastoral Ideal in America* (London: Oxford University Press, 1964), 9.

82. The position of the mechanic in this vignette is reminiscent of the separation of the driver from an immediate understanding of the powerful new mode of transportation available to him in the transition from horse-drawn to steam locomotion carriage. See Wolfgang Schivelbusch, *The Railway Journey: The Industrialization of Time and Space in the 19th Century* (Berkeley, CA: University of California Press, 1986 (1977)), 11–12.

83. Rolt, *Narrow Boat*, 12.

84. Rolt, *Narrow Boat*, 12.

85. Rolt, *Narrow Boat*, 12, 24.

86. The early Inland Waterways Association (IWA) pamphlet entitled *The Future of the Waterways*, which appears to have been written by Rolt just after World War II, concluded that although most boatmen were not ethnically Romani, their traditions had Romani influences.

> The origin of this tenacious tradition [the painting of boats] is lost but, despite the fact that the great majority of canal boatmen are not gypsies and would resent the appellation, it is safe to conjecture that this, like many other customs of boat life, was due in the first instance to Romany influence. ([L.T.C. Rolt?], *The Future of the Waterways*, 8, The Inland Waterways Association, HADFIELD/10/16a, Hadfield Collection [IWA papers], London School of Economic Archives, London, UK)

An early IWA advocate Eric de Maré wrote in his 1950 *The Canals of England*,

> It is often but wrongly believed that canal boating families have gypsy blood, a belief they strongly resent. Though they have been recruited from different sources at different times, many come from English country stock originating around such canal centres as Braunston. (Eric De Maré, *The Canals of England* [London: The Architectural Press, 1950], 71)

Harry Hanson does note that by the late 1960s Rolt was re-visiting his earlier indication that the boat people may have been Romani. (*The Canal Boatmen* (Manchester: Manchester University Press, 1975), 4.) In his recent work, Julian Dutton contrasts Rolt's and Hanson's views and concludes canal boatmen likely had diverse origins. However, he writes, "Rolt's theory is not universally accepted." This is an understatement. Rolt's theory is accepted virtually nowhere today. See (*Water Gypsies* (Cheltenham: The History Press, 2021), 79).

87. Despite some gaps in evidence, Hanson writes that,

> We can state conclusively that owner-boatmen were numerically unimportant, both at the beginning of the nineteenth century and at the end. . . . Considering all the difficulties under which the small man labored, even before the more meagre years of the later nineteenth century, and considering also the complexities of canal trading, reason, if not firm evidence, must indicate that the role of such men was never anything other than a peripheral one. (Hanson, *The Canal Boatmen*, 116–117)

88. Rolt, *Narrow Boat*, 32.

89. Rolt, *Narrow Boat*, 32–34.

90. Rolt, *Narrow Boat*, 154.

91. Rolt, *Narrow Boat*, 194.

92. Henry Vollam (H.V.) Morton, *I Saw Two Englands* (New York: Dodd, Mead, and Co., 1943), 218.

93. Morton, *I Saw Two Englands*, 233.

94. Wright, *On Living in an Old Country: The National Past in Contemporary Britain*, 84.

95. Flora Thompson, *Lark Rise to Candleford* (with an introduction by H. J. Massingham) (London: Oxford University Press, 1945), 10.

96. Alun Howkins, *The Death of Rural England* (London: Routledge, 2003), 151.

97. Howkins, *The Death of Rural England*, 152.

98. Howkins, *The Death of Rural England*, 152.

99. B. W. Clapp, *An Environmental History of Britain since the Industrial Revolution* (London: Routledge, 2013 (1994)), 91.

100. Howkins, *The Death of Rural England*, 195.

101. Matless, *Landscape and Englishness*, 149. Others among a complex web of writers and critics espousing a "ruralist" and anti-modern perspective were Adrian Bell and Rolf Gardiner. In his call to maintain what he identified as a cultural purity in the countryside, Gardiner expressed sympathies with Nazi claims of racial and national renewal in the 1930s. See K. D. M. Snell, "Bell, Adrian Hanbury

(1901–1980)," in *Oxford Dictionary of National Biography*, ed. Lawrence Goldman, September 23, 2004, accessed September 10, 2024, https://www.oxforddnb.com /view/10.1093/ref:odnb/9780198614128.001.0001/odnb-9780198614128-e-56898, and Trentmann, "Gardiner, (Henry) Rolf (1902–1971)."

102. L. T. C. Rolt, *Landscape with Canals* (London: Allen Lane, 1977), 27.

103. Rolt, *Narrow Boat*, Foreword.

104. Rolt, *Narrow Boat*, Foreword.

105. Rolt, *Narrow Boat*, Foreword.

106. L. T. C. Rolt, "The Future of the Canals," *The Field*, June 10, 1944, IWA Press Cuttings, 1936–1948, BW 195/1/1, Waterways Archive, Ellesmere Port, United Kingdom.

107. Frederick Burgess, "Barge Painting: A Traditional Art," *Country Life*, July 28, 1944, IWA Press Cuttings, 1936–1948, BW 195/1/1, Waterways Archive, Ellesmere Port, United Kingdom.

108. Rolt, "The Future of the Canals."

109. Rolt, *Narrow Boat*, 194–195.

110. *Painted Boats* (1945), Ealing Studios, 63 minutes, directed by Charles Crichton, starring Jenny Laird. Also released as *The Girl of the Canal* in 1947 in the United States.

111. *Painted Boats* (1945).

112. Rolt, *Landscape with Canals*, 79–84.

113. Charles Hadfield and Frank Eyre, *English Rivers and Canals* (London: Collins, 1945).

114. Ian Mackersey, *Tom Rolt and the Cressy Years* (London: M & M Baldwin, 1985), 48. Further insight into the early communications of Rolt and Charles Hadfield may be gained through the Hadfield Collection (IWA papers), HADFIELD/10/16b, London School of Economics Archives, United Kingdom.

115. Mackersey, *Tom Rolt and the Cressy Years*, 48.

116. Robert Aickman, *The River Runs Uphill: A Story of Success and Failure* (Burton-on-Trent: J. M. Pearson, 1986), 23.

117. For affiliation with the Council for the Preservation of Rural England, see IWA, *Bulletin* 2 (February 1947), 4.

118. IWA, *Bulletin* 1 (November 1946), 4.

119. Further insight into the early communications of Rolt and Robert Aickman may be gained through Correspondence and Papers of Robert Aickman, PRO 30/82/1, National Archives, United Kingdom.

120. "Painted Boats," *Illustrated*, December 8, 1945, 19.

121. Rolt, *Landscape with Canals*, 92.

122. Mackersey, *Tom Rolt and the Cressy Years*, 65.

123. Aickman, *The River Runs Uphill*, 31.

124. Press Announcement. n.d. Robert Aickman to L.T.C. Rolt, Correspondence and Papers of Robert Aickman, PRO 30/82/1.

125. Press Announcement. n.d.

126. David Bolton, *Race Against Time: How Britain's Waterways were Saved* (London: Methuen, 1990), 21–23.

127. Inland Waterways Association (IWA) *Bulletin* 1 (November 1946), 2.
128. IWA, *Bulletin* 1, 2.
129. IWA, *Bulletin* 2 (February 1947), 4.
130. Barbara Jones, "The Rose & Castle," *The Architectural Review* 100 (December 1946): 161–168.
131. See Barbara Jones, *The Unsophisticated Arts* (London: The Architectural Press, 1951).
132. Jones, "The Rose & Castle," 162.
133. Jones, "The Rose & Castle," 162.
134. Jones, "The Rose & Castle," 161.
135. See H. J. Massingham, ed., *The Small Farmer* (London: Collins, 1947), and L.T.C. Rolt, *High Horse Riderless* (London: G. Allen & Unwin, 1947).
136. Charles Langbridge Morgan, "Freedom to Create," *The Sunday Times*, April 6, 1947. (See IWA Press Cuttings, 1936–1948, BW 195/1/1, Waterways Archive, Ellesmere Port, United Kingdom.)
137. Morgan, "Freedom to Create."
138. Inland Waterways Association (IWA), *The Future of the Waterways* (London: A. P. Tayler & Co, Ltd., [1950]), 9, Charles Hadfield papers, HADFIELD, 10-16a (Early IWA literature), London School of Economics Archives and Special Collections, United Kingdom.
139. See Hanson, *The Canal Boatmen.*
140. IWA, *The Future of the Waterways*, 10.
141. Rolt, "The Future of the Canals."
142. IWA, *Bulletin* 2 (February 1947), 1.
143. Inland Waterways Association (IWA), Memorandum [1947], Charles Hadfield papers, HADFIELD, 10-16a(Early IWA literature), London School of Economics Archives and Special Collections, United Kingdom.
144. Inland Waterways Association (IWA), Memorandum [1947], 13.
145. IWA, Memorandum [1947], 13–14.
146. "Neglected Canals," *Birmingham Post*, October 15, 1947, IWA Press Cuttings, 1936–1948, BW 195/1/1, Waterways Archive, Ellesmere Port, United Kingdom.
147. "English Primitives," *The Architect and Building News*, October 24, 1947, IWA Press Cuttings, 1936–1948, BW 195/1/1, Waterways Archive, Ellesmere Port, United Kingdom.
148. "Beauty and the Barge," *Evening Standard*, October 25, 1947, IWA Press Cuttings, 1936–1948, BW 195/1/1, Waterways Archive, Ellesmere Port, United Kingdom.
149. Eric de Maré, *Canals: A Special Number of The Architectural Review for July, 1949* (London: The Architectural Review, 1949), 4. This special issue was reprinted a year later as a book, Eric de Maré, *The Canals of England* (London: The Architectural Press, 1950).
150. De Maré, *Canals: A Special Number of The Architectural Review for July, 1949*, 14.
151. De Maré, *Canals: A Special Number of The Architectural Review for July, 1949*, 34, 37.

152. De Maré, *Canals: A Special Number of The Architectural Review for July, 1949*, 8.

153. Aickman, *The River Runs Uphill,* 61. See also, Bolton, *Race Against Time*, 33–35. See also articles from the *Daily Mirror, Daily Mail, News Chronicle* and *Birmingham Mail,* among others. IWA Press Cuttings, 1936–1948, BW 195/1/1, Waterways Archive, Ellesmere Port, United Kingdom.

154. John Dower, *The Case for National Parks in Great Britain* (London: Council for the Preservation of Rural England, Standing Committee on National Parks, 1938).

155. John Sheail, "Dower, John Gordon (1900–1947), architect and promoter of national parks"in *Oxford Dictionary of National Biography*, accessed September 10, 2024, May 25, 2006, https://www.oxforddnb.com/view/10.1093/ref:odnb/9780198614128.001.0001/odnb-9780198614128-e-94011.

156. John Dower, "Holiday Use of Countryside & Coastline (An Address Given at the R.I.B.A. on 31 March, as the Fourth in the Series of Six Lectures on Town and Country Planning Arranged by the R.I.B.A.)," *Journal of the Royal Institute of British Architects* 50, no. 8 (June 1943): 181.

157. Dower, "Holiday Use of Countryside & Coastline," 182.

158. Dower, "Holiday Use of Countryside & Coastline," 182.

159. In David Matless' assessment, this work places him within the "planner-preservationist" vision of Britain, which ultimately triumphed over the organicist vision in the years just after World War II. See Matless, *Landscape and Englishness*, 337–342.

160. Dower, "Holiday Use of Countryside & Coastline," 184.

161. New Towns Act, 1946, 9 & 10 Geo. 6, *c*. 68, 9.

162. National Parks and Access to the Countryside Act, 1949, 12, 13 & 14 Geo. 6., *c*. 97.

163. I.G. Simmons, *An Environmental History of Great Britain* (Edinburgh: Edinburgh University Press, 2001), 228.

164. Town and Country Planning Act, 1947, 10 & 11 Geo. 6, *c*. 51.

165. Transport Act, 1947, 10 & 11 Geo. 6, *c*. 49.

166. The non-railway entities initially brought under the authority of the Transport Commission were the Undertakers of the Aire and Calder Navigation, Sheffield and South Yorkshire Navigation Company, Aire and Calder and River Dun Navigations Joint Committee, Company of Proprietors of the Birmingham Canal Navigations, Company of Proprietors of the Calder and Hebble Navigation, Company of Proprietors of the Coventry Canal Navigation, Grand Union Canal Company, Leeds and Liverpool Canal Company, Lee Conservancy Board, Lord Mayor, Aldermen and Citizens of the City of Nottingham (in respect of the Trent Navigation undertaking), Oxford Canal Company, Severn Commissioners, Sharpness Docks and Gloucester and Birmingham Navigation Company, Staffordshire and Worcestershire Canal Company, Company of Proprietors of the Stourbridge Navigation, Trent Navigation Company, Weaver Navigation Trustees, and Company of Proprietors of the Herefordshire and Gloucestershire Canal Navigation (see Transport Act, 1947, 10 & 11 Geo. 6, *c*. 49). In addition to some canal operations that chose to remain independent, components of Britain's inland navigation system exempted from the control of the British Transport Commission were the Manchester Ship Canal, the Bridgwater Navigation,

and the Thames and East Anglia watercourses, the latter of which were overseen by drainage authorities. Sir Reginald Hill, "The Future of Our Inland Waterways," *The Journal of the Institute of Transport*, 25, no. 9 (March 1954): 344. Approximately one-fifth of the watercraft used for carrying on the canal network were taken over by the BTC. Hill, "The Future of Our Inland Waterways," 344.

167. IWA *Bulletin* 18 (February 1949), 24.

168. Robert Aickman, "Nationalization and the Canal Boatman," *Time and Tide*, March 26, 1949, IWA Press Cuttings, 1948–1949, BW 195/1/2, Waterways Archive, National Waterways Museum, Ellesmere Port, United Kingdom.

169. Enid Marx, "Canal Barge Painting [Letter to the Editor]," *Time and Tide*, April 9, 1949, IWA Press Cuttings, 1948–1949, BW 195/1/2, Waterways Archive, National Waterways Museum, Ellesmere Port, United Kingdom.

170. IWA, *Bulletin* 11 (January 1948), 6.

171. "Holidays Afloat," *The Times*, March 4, 1948, IWA Press Cuttings, 1936–1948, BW 195/1/1, Waterways Archive, Ellesmere Port, United Kingdom.

172. "Holidays Afloat," *The Times*, March 4, 1948.

173. IWA, *Bulletin* 13 (April 1948?), 7.

174. IWA, *Bulletin* 13 (April 1948?), 7.

175. De Maré, *Canals: A Special Number of The Architectural Review for July, 1949*, 4.

176. De Maré, *Canals: A Special Number of The Architectural Review for July, 1949*, 4.

177. De Maré, *Canals: A Special Number of The Architectural Review for July, 1949*, 4.

178. David Chapman, *Holidays and the State* (London: Fabian Publications, Ltd., 1949), 6.

179. Dower, "Holiday Use of Countryside & Coastline," 181.

180. Chapman, *Holidays and the State*, 7.

181. Chapman, *Holidays and the State*, 9–10.

182. Chapman, *Holidays and the State*, 18.

183. IWA, *Bulletin*, 18 (February 1949), 16–17.

184. See, in particular, the "Water Gardens" designed in the late 1950s for Hemel Hempstead and the "Jellicoe Canal" designed for the Royal Horticultural Society in Wisley.

185. Thomas Sharp, *The Anatomy of the Village* (Harmondsworth: Penguin Books, 1946).

186. IWA, *Bulletin* 18 (February 1949), 16–17.

187. IWA, *Bulletin* 18 (February 1949), 17.

188. "Canal Cruise May be New Holiday," *Manchester Daily Dispatch*, September 2, 1948, IWA Press Cuttings, 1936–1948, BW 195/1/1, Waterways Archive, Ellesmere Port, United Kingdom. Also, IWA *Bulletin* 16 (September 1948?), 1.

189. IWA, *Bulletin* 16 (September 1948?), 17.

190. Aickman, *The River Runs Uphill*, 145.

191. Aickman, *The River Runs Uphill*, 71. As an IWA *Bulletin* response to BTC actions, likely written by Aickman, suggested in the summer of 1955, "There is

always a proper place for local pride; but the biggest single error our Members can make is to be led, as our opponents always try to lead, into arguing that one waterway should be preserved and another scrapped. Members who suggest such a thing are, however unintentionally, betraying our cause—and without the possibility of advantage even to the waterways they uphold. A cartoon was recently published which depicted a boat in a lock, and the boatman as saying 'It is always our waterway that is uneconomic'." See, "The Board of Survey Report," IWA, *Bulletin* 47 (Summer 1955), 10.

192. L. T. C. Rolt, *The Inland Waterways of England* (London: George Allen & Unwin, 1950), 11–12.

193. Rolt, *The Inland Waterways of England*, 84–85.

194. Rolt, *The Inland Waterways of England*, 98.

195. Rolt, *The Inland Waterways of England*, 133.

196. Rolt, *The Inland Waterways of England*, 138.

197. Rolt, *The Inland Waterways of England*, 176–177.

198. Rolt, *The Inland Waterways of England*, 177.

199. Rolt, *The Inland Waterways of England*, 207.

200. The Market Harborough Festival & Rally of Boats: Program of Attractions, Charles Hadfield papers, HADFIELD, 10-16b (1945–1955), London School of Economics Archives and Special Collections, United Kingdom.

201. The Market Harborough Festival & Rally of Boats: Program of Attractions, 4.

202. The Market Harborough Festival & Rally of Boats: Program of Attractions, 7.

203. See the argument of Joseph Boughey, "From Transport's Golden Ages to an Age of Tourism: L. T. C. Rolt, Waterway Revival and Railway Preservation in Britain, 1944-1954," *The Journal of Transport History* 34, no. 1 (June 2013): 33, and L.T.C. Rolt, *Railway Adventure* (Stroud: The History Press, 1993 (1953)), 130–134.

204. Rolt, *Landscape with Canals*, 170.

Chapter 5

Revival and Regeneration

Canals and a Transforming Britain, 1951–1968

Intermittent assessments of canals' industrial potential through the 1950s kept ideas of leisure and recreation at a distance while generally concluding that at least some needed to be closed. As political assessments came to recognize canals' potential for fishing, boating, and walking, so too would wider cultural conversations note their significance in understanding links between urban and rural spaces and the interaction of nature and technology. Transformations of canals from beleaguered relics of the industrial age to sites of multi-faceted leisure activities in the late 1960s marked the emergence of their contemporary place in the British landscape. Their legacies in fusing technical innovation with natural flows of water and fostering unique living and working communities, as well as their straddling of the pre-industrial and industrial worlds, shaped the beginnings of their physical renewal. Moreover, for many travelers, they offered a prism through which to witness evolving understandings of Britain's industrial past and relationship with ideas of the countryside, on the cusp of the environmental movement of the late 1960s. Just over 200 years after their emergence, canals and inland waterways had returned as symbols of innovation and had become sites for the renegotiation of the nation's relationship with its industrial past and the natural world.

By the early 1950s, within what was now a Conservative government, there was significant concern regarding the dilemmas of a system whose economic rationale seemed to have departed. Sir Reginald Hill, Chairman of the Docks and Inland Waterways Executive (operating under the British Transport Commission (BTC)), remarked in the journal *Public Service*, "If this industry is rationalised by the shedding of responsibility for the maintaining of hundreds of miles of unused canals, it should continue to play a useful if minor role in the national transport system."[1] The lopsided nature of Britain's canal system was abundantly clear in statistics Hill laid out in 1954.

In a speech before the Institute of Transport early that year, he noted that approximately 1,200 miles of waterways accounted for 98 percent of inland traffic, while only 2 percent was carried on the remaining 800 miles.[2] Noting that these 800 miles still needed to be maintained, he argued,

> You cannot safely leave an artificial water course entirely unattended, and this maintenance costs the Commission on an average between £300 and £400 a mile. In terms of net receipts some 850 miles of the Commission's waterways earn a profit of about £300,000 a year, while the remainder lose £400,000. Obviously the ship is waterlogged. If it is to recover buoyancy it has to be relieved of much of the dead weight it is carrying.[3]

Hill went on to describe strategies for making the canals which were capable of carrying significant loads efficient and profitable. But, in what was almost an afterthought, he identified the emerging significance of recreational pursuits:

> One other aspect of inland navigations to which reference should be made is the growing interest in the idea of touring the canals with their mellowed architecture and natural beauty. In this crowded land the preservation of beauty is a social duty, and beauty must not be a mere by-product of decay and I do not regard its creation or retention as something wholly extraneous to our economic life. It is not too much to hope therefore that those more attractive parts of our waterways which are no longer used or needed for commercial navigation will be preserved as amenities.[4]

In 1954, the government appointed Robert Palmer, the first Baron Rusholme and a member of the BTC, to take charge of a three-person Board of Survey to assess its canals. The survey was to be conducted to determine "whether all possible steps" were being taken to ensure that the greatest economic advantage was being drawn from canals and to recommend the next steps to be taken in cases where canals could "no longer be put to economic commercial use."[5] The Board collected the views of various transport and business concerns as well as the Inland Waterways Association (IWA), assessed reasons why transport by canal had declined or ended at specific locations, and reviewed the regional concerns of the canal system. Reasons identified for the loss of traffic included the difficulty of getting quotations on tolls, the increasing efficiency of roads, more competitive rates offered by railways, and the difficulties of ensuring efficient loading and unloading.[6] In its observations, the IWA was steadfast in its contention that there "should be no more abandonment of navigable waterways," and argued the waterways should be publicized for both commercial and recreational possibilities.[7] Moreover, it called for a separate waterways authority to handle canals.

The Board of Survey called for the redevelopment of 336 miles of inland waterways in a first group, with a further 994 to be kept serviceable in a second group and, most significantly, suggested that for 771 miles in a third group, there were "insufficient commercial prospects to justify their retention for navigation."[8] The report recognized an immediate need for the BTC to be relieved of waterways which were taking a toll on finances and administration. Noting that fishing and pleasure boating were among the possible uses of canals, the Board argued it was still unconvinced by the IWA.

> The existence of these ancillary uses has in the past proved an obstacle to physical abandonment, and, as we have explained, many hundred miles of waterways are now being maintained by the Commission at heavy cost for purposes wholly unconnected with transport.

It found the IWA's suggestions for non-industrial use "unrealistic and impracticable."[9] The Rusholme report called for the establishment of a waterways administration separate from the Docks and Inland Waterways Executive that would be placed under the authority of a general manager. This was accomplished at the beginning of 1955 with the appointment of Sir Reginald Kerr as chairman and manager of British Transport Waterways (BTW). Previously, after a distinguished military career, Kerr had been a Divisional Manager of the Midland Division of British Road Services, the national road transport company formed as a result of nationalization.[10] Soon after Kerr's appointment, a £5.5 million investment plan was begun focused upon the Aire and Calder Navigation, the Sheffield & South Yorkshire Navigation, the River Trent, the River Weaver, portions of the Grand Union Canal and River Lee, and the River Severn (to which was attached the Gloucester & Berkeley Canal).[11] This effort, expended on those river navigations and canals considered the most remunerative, did not include those in the Rusholme report's second group. The IWA, in turn, pressed its case for holding on to all the canals in the face of what it indicated was the process of piecemeal destruction undertaken by the BTC.

The IWA's efforts were infused with the grinding labor of convincing a sometimes skeptical public of the value of canals in a context where they became increasingly identified as artifacts of industrial decline and decay—all the while road construction accelerated across the nation. Linked with stark descriptions of trash, collapsing locks, and subsiding canal banks were news stories on the drowning deaths of young children in derelict canals. A February 1954 article in the *Lancashire Evening Post* suggested it was time to "do away with these deadly waters," among which the author included not only canals but disused quarries. The prospect that arteries of trade in the industrial era had turned into "child-traps" was unsettling. The correspondent wrote, "They are all man-made pieces in the mosaic of Britain's golden age of prosperity. . . . Neither cruel sea nor our inland waters furnished by nature can rival them for their fatal lure." The author of the article suggested a cruel irony in that in addition to "gruesome, ugly memorials to the days of rugged individuality," many of the canals now remained as "stagnant, evil, neglected pools."[12] Linked with disused waterways and scarring the landscape of South Lancashire were old clay pits and mines, as well. The article continued,

> The mill lodges, canals, mining subsidences and, in some measure, the disused quarries, are so linked with industry that they often are right on the doorsteps of the homes of the families who have worked or are working in the trades that have created them.[13]

Though drowning deaths in canals were not new, their ascent into the headlines in both rural and urban areas after 1945 became a significant point of concern for the IWA. On April 4, 1954, F. J. Yeates, Secretary for an organization identified as Birmingham Children's Safety (Canals), wrote to the *Sunday Mirror* seeking contact with others across the nation regarding the dangers to children from canals and suggesting that in the last seven months, twenty-one children in the Midlands had drowned in canals.[14] On July 12, 1954, Nottingham police began a frantic search for a seven-year-old boy who was reported to have fallen into the Wollaton locks on the Nottingham Canal. Drawing significant numbers of police and firefighters, as well as onlookers, and involving the draining of one lock, the search for the boy went on for five hours with no body found. A check of local young boys during the search appeared to indicate that all were accounted for, and no further news story emerged.[15] The same issue of the newspaper reporting this story contained a letter to the editor complaining of poor drainage and maintenance near the canal.[16]

Two years later, the IWA *Bulletin* reported on a decision of the urban district council of Stourport, "to take no action in support of the retention of the country's canal system," adding that one councilor had described the Staffordshire and Worcestershire Canal as "nothing but a disease carrier. You can see dogs and cats disintegrating in the water."[17] An amendment arguing that the Staffordshire and Worcestershire ought to be maintained as a commercial waterway failed the council. The *Bulletin* also noted that in the summer of 1956, a councilor J. H. Wardle of Wolverley, just to the north of Kidderminster and also near the Staffordshire and Worcestershire, said that

> many of the canals are nothing more than stinking ditches. If five and a half million pounds can be spent in keeping these dirty, stinking ditches in being,

> where rats thrive and cats are killed—and where even children are drowned—it is absolutely disgusting.[18]

Wardle's proposed solution for canals was turning them into roads.[19] Well into the mid-1960s, concerns were still emphasized within Midlands papers. On August 18, 1966, a petition with 300 signatures was delivered to the Stoke-on-Trent Corporation arguing that local government must make safe a stretch of the Caldon Canal in which three children had drowned within the past year. One of the organizers of the petition stated that, "The canal is now a death trap and every parent in the area is petrified of another tragedy. We want the canal fenced off immediately and, if possible filled in."[20] The general response of IWA members and their supporters was that derelict and dangerous canals should be restored for industrial use whenever possible, and in other cases restored for their amenity value. They contended that children who drowned most often did so on canals that were disused. The IWA argued a second step was instruction in water safety. It emphasized that the number of deaths linked to canals paled in comparison to the number of lives lost on the roads, the latter of which its *Bulletin* identified as a "holocaust."[21]

In contrast to those concerned about drowning, the IWA aligned with protests against automobiles, protests which were increasingly a part of local and national forums. As a column in *The Birmingham Post* of April 1955 asked,

> What will happen when, within very few years, the volume of road transport—freight and passenger—far exceeds the roads' capacity? How vehement will be traders' denunciations of State incompetence, and how peevish the cries of all the voters who have newly become car-owners! M.P.s will hear much from constituents who thought a little car would speed them into the Garden of Eden, but have found that it only gives them a good view of red traffic lights.[22]

In June, 1955, the architecture critic Ian Nairn published his influential article "Outrage" in *The Architectural Review*, arguing state-led and state-planned development, in the name of serving a more affluent population with modern services, was on the verge of destroying the boundaries between urban and rural and creating a monotonous and soul-deadening suburbia. In the process, he coined the term "subtopia." Nairn's extended essay warned of psychological effects of a leveling of human experience and suggested there was a need to escape from bricks and mortar and the psychology of mass society.[23] His emphasis on what he saw as the oppressive impact of centralized planners' broad remit to design suburban development—undoing the individual history and character of location—intersected with others' assessments of canals. In one of only a couple of brief references to canals, Nairn suggested that although canals might follow some practices of standardization, such standardization did not dominate.[24]

The IWA not only decried the actions of railways it believed had undone the initial industrial potential of canals (as had the BTC), but cried foul over the management of a number of canals still under private control and waterways supervised by local and regional governments. One solution emerging from the IWA was for the waterways to be supervised by a group that might operate like the National Trust. Its leader, Robert Aickman, set forth his plan in an article in the *Journal of the Royal Society of St. George* in 1956.[25] Central to his argument was the rationalization of a host of authorities governing waterways into an organization similar to the Thames Conservancy. In the nineteenth century, the Thames Conservancy had offered a means of coordinating interests such as land drainage, navigation, water supply, and fishing and boating on England's premier waterway. As the Land Drainage Act of 1930 and the River Boards Act of 1948 had reorganized and specified responsibilities for the public bodies connected with a host of natural waterways in Britain, the Thames Conservancy remained largely unchanged. In his article, Aickman laid the groundwork for wider unified administration of rivers and canals, writing that authorities managing many of Britain's waterways had done little to prepare for a multi-use future, which would include pleasure boating and fishing.[26] A broad-based national waterways conservancy, he said,

> should be given authority . . . over the whole waterways system as it survives today; should be required to restore and maintain it, and to develop it in all its functions and potential functions; and should be representative of all the interests affected. If desired, the BTC could be given a lease of the few waterways in which they are interested.[27]

The IWA concluded that a national waterways conservancy

> must not be, an antiquarian body with its gaze fixed upon the past. . . . The rivers and canals of the country are properly to be seen as entering upon a new era of usefulness, and much more varied than ever before, more valuable to many more people, and no less profitable, when the true balance is struck, than in the palmiest days of the late eighteenth century.[28]

The IWA stated it should be the goal of the conservancy to "restore as soon as possible to statutory condition, and to exploit in all possible ways, the existing system of statutory navigations."[29] The question of statutory responsibility in navigation was one that had bedeviled the BTC as more and more canals fell into disrepair. Rights of access to waterways were widely recognized, with some like the rivers Thames, Severn, and Trent considered public highways for centuries. The acts of Parliament that established the canals generally included clauses allowing the public access to the new waterways being created, so long as tolls were paid. The obligations of these clauses were passed to railways that acquired canals, with the Regulation of Railways Act of 1873 contending that the canals and associated infrastructure acquired with them were to be kept in good working order with supplies of water unhindered, "so that the whole of such canal or part may be at all times kept open and navigable" for those seeking to use it.[30] Assuming responsibility for the rail canals (and others, as well), the BTC had acknowledged that the act applied to them. The right of access applicable to canals was to operate at all times and be precluded only when repairs were actively underway. In the course of the 1950s, a variety of travelers on canals had found themselves warned away from traveling on underused or disused canals by BTC signage. In some cases, canal advocates had sought and received assistance from the BTC in making journeys possible, but in other cases, had acted to remove barriers to their transit themselves, which they had been advised was legally defensible.[31]

In the 1950s, BTC officials had signaled their desire to obtain legal sanction for ending navigation on at least some canals, arguing that the financial loss on them had become too great in an era of diminished traffic. Provision for the closure of canals was offered by the Railway and Canal Traffic Act of 1888, which had allowed applications for the closure of a canal (and thus the ending of maintenance for navigation) to be made to the Board of Trade if the canal had had no traffic for three years, or if it had become "unfit for navigation" due to the "default" of its proprietors, or if adjoining lands had suffered injury due to an escape of water from the canal (and its owners declined or were unable to effect repairs).[32] In the midst of questionable BTC support, the IWA and others made known their commitment to protest, often by ensuring that boats continued to pass on canals considered at risk. The IWA observed that canal locks, once freely accessible, were increasingly found to be padlocked and argued,

> Most of these padlockings we believe simply to be illegal. Nor can pleas that they are necessitated by various kinds of bad maintenance be accepted, for the bad maintenance itself is usually a breach of statute. Members should (at the least) protest against these padlockings when they encounter them, or soon the whole system will be in chains.[33]

The IWA campaign for a conservancy gained traction as government signals on the future of canals remained mixed. As a result of criticism of the Board of Survey of 1954 (Rusholme Committee), the BTC launched a more wide-ranging study of canals and navigable waterways led by Leslie Bowes, Managing Director of the Pacific Steam Navigation Company. Its preliminary discussion addressed the Land Drainage Act of 1930 and the River Boards Act of 1948, which gave drainage boards and river authorities the right to apply for control over navigable waterways where navigation rights were controlled by another entity but not being activity used.[34] The committee reported an overall operating loss for the inland waterway system of just under £500,000 and reviewed the problems of the aging infrastructure found in locks and bridges. These challenges were endemic, despite what had been a £5.5 million program of waterway improvement begun since the Board of Survey's report.[35] Ranging beyond the simply functional, however, the Bowes Committee began to suggest inland waterways as a whole were a resource with aesthetic, cultural, and heritage value. It stated that disuse of canals could lead to an increase in weeds and malfunctioning locks. It argued,

> The presence of weed on the surface, overgrown towpaths, leaking lock gates, rotting hulls of sunken craft, deserted premises and the absence of traffic can make a waterway appear more decayed than it really is. A derelict waterway in rural surroundings may be inoffensive and even picturesque; but in an urban area, and particularly an industrial area, it is dismal, offensive and sometimes dangerous.[36]

Alongside this, however, a recognition of the aesthetic impact of canals was combined with what may have been one of the first extended governmental statements of their value as a landscape resource:

> Many parts of the waterways, even those which traffic has largely forsaken, have much to commend them. They often traverse attractive country, and sometimes enhance it, bringing water, and associated vegetation and wild life, into otherwise unwatered areas. Our society is increasingly urban and industrial. Both the population and the means to enjoy leisure have grown. These trends increase the value of pleasant surroundings and of opportunities to sail, row, fish, ramble or study natural history, whilst the areas available for these pursuits diminish. The destruction of amenities which can be preserved at any reasonable cost cannot be taken lightly.
>
> Some canals were the work of engineers of acknowledged genius, and include aqueducts and other structures impressive in themselves and reflecting the high standards of design and craftsmanship characteristic of their time. On some of the canals, moreover, there are works which are of considerable interest as examples of the mechanical achievements of the early industrial revolution. Avoidable destruction of such things would be deplorable.[37]

The committee's primary administrative conclusion was similar to that of the Rusholme survey, though it divided waterways into two main categories for preservation: a select set of Class A waterways considered worthy of further economic development with expectations that they could be made remunerative, and a set of Class B waterways which were often narrow but considered worthwhile candidates for upkeep because of their remaining canal traffic, usefulness for drainage and water supply, or potential recreational value. Significantly, the Bowes committee departed from the Board of Survey report in arguing that a commitment of 25 years needed to be made to sustain Class B canals, to ensure that potential traders would be willing to make investments which might bring them up to their full potential. In addition, it recommended the replacement of a complex set of tolls with a fee-based licensing system. A third grouping of waterways was marked out simply for redevelopment. The government eventually rejected a twenty-five-year support plan for Class B canals, generating disappointment and distrust among canal advocates.[38] In response to the Bowes committee recommendations, the Government appointed an Inland Waterways Redevelopment Committee (IWRC) under the chairmanship of recently retired Admiral Sir Frederick Parham. The IWRC, made up of local authorities as well as boat owners and operators, had a mandate:

> To assist as necessary in the promotion of schemes for the treatment of inland waterways which cannot economically be maintained for commercial transport, to consider such schemes when formulated, to make recommendations upon such to Ministers concerned; and to advise Ministers upon any general matters connected with the redevelopment of those waterways.[39]

Including in its membership Lionel Munk (a rising leader in the IWA), L.T.C. Rolt, and John Smith (an official of the National Trust), the committee worked to fulfill its mandate until it was disbanded in the early 1960s.[40] Set against a backdrop of continuing declines in the total tonnage of goods—eight percent between 1954 and 1962—the industrial sustainability of canals seemed dire. In 1954, the BTC's loss on canals was approximately £153,000; by 1962, it was £1,068,000.[41]

Though Conservative governments took little direct action in support of reviving the entire waterways system, the IWA was instrumental in shaping public understanding of canals at a national level, in partnership with other organizations. Its *Bulletin* chronicled the activities of local organizations which campaigned for the restoration of individual sections of canal and initiated clean-up and restoration efforts based on volunteer labor.[42] At times, IWA actions brought valued press coverage. In 1957, "Operation Flyboat" aimed to show that a canal boat could make it from London to Birmingham in forty-eight hours, an undertaking ultimately completed by its member Lionel Munk.[43] In 1959, members of the IWA attempted to navigate the Stourbridge Canal and wrote later, in the Spring of 1960, that they had found the BTC had made desperate attempts to alter their intended route, that the waterway had been obstructed by refuse (which had to be removed by volunteers), and that the route had been manned by rude officials.[44] A second "assault" in December of that year offered somewhat greater success.[45] Offering one of what was a generally regular series of updates on canals in the Midlands, the IWA *Bulletin* of July 1959 reflected on the difficulties of the year:

> The Churchbridge Locks have lately been blown up, in the interests of opencast mining. The West Bromwich Council are trying to close the Ridgeacre Branch Canal and the Halford Branch Canal; . . . The Walsall Council is trying to close the Sneyd Branch, because it is said children fall into it. The Tipton Council have already purchased and propose to fill in the Ocker Hill Branch. . . . The British Transport Commission have in mind to close "fairly soon" the last remaining stretch of the Wednesbury Oak Loop Canal. The unique Dudley Tunnel is now officially closed to all traffic. Netherton Tunnel is no longer lighted. Every discouragement is made officially to navigation through the Black Delph Locks, a not unimportant through route; and the future of this line is regarded as in jeopardy. The top gate paddles are being removed on the Stafford and Worcestershire Canal; where the rule that locks must be left empty, has already resulted in bad leakages of the bottom gates. On this waterway also, the beautiful lock cottages are being demolished; and the Basin at Stourport power station, until a few years ago a hive of activity, is being filled with rubbish, no doubt "harmless and inoffensive."
>
> And the process is accelerating; as always happens when once a rot starts. . . . There is no answer to the problems of the immense Midlands system but a revival in waterway carriage; and that is unlikely to be sought under the present administration.[46]

By the late 1950s, as Britain's industrial prowess slipped further into the past, new attention was directed to the artifacts of industrialization. Faced with closures among the workshops, plants, and warehouses associated with the industrial era, campaigns to preserve these spaces as critical to the narrative of British history began.[47] Having left the IWA, but still a member of the IWRC which advised the government on canals, Rolt was a pioneer both in the movement to preserve industrial artifacts and within the discipline of industrial archaeology. In 1957, he published his biography of the nineteenth-century industrial engineer Isambard Kingdom Brunel, soon to be followed by those of others including Thomas Telford, George and Robert Stephenson, and Thomas Newcomen.[48] Indeed, in recounting his work on the biographies, Rolt said he felt himself a "latter-day Samuel Smiles," and that as he watched sales of his engineering biographies rise, he was satisfied that they were eclipsing sales of *Narrow Boat* and that he was now introduced as the "engineering historian."[49] The railways, he wrote in 1960,

> represent a colossal outpouring of creative energy unmatched by any other age; the embodiment of the pride, the hope and the aspiration of a pioneer generation expressed in the cyclopean masonry of arch and pier, and in the smoke-blackened architectural splendour of great stations.[50]

However, as his writing reveals, this was a nuanced engagement with the early industrial era, an engagement that rejected many of the ultimate effects of the industrial world. Rolt contended in his survey of Victorian engineering published in 1970 that the nineteenth century could be divided into two epochs marked by a divergence around 1860. The first (despite Peterloo and the "Hungry Forties") he found to be one of optimism about the ultimate effects of material prosperity, and the second one of pessimism, "when many of the more acute and perceptive minds were deeply disturbed and bewildered." [51] He remarked that a man born in 1800 would have been born into an "ordered" world where, "Despite the new roads and canals, life went on much as it had done for un-numbered generations."[52] Yet, he added,

> To stand upon the walls of Dudley Castle and to gaze down at the landscape below, ringed by fires and under a perpetual pall of smoke, was like looking into the mouth of hell itself. No wonder that all persons of sensibility tried to forget the existence of the Black Country.[53]

For Rolt, the mid-nineteenth century marked not only the passing of the heroic generation of engineers, but both the sublimation of individual effort in wider projects and an overall loss of confidence in the engineer's skills. He wrote:

> Increasingly, teams of more or less anonymous experts became responsible for new developments in engineering rather than a single brilliant individual. . . . No longer bestriding his profession with such easy assurance, he [the individual engineer] became the servant instead of the master of the boardroom and the accountant. . . . That mood of uncritical reverence for material progress which had made the engineer the hero of the hour did not long outlive the Great Exhibition [1851]. . . . Just as the engineer had shrunk the physical world and brought about a major social revolution by the power of steam, so it appeared as if the scientists were deliberately undermining the whole structure of thought and belief.[54]

Rolt's conflicted attitudes toward the coming of industry reflected those of others who read his work and found that, among other powerful forces of industrialization, canals represented both a balance of innovation and change and enduring links with the natural landscape. Rolt continued to work through the 1960s and early 1970s. During this time, he was instrumental in establishing the field of industrial archaeology, which had taken an important step forward with his attendance at the Industrial Archaeology Conference convened by the Council for British Archaeology in December of 1959. Here, Rolt joined the Industrial Archaeology Research Committee, which saw one of its primary tasks as pressing for the "Industrial Monuments Survey."[55] An industrial monument was defined as

> any building or other fixed structure—especially of the period of the Industrial Revolution—which either alone or in association with plant and equipment, illustrates or is significantly connected with the beginnings and evolution of industrial and technological processes, including means of communications."[56]

Rex Wailes, a past president of the Newcomen Society who led the project, added in simplified form that "an industrial monument is something of industrial interest which could not be preserved in a museum without first either destroying its housing or uprooting it from its foundations."[57] Emphasis upon built artifacts as a lens through which to view Britain's industrial past helped to set the stage for the use of canals as a route of transit to that past. Once initiated, Rolt recalled the survey as particularly successful, eventually becoming "a popular nationwide pursuit engaging people from every walk of life."[58] For the many who became involved, he said, time seemed to be of the essence, remarking, "All too frequently it was a case of here today but gone tomorrow."[59] Soon after, he became involved with an effort that eventually resulted in the York Railway Museum, which opened in 1975.[60] However, though still emphasizing the loss of a pre-industrial world, Rolt in his later life acknowledged he was helping to construct a particular vision of the English past, one that might indeed be based on one's perception of change:

> Nowadays, when surviving examples of the agricultural traction engine, heavy-breathing, brasswork a-twinkle, lumber round the show ring, we regard them with affectionate nostalgia. Like the shire horse and the gaily painted harvest wagon they seem to have become a part of the romantic past of rural England as the townsman conceives it. But one has only to read Thomas Hardy's description of the steam threshing scene in *Tess of the D'Urbervilles* to be made aware of the true impact of steam machinery on Victorian country life. Here the steam engine and its overalled attendant are portrayed as black and alien invaders from another world, imposing upon an age-old rhythm of field labour the relentless and untiring pace of the machine. For us, the Golden Age is always yesterday. Perhaps tractor and combine harvester will one day be regarded in the same romantic light.[61]

The idea of a "golden age" in canals seemed easy to picture by the mid-twentieth century for Rolt and other canal enthusiasts. The canals had been built by workers drawn largely from rural areas in an age before the application of large-scale, steam-driven machines to move earth. Their supporting structures had been built with simplicity in mind, often using local materials. If their construction had been driven by capitalist motives, they had nonetheless been built with a diversity of engineering solutions responding to the landscapes they inhabited. The presence of canal boatmen in both rural and urban areas and canals' association with crime, immorality, and danger had created significant anxiety in the nineteenth century; but for explorers on the canals in the early and mid-twentieth century who encountered boatmen who appeared to exude honest simplicity (if a lack of formal education), it seemed easy to attribute such stereotypes to Victorian social condescension. At the height of Britain's commercial and industrial achievement in the mid-nineteenth century, when railways dependent on coal took center stage, canals had retired to the background, continuing to depend upon human and animal power and boat crews who, if not meeting the definition of independent entrepreneurs as Rolt imagined, were still free of the factory system. The failure to invest in or upgrade existing engineering, undertake major new projects (other than the massive effort of the Manchester Ship Canal), or to consolidate under a centralized authority meant that individual canals retained their own local or regional characteristics and avoided being implicated in the cultural homogenization many critics derided by the early twentieth century.

In addition, canals seemed the very antithesis of the speed, destruction, and carbon-based pollution increasingly associated with automobiles. Enthusiasts had long touted the waterways' capacity to transport bulk goods at a significant cost-per-mile savings (admittedly slowly and with potential costs of transshipment to other modes of delivery). However, by the early 1960s, this argument for waterborne trade was conjoined with complaints regarding new highways that crossed urban areas at significant expense and the massive loss of life from car accidents. In 1945, there were approximately 2.5 million vehicles on British roads (approximately 58 percent of which were cars).[62] This increased to nearly 4.5 million in 1950 (approximately 51 percent of which were cars) and nearly 9.5 million in 1960 (of which 59 percent were cars).[63] An October 1959 article in the *Architects' Journal* entitled, "Metropolis" chronicled an unsettling toll for automobiles on society. Building upon the work of the Road Research Laboratory, the *Architects' Journal* calculated the cost of road congestion at £200 million, with an increase to £500 million when allowing for traffic delays outside of working hours.[64] The financial burden of road accidents was placed at £130 million, with the death toll estimated to be 72,000 since the end of World War II.[65] Indeed, the estimates came at a time of rapidly rising road fatalities in the United Kingdom. Declining from a peak of over 9,000 per year in 1940 (a number largely ascribed to the effects of the blackout), fatalities had declined to under 5,000 by war's end. They then began a steady rise, reaching a peak of just under 8,000 in 1966, when more stringent drunk-driving laws were instituted.[66] The *Architects' Journal* highlighted the United States, and in particular Los Angeles, as the arena in which the terrifying impact of the automobile was most acutely felt. The city had made manifest the subtopian future Nairn had roiled against, under a layer of smog. "The conclusion to be drawn from American experience seems to be that road building by itself can prove a bottomless pit, and that no matter how much money is sunk in it no solution can ever be found so long as other measures are not taken to bring the volume of traffic within manageable limits," the IWA *Bulletin* noted in recounting the *Architects' Journal* article.[67]

Rolt continued to face difficult decisions regarding canal closures during his time on the IWRC and learned the significance of leisure from his time on the Talyllyn narrow-gauge railway [see chapter 4]. Despite his best efforts to keep it operating as an authentic railway, he was forced to acknowledge its reliance on tourists.[68] Step by step, national policy responded to the demands that were now manifesting themselves. In 1959, the Labour Party published *Leisure for Living*, a document circulated for consideration at the party's annual conference at Blackpool in October. The report stressed the significant work yet to be done by the party in planning for and fostering artistic and cultural institutions, as well as sporting events and outdoor pursuits. The statement emphasized such things as musical appreciation, the importance of public support for art (and initiatives to expand its accessibility for the masses), and the significance of architecture that represented contemporary and innovative design. The report's authors emphasized that nearly fifteen years after the close of World War II, postwar planning could turn its attention to the hours when workers would not be laboring:

> The [first] post-war Labour Government proved that, in a properly planned society, it is possible to guarantee full employment; and, as automation spreads, it will also become possible, while maintaining full employment, steadily to lessen the number of hours that most people have to work. . . . Once full employment is again secured, the emphasis will increasingly be not on jobs for all but on leisure for all—leisure *and how to use it.*[69]

The potential of canals to be a part of the overall experience of the countryside was limited to a very small portion of the report, yet its authors noted that they welcomed the proposals of the Bowes Committee for the redevelopment of some commercially unviable canals, "many of which pass through beautiful and unspoiled country."[70] They wrote that canals and their towpaths created opportunities for "cruising, boating, fishing, and walking."[71]

Though brief, the discussion of canals in *Leisure for Living* pointed the way toward greater consideration of the routes as natural amenities. Generalized provision for the enjoyment of outdoor amenities on a national basis had been marked by the passage of the National Parks and Access to the Countryside Act of 1949. In the 1950s, the legislation allowed for the designation of ten broad regions within Britain which were to emphasize conservation and public access but to have day-to-day administration that remained largely with local and regional authorities.[72] Among *Leisure for Living*'s recommendations was that the scope of the National Parks Commission be expanded to "become the chief agency for the use of public funds for preserving and promoting enjoyment of the beauty of the countryside, functioning in much the same way as the Arts Council and the proposed Sports Council."[73] Moreover, the National Parks Commission's purview should be extended to waterways. The report concluded,

> We recognise that some financial support may be needed to make up for the long years of neglect. It will probably be necessary for stretches of suitable canals to be taken over by the National Parks through which they run, by local authorities, or by the National Trust (which has shown itself a worthy custodian of many other good things).[74]

The focus of the BTC, however, remained on upgrading only specific portions of the canal system. As part of its remit, the IWRC investigated the second and third groupings of canals identified by the Bowes Committee, compiling information from local authorities, property owners, businesses, and a variety of special interest groups. It also contacted local administrative authorities, leaders of outdoor and recreational groups, and industrial leaders whose organizations still made use of individual canals. Considerations were given to elements of aesthetic appeal in the surrounding environment, physical degradation, costs of restoring navigation, and the uses to which individual portions of canals (or their entire lengths) might be put if restoration efforts got underway. Many waterways were dealt with as individual navigations rooted in local concerns and not necessarily a part of a wider waterway network.[75] For other canal enthusiasts, however, the record of the IWRC was mixed. After initially rejecting a BTC proposal for the closure of five sections of the Birmingham Canal Navigations (BCN), the IWRC later agreed to four of the five closures, saving the Dudley Tunnel, a 1.8-mile passage on the Dudley Canal. The IWA contended the reprieve was due to a "Protest Cruise" organized by its Midlands branch which succeeded in garnering national media attention.[76]

The IWA also suggested that a protest cruise had been instrumental in securing the future of the Stratford-upon-Avon Canal, most specifically its southern portion. Indeed, the subject of Stratford-upon-Avon was one in which the industrial history of the waterways and a future based upon amenity could be amalgamated. Already established as a significant inland port due to the River

Avon, Stratford's significance had increased with the opening of the Stratford-upon-Avon Canal in 1816. Never marked by heavy industry, the town was well-established as a tourist escape focusing on its Shakespearean legacy by the mid-twentieth century. The southern portion of the Stratford-upon-Avon Canal ran for about 13 miles from its northern end near Kingswood Junction—where it was connected with the northern section of the Stratford Canal, as well as the Grand Union Canal—to the River Avon at Stratford-upon-Avon.[77] Discussions regarding the transfer of the waterway to the National Trust, for restoration and use as an amenity, had begun in the 1950s, though it was not until after a positive IWRC review that the Trust in 1960 took control. A complex history of control on the Stratford spoke to conditions which affected a number of other canals. Once under railway control, a portion of the canal had been leased to the Stratford-upon-Avon Borough Council, though maintenance was not kept to navigable standards and the waterway's primary purpose had been to provide water for agricultural purposes. In 1958, the Warwickshire County Council had sought a warrant declaring the abandonment of the canal in order to construct a fixed bridge over it, which could be constructed at much less cost than one preserving navigability. The action was blocked as evidence emerged that the canal had been used for navigation within the past three years. Subsequently, a lease agreement was reached between the BTC and the National Trust, encouraged by John Smith, a member of the National Trust's executive committee who was committed to canal restoration.[78]

The financing and costs of the Stratford project incorporated volunteer labor and the continuing participation of the BTC and the Warwickshire Council in addressing issues of roads and bridges. After securing control, the National Trust, in an annual report, echoed the emerging interest in recreation and the canals.

> The transfer of canals represents a new departure; it is not to be expected that the legal and financial arrangements will be simple or rapidly completed. But the council are aware of the pleasures—not the least an escape from the bedlam of the roads—which can be given to a wide cross-section of the community by the preservation, rather than the destruction, of these long-neglected amenities.[79]

By 1962, £42,000 had been raised by the National Trust for restoration.[80] In his full-time work alongside the National Trust, IWA member David Hutchings, chair of the Midlands branch of the IWA, contextualized such expenditures, contending it was costing the BTC £3,000 [a year] to maintain the canal (maintaining locks and the banks, to prevent flooding) and that an estimated £120,000 would have been required in construction/de-construction costs and then infilling to eliminate the canal and its risks.[81]

In addition to the southern section of the Stratford-upon-Avon Canal, one of the notable projects where IWRC actions helped to stem canal closures was the Kennet and Avon Canal. The Kennet and Avon was 75 miles long and linked the two rivers in its name, providing a navigable waterway across southern England. The BTC sought permission to end navigation in the Transport bill introduced in 1956, drawing immediate opposition from the IWA and its affiliate, the Kennet and Avon Association.[82] The clause supporting the abandonment of the canal was eventually thrown out, providing a platform for an accelerated voluntary movement to restore the waterway. The Bowes Committee of 1958 argued that it was "pre-eminently a case for re-development," which could take into account "amenity and recreational values."[83] The IWRC concluded that although it could not immediately recommend authorization of the "enormous expense" of restoring the entire canal, it might still be possible to harness enthusiasm and begin projects that would prevent its further deterioration.[84] It also indicated the majority of its members believed that no precipitate action should be taken that would "damn for ever as a navigation a waterway in the future of which there is more public interest than of any other in the country."[85] Efforts to secure the future of the Kennet and Avon continued, eventually coming under the authority of the Kennet and Avon Canal Association (now the Kennet and Avon Trust).

IWA rhetoric, which until the mid-1960s remained very closely associated with Aickman himself, argued that the organization's actions were critical to canals' survival. The *Bulletin* stated in 1961,

> It is important to emphasize also how much a Protest Cruise can achieve. . . . Without the two Protest Cruises on the Stratford Canal, that waterway would never have been saved; and the Dudley Tunnel line has been saved *only* by the Protest Cruise.[86]

The organization's determined tone was reiterated in a concurrent campaign for the Ashton Canal in Manchester, where the Manchester Corporation and other local authorities, citing the drowning deaths of children and a lack of traffic, had called for its closure and potential transformation into a series of lagoons. The IWA stated that "blind prejudice" drove the argument for filling in the canal rather than restoring it.[87] To its members, it wrote, "If soundness lies, as it does, in upholding the pessimism, at once tepid and desperate, of the assembled local authorities, then let us please be rebels."[88]

Amidst wider public discussion of canals' recreational potential, the BTC, and BTW which it supervised, had not been inactive. However, the era of pleasure boating arrived only via a series of conservative and intermittent steps adjunct to BTW's continuing focus on industrial transport. Between 1954 and 1960, BTW reported the number of "cruising licenses" issued for travel independent of tolls on the waterways had risen 77 percent from 4,716 to 8,342 per year.[89] Moreover, some initial steps to recognize the recreational potential of waterways had been taken by stepping into the new realm of passenger transport. Though nascent independent rental agencies would eventually be the most successful in advertising "get-away-from-it all" holidays on the waterways, BTW indicated that for its own part it was now providing twenty-eight two-berth cabin cruisers ("Water Baby" class), as well as eighteen "family" cruisers, which were converted narrow boats with four to six berths each.[90] Moreover, it reported on its own tours including two which were five-day "Heart of England" cruises taking passengers through "some of Britain's finest countryside."[91] A further one was a "sleep aboard" holiday on the "Water Wanderer" offered between Nottingham and Boston, and a final "through Shakespeare land" on the "Water Rambler" between Oxford and Birmingham.[92] Day-cruising vessels had also been increased from two to eleven since 1955, with each carrying sixty passengers. In addition to a passenger service established to the London Zoo, a final, luxurious day cruiser known as the "Fair Lady" was provided in the London area for conferences and cocktail parties and had even been used to host a wedding reception.[93] Yet, the prospect of financial sustainability was blunted as by April of 1962, the "Heart of England" cruises on the vessel "Water Rambler" had been ended while those on the "Water Wanderer" had been truncated.

One of the most significant steps in establishing new canal oversight in the early 1960s was the Transport Act of 1962, which reorganized the responsibilities of the BTC under a set of separate public authorities for Railways, London Transport, the Docks, and the Waterways. A new British Waterways Board (BWB) replaced BTW and was governed by a chairman and vice-chairman and assisted by a board selected by the Minister of Transport.[94] The Minister was empowered, subject to the authority of Parliament, to make grants to the BWB to meet deficits.[95] Critically, however, the act offered power for the BWB to divest itself of control over waterways no longer considered viable,

> It shall be the duty of the Board to review the manner in which the inland waterways owned or managed by them, so far as not required for the discharge of their duty . . . may be put to the best use, to formulate proposals with the object of putting them to the best use and to take all steps open to them to achieve that object whether by developing or converting any inland waterway themselves or by selling or leasing to other persons.[96]

This provision reinforced a divide that had opened between the government and the IWA as the selection of members for the BWB was made with deliberate effort to keep out those seen as extreme members of the IWA. With the board chaired by Parham (with Sir John Hawton acting as deputy chairman) and with representation from the National Parks Commission and trade unions, governmental representatives also selected the noted canal writer Charles Hadfield as a member, with one observer noting he was, "a well-known writer on canals but not a fanatic."[97] This last decision sharpened the differences between Aickman, who was to continue to exert significant power over the IWA, and others interested in canals. In his autobiography several years later, Aickman wrote that he found Hadfield's position of establishing priorities among the canals to be saved detrimental to the forming of a national organization:

> In his books, he not infrequently suggests that this or that canal should never have been built at all. It was not an attitude that was likely to appeal to me, or that seemed to me conducive to the advancement of the Association's extremely difficult struggle. . . . It would be difficult to run a national campaign while admitting to many enquirers that it was not proposed to do as much for their local waterway as for various others. Most enquirers are naturally interested in their local waterways first. Any attempt at discrimination would have resulted in the weakening of the national campaign (anyway at that early stage) by the atomisation of the single body into numerous merely local societies, all more or less powerless; much to the gratification of opposing authority.[98]

As the BWB initiated a major survey of its holdings in 1963, many of the last vestiges of the cultural world Rolt had done so much to publicize with *Narrow Boat* were vanishing. Published accounts of efforts to travel the canals as he had, and even to revive commercial use, became central to the movement to save the waterways, even though they relied on an element of nostalgia. In 1958, Hugh Malet, who had purchased a small wooden boat (a "dory") for £35, spent several weeks making his way from London to Liverpool. Writing in a published account that canals offered an immediate escape route for the "wage-slaves of our expanding industrial economy," he ruminated about why so much effort was spent on saving the less important of Britain's country houses while "several of the loveliest waterways are allowed to turn into costly and evil-smelling ditches."[99] He recorded his impressions of his journey through Birmingham, in which the contemporary city, imprisoned in the noise and dirt of industry, had left the canals to decay:

> There was an extraordinary contrast between the streets of Birmingham, with their belching diesel lorries loaded down with merchandise, and the hurrying cars and people, and the deserted canals beneath them. Parts of the centre of the city had a strange resemblance to the less fashionable backwaters of Venice, and on the canal it seemed almost as though an atom bomb had robbed all but a few of the citizens of their past lives.[100]

For some, the experience of travel on the canals continued to offer an appeal not unlike that which had emerged at the close of the nineteenth century. In the early 1950s, one of the last investments in narrowboat carrying came with the founding of the Willow Wren Company by Captain Vivian Bulkeley-Johnson, later a member of the Advisory Committee of the IWA.[101] In the early 1960s, David Blagrove, who had been inspired by the IWA when first being introduced to the organization in the mid-1950s, took on a series of contracts with the Willow Wren Company in his narrowboat "Enterprise," traveling from London to Birmingham and back primarily by way of the Grand Union and Oxford canals, as well as the Birmingham Canal Navigations. Operations at Willow Wren were headed by Leslie Morton, once a manager with the Grand Union Canal Carrying Company. Narrowboat trading could be profitable, Morton argued, if licensing rather than tolls were introduced for trading boats, as had been recommended in the Bowes report.[102] The Willow Wren operation turned the remaining canal boat workers into independent agents, offering the boats for hire to them as they pursued their remaining contracts on the waterways. Blagrove's effort was one carried out in the twilight of the industrial era, but it was one in which he was able to capture the voices of the last of the canal boatmen he met and was able to reflect upon what seemed an inevitable demise of industrial traffic, particularly in the process of loading and unloading of craft, whose inefficiency put the canals at a great disadvantage.[103] Recounting the experience from the perspective of an account written in the early 1980s, Blagrove was able to remark that the world of the pleasure boater had not arrived in the early 1960s and that the canals still seemed a convenient place for trash. "To officialdom it was cheaper to pour the sewage into the canal than reclaim it. The environmental bandwagon had not then started to roll."[104]

Blagrove noted an appreciation both of changes in the landscape canals had brought and of their old industrial routine. He even managed to catch a glimpse of Sister Mary Ward, who had worked the canals as a nurse to boating families.[105] He lauded Bulkeley-Johnson, whom he believed had been motivated in his business efforts by the decision in the 1950s to standardize the livery of BTC boats, writing, "This final straw made him resolve to put his own fleet on the canals; a fleet painted in traditional style and worked by skilled boaters whose culture and way of life would be thus retained."[106] Nonetheless, in recounting a conversation with Morton, he noted that the latter only saw narrowboats surviving if a very favorable set of policies were introduced, including licensing rather than the paying of tolls.[107] In his travels in the early 1960s, Blagrove noted faint signs of revival on the canals in some boatyards, which were now beginning to cater to emergent pleasure boating, but found himself most drawn to the world of the industrial narrowboat, which he believed he was making a desperate effort to save. Describing the care given to the maintenance of cabins by remaining boatmen, he wrote,

> They irresistibly reminded me of the colors of a doomed regiment. . . . The heartbreak of the dying canal system was manfully concealed by such symbols and we determined to have them, not merely to attract custom but to identify ourselves with our new-found friends.[108]

Similar to other writers on canals by the mid to late twentieth century, Blagrove remarked upon recorded canal journeys of the nineteenth century, drawing comparisons and contrasts in the experience of the countryside of both eras. John Hollingshead's journey in 1858 had seen Leamington Spa poised on the edge of the industrial regions of the Midlands, but still attractive. For his part, Blagrove noted a new impact of civilization that undermined his vision of the rural ideal:

> Opposite Leamington Station we entered a cutting with a housing estate at the top. The canal and the cutting sides were liberally sprinkled with old bicycles, prams, filthy mattresses and other debris. The contrast between the urban squalor of Leamington and the pleasant rolling country of the Fosse, a mere three or so miles back, emphasised forcibly the foulness of industrial urbanisation. The humbler inhabitants of the rural communities are doubtless just as poor, deprived and disadvantaged as their urban cousins[;] many years of rural life and work have left me with few illusions about the countryside. But in the poorest and meanest quarters of rural habitation I have never seen the filth and general untidiness of some of our big, planned estates.[109]

Much of Blagrove's recounting, which was published two decades later in the 1980s, was infused with his take on the values of an early industrial England he believed to be passing away amidst the emergence of the welfare state. It may also have been influenced by the social and economic tensions of the Thatcher era. He remarked upon his brief visit with Joe and Rose Skinner (who were to be highlighted by many canal supporters as among the last true boat people) at a canal-side pub known as the "Greyhound" located near the intersection of the Grand Union and Coventry canals at Hawkesbury Junction.[110] Blagrove linked them with an "ethos" of a different era.

> They were born and grew up into a harder social environment than Welfare State Britain and had survived wars and hardships bruised but not broken. Their nineteenth century world still admired and respected individualism both in initiative and responsibility and, for all their lack of wealth and position, they were people of spirit and character.[111]

Yet, the ability of the Skinners, Blagrove, or any others to survive on the narrowboat trade was drawing swiftly to a close. In the Fall of 1962, Blagrove recounted the reinforcing decisions of Bulkeley-Johnson to pull his investments from the canals and the decision of the BWB to halt its own commercial fleet. And, immediately following, came the plummeting temperatures of the winter of 1962–1963.[112] As the "Big Freeze" engulfed much of Britain and Europe, canal traffic came to a near standstill. Blagrove left his canal work in the spring of 1963, noting the ineffective bureaucracy which he believed had undermined narrowboat trading, and the steady decline of the boats controlled by Willow Wren. Between 1963 and 1964, Willow Wren did take over the remaining Northwestern Fleet of the BWB (about thirty boats), offering itself as one of the last major carriers connected to narrowboat transportation.[113] Morton later indicated that in the first full year of working the former BWB boats, he had been able to erase what had been a deficit under the BWB.[114] He remained adamant regarding the commercial narrowboats, arguing that one of his most significant difficulties was the inability to efficiently transfer loads between narrowboats and roads.[115] By the close of 1968, however, after his death, Willow Wren went into liquidation.[116]

While Blagrove was among the last to seek out and participate in a life on the canals that retained echoes of the nineteenth century, the tradition of contrasting canals with the understood ills of modern Britain continued to evolve. In July 1965, the young journalist Nicholas Tomalin offered an embrace of the canals as he recounted a five-day trip on canals in and around London, finding within them "a most magical nostalgia cure."[117] He asked, "Why thousands and millions of people . . . don't spend their free time compulsively sailing the canals is to me inexplicable. And why the British Waterways Board, cherishing a wasting industrial asset, doesn't spend more time and money wooing flotillas, squadrons of paying pleasure customers seems to me a scandalous waste of undoubtedly the most beautiful mysterious design in Britain."[118] Citing what seemed to be prime areas for re-development, including Paddington Basin and City Basin, he highlighted an apparent BWB reluctance to support projects. Yet, at the same time, he appeared to have discovered in the world of London canals a netherworld that invited him to explore. The Docklands, already in the process of being abandoned, had become a mysterious landscape of de-industrialization through which canals offered a fascinating path. He wrote, "The Regent's Canal from Camden Town down to the London Docks at Limehouse is utterly unknown, unexploited, and sensational. You need a liking for Dark Satanic Mills to appreciate its dank, foul-smelling charm, an appreciation of Pranesi engravings . . . a cross-section of industrial revolution London, unchanged by contemporary development."[119] It was the ability to see the back of an older industrial London unadorned by contemporary fixtures that seemed key, "Not a solitary lemon yellow door, neon sign, steak-house, coffee bar, or picture window disturbs the tranquil period charm of the back of the ABC Bread Bakery, the Egyptian extravagance of some 1910 furniture factory, the mock-Gothic backs of Finsbury houses, the half-mile long Islington tunnel, scarcely wider than a tube train, sulphurous, black, more dramatic than the blue grotto in Capri, the timber wharves, the mysterious fires which burn on the towpath all day."[120] There was a "mysterious magic," he said, as one chugged on the Regent's Canal, "infinitely slowly across a crumbly, dripping aqueduct right over the high-speed dual-carriageway vulgarity of the North Circular Road." He said canal builders had built "beautifully. All canal shapes are simple, solid, rounded and elegant."[121] Canals were "immune from the nastiness of fashion," he said, adding in the vein of Rolt, "They're neither oppressively natural, nor artificial, but a product of the struggle of man's ingenuity with nature."[122] Finally, he concluded with a slight against contemporary leisure:

> Of those who are in love with water, let the young, with-it obsessed and over-sexed go surfing, or skin-diving or dinghy sailing. Let the sea-sickness worshippers, the compulsively competitive, go ocean-racing. And let the flabby tycoons, with blondes on their foredecks overcrowd the Thames or the Broads.
>
> For the mature, the content, the nostalgic, and the well-adjusted, the canals are indubitably the best kind of messing about with boats.[123]

In the waning days of Conservative rule in early 1964, the chairman of the BWB, Sir John Hawton, issued the Board's interim report entitled *The Future of the Waterways* and offered significant support for a new "multi-use" philosophy of the system. The report outlined what were to be the critical challenges of the 1960s, which were focused on melding an emerging understanding of amenity with the diminishing chances for industrial transportation and attempting to provide some insight into the relative costs officials could expect in the future. Among the different policy tracks that could be followed, the Board gave its support to "a positive policy of cohesion and development of the waterways as an organised entity, in light of all current social and economic conditions."[124] The attitude we see emerging, therefore, is one of a general policy of "multi-user." . . . We think it can best be achieved by treating the waterways as a single undertaking capable of fulfilling a variety of combined national purposes, ranging from strictly commercial transport to more general social amenity.[125] The authors went so far as to map out what they believed might function as a "basic pleasure cruising network," including a series of "touring rings" which might allow pleasure boaters to depart from one location and return to the same location again without back-tracking.[126] Subsequent to their discussion of pleasure boating, they highlighted existing "bankside facilities" already under the control of the

Board which they believed might support the future of pleasure boating and the roles that fishing and general amenity appreciation of canals might play in the future. Although a significant engineering review of the canals had not yet been undertaken, the authors included maps that they believed spelled out the potential future for recreation and amenity on the canals. The IWA *Bulletin*, reflecting what was an emerging moderation in its pronouncements, indicated it felt there were substantive signs of progress in the BWB publication, noting the significance of the Board's recognition that it was much more costly to close canals down than to maintain them.[127] Aickman and Munk, the latter now chair of the IWA, met with members of the BWB in May 1964, and afterward, the BWB implemented a trial of licensing for two industrial carrying firms (of which one was Willow Wren). In his annual report for 1963–1964, Munk wrote that Hawton "had a refreshingly new approach" to the problems of the waterways.[128]

Yet, where new approaches might mean centralized planning, the IWA remained skeptical. In an *Architectural Review* article entitled "Do-it-Yourself" reprinted in the *Bulletin*, Nairn applauded the commitment of volunteers who had offered their time, labor, and money to the ongoing restoration of the Stratford-upon-Avon Canal.[129] Public subscription to the costs and contributions of labor by volunteers, including military service groups, meant that headway was being made on projects that might otherwise bog down under a "Whitehall mentality."[130] He wrote that such a "do-it-yourself" mentality was characteristic of the English and one that had been rejected by both political parties, which seemed more and more intent on "do-it-for-you."[131] The "do-it-yourself" mentality, he said, could take over large sections of the economy, perhaps those that had become economically unhealthy. Local planning could perhaps be left in the hands of local people themselves. "Footpaths, walls, hedges, signs and planting could all be decided by local option, . . . It would not be anarchy, but the principle of freedom of parts within a guiding pattern."[132] He wrote,

> Give it some freedom, freedom really to concern itself with the environment, and it will not be abused, because in this respect the English are deeply civilised, though every year bureaucratic obstruction whittles down the inherited store of good nature and fair play.[133]

An emphasis on action from below rather than reliance upon administrative action from above had been embedded in the workings of the IWA over the past 18 years. In July 1964, the restored southern portion of the Stratford-upon-Avon Canal—so much the product of volunteer labor—received a grand new opening presided over by Queen Elizabeth, the Queen Mother. In October of that year, issues of both canals and wider opportunities for recreation and access to the countryside would see important changes as Labour stepped into power.

In January of 1965, Aickman, who had been founder and vice-president of the IWA for 18 years, stepped down from both a leading role in the organization and the editorship of the *Bulletin*. In announcing his resignation, he said that work and responsibility on behalf of the IWA had come to "overwhelm" his life. Yet, the door may have been held open for him as he indicated that many people in the organization now believed a policy of cooperation with the BWB was advisable. Hugh McKnight was appointed editor of the *Bulletin*, taking over much of the work from Aickman. For his part, Munk, also chairman of the Kennet and Avon Canal Trust, chairman of the Association of Pleasure Craft Operators, director of the Thames Hire Cruiser Association, and operator of Maid Line, a hire cruiser company operating on the waterways, was heavily invested in the future of the canals. He had also been a conservator of the River Thames since October of 1963.[134] Aickman would continue to serve on the Council of the IWA. Though Munk believed the BWB had now expressed its goodwill and that the IWA's attitude must be "completely flexible," Aickman cast a long shadow.[135] In April of 1965, Parliamentary Secretary Stephen Swingler chaired a meeting that included Munk, other members of the IWA, and representatives of the Ministry of Land and Natural Resources, the Department of Education and Science, and the Ministry of Transport.[136] A brief prepared for Swingler in advance of the meeting signaled enduring distrust regarding the IWA. It argued,

> Its propaganda has often been extreme and sometimes unscrupulous, and it is probably fair to say that some of its activities have done more harm than good to its own cause. It has been a bitter critic of the Ministry and British Waterways and has not refrained from personal abuse.[137]

The brief highlighted the "autocratic" hold which it said Aickman had had on the IWA, particularly in formulating the association's more "extreme" policies.[138] Placing Aickman's "retirement" in quotation marks, it highlighted his continuing position on the council of the IWA and influence over the organization.[139]

In July 1965, the Parliamentary Inland Waterways Group (a political grouping offering support to the canals in Westminster), the IWA, and the National Trust submitted a pamphlet entitled *Proposals for a Waterways Trust* to the Minister of Transport, again reviving the idea of a national organization that would take control of non-commercial waterways. Reflecting an ongoing IWA position, the three groups argued that the BWB's estimates for the elimination of canals were low and that a total of at least £22 million would be needed.[140] Reflecting the philosophy of multi-faceted use of the canals, the three groups indicated that a non-taxable Trust should be established by Act of Parliament, to be guided by a council that might take into account the interests of groups thus far associated with canals, ranging from the Angling Association to the National Parks Commission.[141] The proposal for the trust was given greater impetus as later that year, the BWB issued a pamphlet entitled *Facts About the Waterways,* which expounded upon the pressures facing the canals—noting debts, interest, maintenance concerns, and declining tolls—while also identifying nascent efforts at redevelopment for leisure and speculating on costs to close or maintain canals for various purposes. The BWB's deficit from 1954 to the end of 1962 grew more than seven times, comprising 40 percent of gross revenue in 1962.[142] The report's authors documented the anticipated costs of eliminating canals (or maintaining them with certain specifications) and predicted that a minimum total of £600,000 per year would be needed to pursue the most economical option for the entire network—whether it be engaging in minimum maintenance or moving toward complete elimination.[143]

In the course of the next four years, however, conditions changed significantly as postwar interest in leisure and access to the countryside expanded. Initially, those grappling with large numbers of holiday travelers paid canals little heed. In early 1965, however, in a report later entitled *The Challenge of Leisure*, Michael Dower published the results of a survey for the Civic Trust on the problems of congestion, inadequate urban parks, overburdened sites of natural beauty and public recreation, and the infrastructure changes needed to accommodate a public with more money and leisure time. The impact of leisure on the countryside, he remarked, amounted to a "fourth wave" that followed upon the dramatic changes that had come with the advent of industry, the expansion of the railways, and the expansion of new automobile-centered suburbs. Leisure brought with it new demands ranging from homes that accommodated recreational interests to sporting facilities and parks suitable for multiple uses. While cities had to grapple with the need to reconsider space within their own borders, regional and national planning were necessary to ensure that the nation could handle the multitudes who now flooded out of cities on weekends. Interest in transport other than in automobiles was significant, as Dower remarked upon the need to provide for "the growing desire for movement by foot, cycle, horse and even boat between various open spaces without entanglement with motor cars."[144] Yet, although Dower cited the significance of coastal areas and some inland bodies of water such as reservoirs, canals received very little attention.

Nonetheless, hundreds of miles of canal crisscrossed rural areas of Britain, offering an alternative to the building of new infrastructure, most specifically that related to the automobile. In the Lake District and Pembrokeshire Coast national parks, authorities indicated, "the weight of numbers of people and cars crowding into those areas threatens to kill what they come to enjoy."[145] The Central Statistical Office indicated there would be a one-third increase in the population by 2000 and that the number of cars would rise from 6.5 million to 18 million by 1980.[146] There was a growing consensus among politicians and policymakers that the Countryside Act of 1949 was no longer adequate to manage recent increases in recreation.[147] Such a conclusion had already been highlighted in a series of meetings that since 1963 had brought various groups interested in the development and/or conservation of rural resources together under the umbrella of the "Countryside in 1970."[148] In a second meeting of the "Countryside in 1970" conference held in November 1965, Frederick T. Willey, head of the Ministry of Land and Natural Resources (MLNR), called for a new Countryside Commission that would take over from the National Parks Commission and have "a new function of encouraging the provision of opportunities for the enjoyment of the countryside generally."[149] In early 1966, the MLNR offered an update of the Countryside Planning Act of 1949 in its white paper *Leisure in the Countryside*. It called for new initiatives to provide wider recreational amenities, particularly Country Parks, to buffer the demands placed upon National Parks.[150]

Yet, if the potential for amenity waterways seemed to reach new heights within the BWB and organizations actively promoting their future, it had yet to be entrenched at a ministerial level. In a brief for Minister of Transport Barbara Castle, written in preparation for a meeting with waterways representatives, R. D. Poland, Undersecretary at the Ministry of Transport, noted that given the costs associated with keeping some waterways open to transit by powered craft, the minister would "not want to sound too enthusiastic about the future role of the waterways for pleasure cruising."[151] In the meeting, which involved members of the Parliamentary Inland Waterways Group, as well as the IWA and National Trust, Castle said little regarding the idea of a trust for the waterways.[152] However, in July 1966, Castle presented the white paper entitled *Transport Policy* to Parliament. The report reinforced a sense of impending crisis regarding the effects of increasing road traffic—though the commitment of the government to roads seemed not in doubt. In contrast, waterways were considered a nearly defunct component of commercial infrastructure and given very little attention. Their use for activities such as canoeing and fishing was estimated to mean little to no further expenditure. Pleasure boating, however, meant further necessary investment but uncertain returns, thus posing a problem for the report's authors.[153] The white paper indicated that the future maintenance of waterways for pleasure boating was expected to be decided through discussion between the BWB, the Economic Planning Councils, and other groups, including those representing users.[154] Moreover:

> On the basis of these discussions new legislation will be enacted to take the place of the present statutory provisions which expire at the end of 1968. The

> network which the Government decides should be maintained for use by powered pleasure boats following these discussions will again be reviewed after five years, and thereafter at regular intervals, in the light of the use made of it, of the success achieved in attracting more custom and in exploring the possibility of increasing charges and other sources of revenue, and of prospects for the future. The Government must, however, reserve the right during the period between these reviews, exceptionally, to close a waterway to use by powered pleasure boats by statutory order if the cost of keeping it open proves to be out of proportion to the benefit from it.[155]

Although the white paper gave support to the idea of cooperation between the BWB and voluntary societies for work on canals, it eschewed the idea of a trust. Poland had noted earlier in the month,

> There is little doubt that when the White Paper is published the waterways section will be looked upon by most of the enthusiasts as a betrayal. . . . Everything that follows (including the passage of legislation through Parliament) will be fought by a most vigorous, vociferous and effective lobby.[156]

Indeed, he was right. Munk was unnerved by the prospective decision to continue to review waterways for possible closure every five years, making it nearly impossible for those seeking to invest in transport to secure any financial backing. "Such wicked nonsense must effectively stifle the development of all transport and pleasure boat trade on the canals and even the purchase of boats by private individuals."[157] He found the rejection of a Waterways Trust particularly galling, noting that all the "advantages" of a trust, which could include voluntary efforts, memberships, covenants, and the use of prisoners for labor, were all lost. However, he cited local examples of canal revitalization schemes and concluded,

> Let this present Government know that now they really have a fight on their hands and we shall never give up that fight until we have finally achieved the aims of our Association—the guaranteed retention and development of the whole canal system.[158]

The IWA kicked off a campaign focused on a membership drive as well as prompting members to organize local protests and pressure MPs.[159] The scope of the IWA's efforts to widen the appeal and accessibility of canal journeys was evident in the 1966 publication, *The Canals Book, Navigator Guide Directory*. It depicted a boating culture on both canals and rivers that was vibrant and within reach of individuals and families who sought to experience history, architecture, and nature in both the countryside and cities. In his introduction to the book, McKnight encapsulated much of the IWA's program by asking,

> Do you like boats? Peace and quiet? Wonderful countryside with breath-taking views? Fine architecture? Independence and a mild challenge of adventure? Do you want to explore a way of life virtually unchanged since the eighteenth century? You'll find all that and more on the canals and rivers of England.[160]

The opportunity to escape the rhythms of twentieth-century life, if not outright reject them, was evident in the book's derision of what automobile and road travel had brought. The ideas outlined by McKnight returned to the ideas at the heart of the postwar movement as founded by Aickman and Rolt:

> Today, who would choose to tour England by road, for pleasure? The speed of travel and the boring similarity of the landscape as seen from a motor car make for little enjoyment. The roads lead you from one tasteless petrol station to the next: from one drab suburb to another. In contrast, canals seem deliberately to avoid towns, ribbon development and the sadness that the twentieth century has brought to Britain. Rather more than 90% of the inland waterways wind through remote countryside: the exact opposite of their generally accepted image. Let it be stated once and for all that the dead-straight canal, connecting gas works and glue factory, is a rare exception.[161]

McKnight did offer a nod to Birmingham, where he noted that one was able to gain insight into industrial England of a century ago and where the mileage of canals was greater than Venice. Though there was a "less uplifting panorama of architecture," he noted that one could find overnight mooring at Gas Street Basin, where there was "something of the decayed charm of Bruges" and access to shops, the theater, and art galleries.[162] Yet, most waterways "avoid" towns and were unaffected by the modern world and its advancements, he said.[163]

In early January, 1967, the IWA prepared a memorandum entitled *The Way Ahead for the Amenity Waterways* for the Ministry of Transport in an attempt to outline what it believed would be a responsible approach to the non-commercial waterways. Citing the degree to which both the numbers of hired boat cruisers and holiday-makers on canals had increased since the end of World War II, *The Way Ahead* emphasized the new layer of experience of the countryside that now presented itself. Rather than simply artifacts of the industrial era, canals had now become green. *The Way Ahead* remarked, "The canals are in general unspoilt and undeveloped and convey the peaceful and graceful atmosphere of a hundred and fifty years ago. . . . Waterways stretch green fingers into the Black Country. It is indeed surprising how quickly one proceeds from the center of an industrial town into pleasant country on the waterside."[164] The idea of a linear park, according to the memorandum's authors, provided not only a holiday, but a markedly different way of seeing Britain:

> Our waterways form a wonderful national park of a particularly valuable sort in that it is linear and occupies little space. To create such a water paradise from scratch today would cost many millions. Inland waterway boating is cheap, classless and healthy. It offers mild exercise and it takes people off the roads. It takes the businessman away from his telephone and it provides relaxation and tranquility. The system enables one to cruise all over the country—from the Thames to the Mersey, from the Humber to the Severn, to the mysterious waterways of the Fens and Middle Level and into the Welsh mountains, and the exciting crossing of the Pennines and to the lovely Lancaster Canal extending up towards the Lake District. One cruises at such a speed as to enable one really to see the countryside, to see the waterside flowers and wild life. There is the interest of eighteenth century architecture, thrills of navigating through tunnels and across aqueducts and embankments.[165]

The water within the canals had now become a recreational commodity itself, the report stated, noting that in the "clear, clean water" of a maintained waterway lay opportunities for

> canoeing and rowing, for skating in the winter or walking and pony-trekking, for cruising, fishing and swimming, for bird-watching, and their reservoirs are available for sailing and water skiing. If we neglect or abandon our waterways, we shall do a tremendous dis-service to both young and old people of the future. We shall have cast away a wonderful heritage.[166]

The report indicated the standard working week would be gradually reduced in coming years, noting, "The Government will be faced with making the necessary provisions to occupy that recreational time. The amenity network of waterways is a ready-made solution and a therapy for the ills of the modern life."[167] For Britain, the "ills" of modern life might have been expanded to cover the balance of payments deficit which Britain still confronted in the mid-1960s. The IWA indicated a potential boom or bust for the nation, depending on whether a clear-cut choice was made to preserve or abandon the canals for holiday-makers who might otherwise go abroad. Moreover, there was now an opportunity to attract tourists from the United States, Australia, South Africa, and Germany, who could help make canals a "booming branch of the English tourist industry."[168] The IWA optimistically suggested that the annual increase in the number of boat licenses could range from between seven and 20 percent, offering a substantive way for the BWB to overcome its deficit, in addition to the fuel taxes that could be collected. The position of boats would be enhanced as the number of automobiles in society increased.

> As pressure on open spaces builds up and cars become too numerous to be "a way of getting away from it all" the public will realise even more that canals present a long continuous beauty spot accessible at many points while also giving everyone the reward of seclusion.[169]

The IWA offered several recommendations on fees linked with canals or their waters, perhaps seeking to demonstrate that if implemented as part of a trust, the charges could help to ensure a sound financial basis for management. It also offered suggestions on additional spending that might include improvements to bridges, the provision of water points and toilets, and encouragement for the development of snack bars, restaurants, and pubs on canal routes. It said reviews of canal viability should come after approximately 25 years.[170] Acknowledging that in the past enthusiasts were likely "over-optimistic as to the remaining economic potential of narrow boats," it contended that the BWB had been too hasty in dismissing them.[171] Some economic potential remained, they said, and the boats could both contribute license fees and help keep canal channels navigable. Moreover, their cultural value remained. "One of the attractions of the amenity waterways is their colour, their beauty and their atmosphere, and one of the chief contributors to these qualities is the gay traditional narrow boat."[172]

In January of 1967, McKnight also undertook a much more wide-ranging survey of recent efforts to integrate canals within urban design in a *Bulletin* article. In it, he built upon a case the IWA made before a "Canals Consultative Committee" organized by the Greater London Council (GLC) in which the group had pressed the case of a "multi-functional concept of canals."[173] One of the participants in London, the Regent's Canal Group, McKnight wrote, was composed of "professional people" drawn from the Paddington Waterways Society, the Islington Society, and the St. Pancras Civic Society, along with some IWA representatives.[174] The group, in a draft copy of a report yet to be released, outlined the major facets of a "linear park," which was to connect areas west of metropolitan London with the East End and the then intended Lee Valley Park. The group was considering some commercial use of the waterway in addition to increasing the services for pleasure boating and was placing the canal at the center of a transformed townscape that would include pubs, shops, and restaurants, as well as enlarged parks. Significant here was a plan for the re-working of an urban area that would contrast substantively with rural areas commonly associated with canals. McKnight wrote, "A town waterway can never be changed to look like a rural river, nor should the attempt be made."[175] Indeed, when published later in the year, the report of the Regent's Canal Group highlighted the unique impact the canal corridor could have on the dilapidated areas through which it traveled. The towpath, for example, could offer a "traffic-free walk through north London broken only at the two tunnels," (should appropriate provision be made for maintaining safety and preventing vandalism).[176] Moreover, the geography of the canal, with varying areas of openness and enclosure along its route, offered a "special character and attraction."[177]

The efforts in London, where beautification had already begun in Little Venice and the building of a waterside walk had started in the Borough of Westminster, were being mirrored in Birmingham. Here, reconstruction plans for the City Centre included housing, shops, schools, and light industry bordered on one side by the BCN and Gas Street Basin.[178] Even Banbury, having moved decisively away from the one-time idea of filling in the Oxford Canal basin, was now re-imagining its city center with a marina that would support the growing numbers of pleasure boaters.[179] As the IWA *Bulletin* would remark late in 1968 after a presentation by J. A. Hurrell, Architect/Planner for the Birmingham City Centre Re-development, who had just outlined plans for the BCN:

> It really is most heartening to find that at long last initiative for using canals in the urban landscape is coming from planners as opposed to boating or canal enthusiasts. The next five years should see the completion of many brave schemes in Birmingham and elsewhere, and these will do much to generate similar work in other parts. The concept of amenity value of inland waterways is gathering momentum.[180]

In remarking upon the twentieth anniversary of the IWA, McKnight wrote,

> It is said that it takes twenty years to change public opinion: where the canals are concerned, we managed this in less than fifteen. By the mid-Fifties, no impartial observer would fail to agree with the fundamental claims of the campaign. Now, in 1966, we appear to be on the very threshold of seeing all that we have fought for becoming a reality.[181]

In March 1967, the IWA hosted a conference in Oxford that included several speakers, among whom the most influential was Dower, also a member of the Civic Trust, the Sports Council, and director of the Dartington Amenity Research Trust (and much later, the director-general of the Countryside Commission). In his talk, he was able to combine his own observations with the contributions of others who had participated in the conference. He situated the movement for canals within a transforming Britain, where by 2000, the population was expected to be 70 million, the number of cars at 30 million, the number of people of retirement age at 12 million, and the industrial work week at thirty hours.[182] Along with these changes, transitions in leisure, marked by activities that were increasingly participatory, individualized, and focused on the outdoors.[183] Noting interest in industrial archaeology, he called for the preservation of old canal buildings, stating,

> We cannot condemn the British Waterways Board Staff to live in archaic conditions: but this is one of the few remaining working systems of industrial revolution machinery. Although this may seem a minority interest, it is a very powerful supporting argument, not only for maintaining waterways and their ancillary buildings, but maintaining them at work.[184]

His wrap-up of recommendations would likely have cheered Aickman as he offered his support for maintaining navigation over all 2000 miles of the canal network, integrating those miles with other recreational opportunities.

In September of 1967, as the IWA prepared for its "National Waterways Week" in October, the Ministry of Transport published a further white paper entitled *British Waterways: Recreation and Amenity*.[185] The Ministry of Transport stated that over the past twelve months it had conducted "extensive and exhaustive" consultations with any organization likely to be interested in the "recreational and amenity potential of the waterways."[186] Groups consulted included "regional Economic Planning Councils, Sports Councils, local authorities, trade associations and national societies."[187] Its three main conclusions after these conversations were:

> First, apart from some pruning of disused arms and lengths of canal here and there, all the waterways now open for powered pleasure cruising should remain open. Great stress [by those consulted] was laid on the need to retain waterways forming part of the through or ring routes and it has been urged that it would be wrong to close any navigable routes to pleasure craft at a time when increasing leisure and a rising standard of living are likely to put all our facilities for recreation at a premium. Second, the great majority of those consulted think that to review the network at the end of five years and at intervals thereafter [as proposed in *Transport Policy*] would discourage private commercial investment and development by making the future of individual waterways uncertain. Thirdly, it was strongly represented that some waterways not now navigable should be restored for pleasure cruising.[188]

The Ministry of Transport felt confident in stating that pleasure cruising on canals and rivers was now no longer a pastime for the wealthy and that in the near future more and more people would want to utilize the waterways for relaxation and holidays. The BWB was to assume an expressly "positive" duty.[189] It was, "thus the Government's intention that for the first time, this recreational purpose of the nationalised waterways should be recognised by public Act of Parliament."[190] Gone was the organization of canals into classes, with a third grouping destined for closure to navigation. Henceforth, select waterways were to be maintained for predominantly commercial activities in a first category while those referred to as "cruiseways" were to be in a second category. The Minister was to retain the power to initiate re-classifications of canals, but parliamentary intervention could prevent "sudden or arbitrary closures."[191] Central to the incorporation of a diversity of voices in decision-making regarding canals was an Inland Waterways Advisory Council, which was indicated as a standing committee to be established promptly, with statutory authority as a permanent body to be proposed in the next Transport Act. Chaired by a member of the BWB, it was to both make recommendations to the board, such as for the restoration of certain waterways, and to take up complaints.

To the chagrin of the IWA, when it was introduced, the Ministry's new transport bill revoked the requirement of the Regulation of Railways Act of 1873 to maintain public navigation on specific inland waterways.[192] In addition, waterways that were abandoned by the BWB as unworthy of either commercial or cruising maintenance could be turned over to local authorities who could then enforce components of public health and town planning statutes, a step which had the potential to demolish the canals in the name of removing public nuisances.[193] Munk wrote to Stephen Swingler, the parliamentary secretary to the Minister of Transport,

> We appreciate fully that the Minister and the British Waterways Board last year adopted quite a far-seeing outlook towards the future of our inland waterways but, as watch-dogs of the public interest, we cannot assume that future Ministers and their staff and future British Waterways Boards will take such a generally satisfactory line. The safeguard of the Courts is therefore urgently necessary.[194]

However, despite significant pressure levied by the IWA, particularly in the House of Lords, as passed on October 25, the Transport Act of 1968 did not return legal protection for navigation.[195]

Central to the provisions of the Transport Act, and reinforcing the idea of an integrated understanding of recreation in the countryside, was the creation of the Inland Waterways Amenity Advisory Council (IWAAC). The IWAAC was set up informally in May 1968 but was constituted more formally later that year, with member of the Greater London Council (GLC) and the first Chairman of the London Canals Consultative Committee, Illtyd Harrington as chair. Its task was defined generally as making recommendations to the Board or Minister of Transport regarding the use and development of "cruiseways."[196] The committee drew members from bodies representing rural preservation, planning, outdoor sports (including canoeing and fishing), waterway cruising firms, and local waterway clubs. Among those appointed to the committee with IWA connections were Hadfield and Munk, with

Aickman absent.[197] As it argued for the preservation of rights in navigation, the IWA operated in a context of substantive government action on the democratization of the countryside. As passed in July, the Countryside Act of 1968 sought to expand upon the National Parks Act of 1949 by sketching out responsibilities for a new Countryside Commission and expanding the power of local authorities to establish country parks. At a national level, the new responsibilities of the Countryside Commission included,

> (a) the provision and improvement of facilities for the enjoyment of the countryside, (b) the conservation and enhancement of the natural beauty and amenity of the countryside, and (c) the need to secure public access to the countryside for the purposes of open-air recreation.[198]

At a local level, the act empowered authorities to establish both the boundaries of parks and provide for recreational amenities, which might include sailing, boating, bathing [swimming], and fishing. Canals were included within the act's definition of "open country," and although provisions were made for canals and waterways managed by the BWB, the powers of the act allowed for local authorities to negotiate with other groups for access to them.[199]

Though government officials now clearly favored outdoor experiences of nature as a counterbalance to contemporary urban life, the widened environmentalism of future decades was only just emerging. Phil Macnaghten and John Urry have written that for most of the decade government officials had primarily used vocabularies related to rural preservation and amenity, and that there was a

> tacit but prevailing consensus that concerns over nature were largely irrelevant to public policy development aimed at further modernisation of such areas as agriculture, energy supply, transport infrastructure and nuclear power stations, over and above their impact on landscape and amenity.[200]

By the mid-1960s, however, things began to change. Early in the decade, Rachel Carson's *Silent Spring* had awakened concern regarding the use of pesticides, a theme that had once been related to Massingham's and the Soil Association's concerns regarding the introduction of industrialized farming. In 1966, Lady Eve Balfour, founder of the Soil Association, was instrumental in helping to found the Conservation Society which, while focused primarily on issues of over-population, brought together a coalition of both older conservatives and younger, more radical activists.[201] The impact of the Conservation Society was relatively brief, but it intersected with widened fears of environmental catastrophe. These fears were often argued to be systemic concerns transcending previously local concerns; as such, their advocates believed they needed to be addressed at a global level.[202] B. W. Clapp has written that the index to *The Times* contained no entries for "pollution" or "conservation" in the last quarter of 1946, 1956, or 1966. Only toward the end of 1969 did environmental questions make a substantive appearance, with coverage becoming more regular in 1970.[203]

In its coordination with local groups, the IWA had earlier emphasized the volunteer activism, often among the young, that would become characteristic of the new environmental movement by the late 1960s.[204] The momentum generated by 1969 was reflected in the IWA rally of boats designed to coincide with the bicentenary celebrations of the BCN in Birmingham in July of that year. The lead-up to the rally revealed some of the problems that still confronted the canals. A week before the beginning of the rally, canal supporters organized a clean-up of the Walsall Canal for a smaller rally planned to take place in advance of the major IWA gathering.[205] In the process of their clean-up along a 200-yard stretch of water, they removed "one car bonnet, two bedsteads, an enormous coil of wire, two mattresses, a number of lorry and car tyres, a children's tricycle, bricks and several tons of weed."[206] According to the *Walsall Observer*, the accumulation of trash was an ongoing problem, with a BWB spokesman indicating, "We can clean up the canal one week, and the next week it is just as bad again."[207] Despite fines, catching perpetrators was difficult as much of the dumping took place under cover of night. Yet the Birmingham rally, set to begin on July 25 and highlighted by a floating pub, and tug used to ferry people from one end of the two-mile stretch of moorings to the other, was expected to draw 450 boats and 100,000 spectators.[208] Among those who found himself in the center of Birmingham amidst converted narrowboats and canal enthusiasts was George Radford, one of the few remaining working boatmen who was waiting for a new engine for his boat.

> I was born on a narrowboat. I worked from the time I could walk. I went to school for just seven months. The canals are my way of life. . . . Some of these new canal people are all right. As far as the teenagers are concerned, its best just to ignore them. In the old days it was peaceful. The trouble is now everybody wants to overtake each other.[209]

And, indeed, in his own way, chairman of the BWB Frank Price expended significant effort to show that the BWB was leading the change process. In an address circulated to the press, entitled, "Ditching the industrial image of our canals" in the *Birmingham Post*, he wrote that it was "incredible" that the canals of the BCN should have for so long been neglected as to "now resemble in many parts nothing better than open sewers."[210] The situation was much different beyond Birmingham, and he indicated that he saw no reason why

Birmingham canals "should not be things of beauty as well as providing the citizens with a much needed channel for leisure."[211] Representatives of local and national government from all political parties were coming to understand that the role that waterways could play in the recreational field was "truly enormous."[212] He concluded, "The future looks bright, but it still needs a great deal of patience, a great deal of money and a great deal of very hard work."[213]

As the Birmingham rally of 1969 showed, life afloat changed dramatically in the 18 years between 1951 and 1969, and even more so in the 30 years since Rolt's pivotal voyage in 1939. In 1951, canals faced declining tonnage and a sense, even among those still working the system, that the end of traffic in many areas was near. By 1969, nearly all of the old canal boat workers were gone and their place was taken by pleasure-seekers who traveled in modern or updated craft participating in a new world of experiences. The understandings these advocates carried with them were an amalgam of evolving experiences over the last 200 years, though the years since 1951 had been pivotal. By 1968, the IWA as well as many other organizations with an environmentally linked mission were in the midst of massive expansion. Membership in the National Trust rose from about 200,000 in the 1960s to 1 million by the end of the 1970s. The Royal Society for the Protection of Birds had 25,000 members in 1964 and 300,000 in 1979; in 1969, it had control of 6,500 hectares of land and in 1979, 32,000.[214] For its part, IWA membership reached 17,728 in 1979.[215] In the late 1960s, the results of neglected maintenance were visible, and in some cases getting worse, but both governmental employees and volunteers were ready to participate in the reclamation of the waterways for recreation.

Indeed, canals were able to traverse significant divides in a modernizing Britain between 1951 and 1968. Initially championed by those who inherited and continued to enunciate the precepts of interwar ruralism, including the IWA, they were able to evolve to embrace modern ideas of recreation and leisure by the late 1960s. The first stage in this process had been an early acknowledgment and embrace, of holiday travel. Rolt had only cautiously accepted a tourist presence, and in the late 1940s many canals were seen by private (and soon, governmental) officials as assets sustained by those willing to pay tolls. However, as the decline of industrial traffic necessitated a new economic argument for their maintenance, holiday travel made sense. In the eyes of canal advocates, individuals touring their waters might find on their journeys peace and an appreciation of the world Rolt and others had witnessed. Channels might be kept free of weeds and pressure for government maintenance expanded. Holiday-making undermined Rolt's commitment to the active use of canals in their original form but spoke to the adaptability of the waterways to changes of interest and use, and positioned them for the new era of leisure in the late 1960s.

Initially, most government officials expected holiday travel to represent only a small part of overall revenue. Receipts had fallen on many canals, and use and maintenance had essentially ended on others. Questions regarding their future trended more and more toward how and when disused routes might be eliminated. In 1954, Sir Reginald Hill, Chairman of the Docks and Inland Waterways Executive, found the canal system lopsided as he focused upon its economic returns, a position that was reinforced through the conclusions of the Rusholme Board of Survey. The latter argued for a focus upon only 336 miles for the maintenance of navigation, out of over 2000 miles under the BTC's supervision, finding the IWA's commitment to saving the whole system "unrealistic and impracticable."[216] The proposal of a conservancy forwarded by the IWA and others suggested an importance for the waterways that transcended their industrial use and made them of value by virtue of both their links to the cultural world of the boatmen (then disappearing) as well as to the experience of nature. The more thorough Bowes Committee in the late 1950s reviewed the operating losses of the waterways, the results of the over £5 million that had been invested in them in the mid-1950s, the continuing decline of some infrastructure, and the dangers associated with some disused canals. Although greater interest in a diversity of uses for the canals was signaled with the creation of the IWRC, there was still no full government commitment to saving the waterways and little indication that pleasure cruising could one day financially support the system as a whole.

One of the most significant challenges the IWA had was re-shaping public attitudes toward canals, which tended to dismiss them simply as ramshackle leftovers of the age of industrialization and to focus, rather, on the dangers they held. A century after their eclipse by the railways and after several decades in which no overarching public or private plan for centralization and revivification had been put in place, an understanding of canals as sources of danger and neglect was embedded. However, a new movement—public interest in, and academic recognition of, industrial archaeology—was also on the horizon. Inspired in part by Rolt's new work on the lives and accomplishments of Victorian engineers, industrial archaeology suggested a value for the remnants of industrial life represented in factory buildings and houses, as well as the routes of the canals. This appreciation of industrial heritage amidst a post-industrial society helped to link canals with efforts at heritage preservation. Moreover, many waterways represented earlier stages of industrial transformation in which the pioneering works of industrial engineers had appeared most significant. For Rolt and others, this was the age most in need of preservation. Undertakings such as the "Industrial Monuments Survey" helped to crystallize the importance of preserving industrial history in the minds of the British public.

As the "Industrial Monuments Survey" had suggested, engagement with the history of industrialization could be made democratic. The purchase and restoration of a narrowboat might take significant funds, but canoeing, angling, and walking in the spaces along canal banks did not. Canals were not country houses, the preservation of which appeared in the eyes of some to be an implicit nod to the influence wielded by the upper classes (despite the homes' popularity for tourist visits).[217] In practice, canals had created transit routes that might cut through aristocratic estates, bringing with them evidence of a modern commercial era both on their waters and along their banks. These intrusions, particularly once traffic had begun to decline, could be easily integrated into the natural world, thus leaving intact many of the landscape features and settings associated with pre-industrial Britain. The legacy of canal boat workers also suggested canals as sites of the commemoration of labor. Though some accounts of this labor might bias the role of the independent "Number Ones" as boat captains, they spoke to a population that had once been maligned in the early nineteenth century but was now integrated into the fabric of nostalgic ruralism.

Travel on the canals meant recognizing industrial labor not simply through passive acknowledgment but through active labor. Participating in the labor of re-building the waterways meant replicating the labor of their construction. Significantly, for many canal advocates, such efforts represented an antithesis to modern travel through the countryside. Access by auto had become an important source of concern in the interwar period as the new freedom it offered was countered by noise and air pollution. By the 1950s, the advance of high-traffic auto routes helped to define new treatments of canals. For speed, canals substituted slowness; for bypassing the spaces in between points of interest, canals substituted an appreciation of the spaces in between; for travel upon asphalt and concrete, canals substituted paths upon water through a largely rural environment; for visions of the future, canals substituted experiences of the past. The lives lost on the motorways made their own case, in the eyes of the IWA.

Ultimately, the future of canals would be secured amidst a turn to multi-use. The context was set by two significant steps at the end of the 1950s. The first was the publishing of the Labour Party's *Leisure for Living*, which suggested new possibilities for canals' extended lengths in providing opportunities for a nation with more time on its hands. The second was the National Trust commitment to the Stratford-upon-Avon Canal. In the latter, the National Trust signaled a new significance for canals, as Stratford-upon-Avon was one of England's most significant tourist spaces. The canal's restoration offered a model of multi-party participation in a hybrid project reflecting heritage and recreation. Both actions helped to secure a place for canals in the coming world of leisure. This new world, however, was still linked with the past, as evidenced by the recorded journeys of those who still sought to engage the working life of the waterways. Both Malet and Blagrove were among those who offered a sense of the waterways as a means of escape from modern life and saw within their journeys a means of commenting upon modern life. While Malet and Blagrove were able to witness the last vestiges of the disappearing world of the boatmen, Tomalin's voyage just a few years later suggested that that old world had now ended. However, rather than being at an end itself, the world of the canals had opened a new opportunity to explore an antithesis to what some saw as the superficiality of contemporary life. Canals were "immune from the nastiness of fashion," Tomalin wrote.[218]

The Transport Act of 1962 brought an administrative shift as the waterways came under a separate and dedicated administration in the BWB. Although at least one portion of the puzzle of waterways administration had now been addressed—that of a dedicated and centralized authority—the potential loss of routes via administrative and parliamentary action continued to worry many, particularly the uncompromising IWA leader Aickman. However, having become embedded in the landscape of Britain, canals were now not easily removed. It was with Dower's "fourth wave" that the move toward leisure and recreation appeared to have made great strides. Though Dower's work was at least in part a warning, it cleared a path for the integration of canals into new visions of the countryside. The Ministry of Transport began to act more favorably toward the prospect of canals, though it stopped short of guaranteeing a future for navigation. In its response to the Ministry of Transport's initial white paper of 1967, the IWA embraced the idea of the waterways as green space linking rural with urban space, taking up the idea of linear parks which were now argued by planners as an effective way to provide for a multi-use future, as well as open and democratic access for a wide variety of participants. The practical aspects of linear parks were made evident in plans for canals that emerged in both London and Birmingham. Here, the adaptability of canals to a contemporary world was evident, as canal banks became spaces for shops and restaurants. Though the water channel itself remained separate, it was still central to the new identity being established along canal banks. The Transport Act of 1968 marked a new commitment to a future for the canals based on leisure, and the creation of the IWAAC signaled, in both its title and prescribed responsibilities, a new identity for canals.

By the late 1960s, canals were poised to play a broad new role in urban and rural spaces. Offering green corridors of transit, they often blended into the landscape and mimicked natural flows of water. Continuing to make use of early engineering solutions to water management, they offered engagement with an industrial past that eschewed the heavy pollution of the carbon age. Making use of unique boat types in their restricted spaces, they offered a chance to acknowledge and actively re-create the rhythms of work of the boatmen, whose presence had diminished amidst the changing mechanisms and pace of modern industry. As the twentieth century closed, these qualities—and many others —would continue to drive canals' evolution as essential components of the British landscape.

NOTES

1. David Bolton, *Race Against Time* (London: Methuen, 1990), 106.
2. Reginald Hill, "The Future of Our Inland Waterways," *Journal of the Institute of Transport*, 25, no. 9 (March 1954), 344.
3. Hill, "The Future of Our Inland Waterways," 344.
4. Hill, "The Future of Our Inland Waterways," 344.
5. British Transport Commission (BTC), *Canals and Inland Waterways: Report of the Board of Survey* (London: British Transport Commission, 1955), 8 (Preface).
6. BTC, *Report of the Board of Survey*, 37–42.
7. BTC, *Report of the Board of Survey*, 48.
8. BTC, *Report of the Board of Survey*, 70.
9. BTC, *Report of the Board of Survey*, 65.
10. W. L. Ives, *The Problem of our Inland Waterways* (London: British Transport Commission, 1957), 11.
11. Ives, *The Problem of our Inland Waterways*, 12.
12. "Do Away with These Deadly Waters," *Lancashire Evening Post*, February 3, 1954, 4. British Newspaper Archive (BNA).
13. "Do Away with These Deadly Waters," 4.
14. "Death Traps," *Sunday Mirror*, April 4, 1954, 12. BNA.
15. "Vain Search in Canal," *Nottingham Evening Post*, July 12, 1954, 5. BNA.
16. "Whose Pigeon?," *Nottingham Evening Post*, July 12, 1954, 4. BNA.
17. "The Town that Voted the Earth was Flat?," *Bulletin* 52 (October 1956): 22–23.
18. "The Town that Voted the Earth was Flat?," 22–23.
19. "The Town that Voted the Earth was Flat?," 22–23.
20. "Petition on 'Death Trap' Canal," *The Birmingham Post*, August 18, 1966, 5. BNA.
21. IWA, *Bulletin* 43 (July 1954), 15–16. See also, *Bulletin* 41 (February 1954), 10–11. In the course of 1957, the *Bulletin* contended 5,550 people had been killed by automobiles on British roads, 63,706 sustained serious injury, and 204,602 found themselves slightly injured.
22. C. V. Hancock, "Rod and River: Towpath and Warpath," *The Birmingham Post*, April 22, 1955, 4. BNA.
23. Ian Nairn, *Outrage* (London: The Architectural Press, 1955), 366.
24. Nairn, *Outrage*, 452. While Nairn emphasized the preservation and enhancement of visual design in planning and spoke directly to those charged with the design of urban landscapes, the poet and advocate of Victorian preservation John Betjeman likely spoke to a wider swath of the English population. He was one of the most

well known among a variety of voices who argued postwar growth was destroying an essential English identity, particularly in the countryside. His membership in the IWA and interest in the restoration of canals intersected with his interest in landscape design and Victorian engineering. In 1955, he contributed to a BBC "Panorama" television broadcast on inland waterways and on July 21 wrote an article "Unfair to the Canals" for the *Daily Telegraph*. (See, IWA, *Bulletin*, 48 (Autumn 1955), 43.) In the column "City and Suburban" in the *Spectator*, he criticized the Rusholme Committee, suggesting that it simply represented the views of the BTC itself and had arrived at its conclusions far too quickly. (See "City and Suburban: Ditching the Canals," *The Spectator*, July 22, 1955, 11.)

25. Robert Aickman, "The Case for a National Waterways Conservancy," *Journal of the Royal Society of St. George* (Summer 1956): 9–12.

26. Aickman, "The Case for a National Waterways Conservancy," 11.

27. Aickman, "The Case for a National Waterways Conservancy," 12.

28. IWA, *A National Waterways Conservancy: Proposals by the Inland Waterways Association Limited* (London: W.S. Cowell, 1958), 7.

29. IWA, *A National Waterways Conservancy*, 24.

30. Regulation of Railways Act 1873, *c*. 48. See also, "A Barrister," "The Removal of Obstructions to the Right of Navigation," IWA, *Bulletin* 56 (July 1958), 3.

31. See "A Barrister," "The Removal of Obstructions to the Right of Navigation," 4.

32. Railway and Canal Traffic Act 1888, *c*. 25.

33. "Creeping Restrictions," IWA, *Bulletin* 58 (April 1959), 14.

34. UK Parliament, *Report of the Committee of Inquiry into Inland Waterways* (London: HMSO, 1958), 9–10.

35. UK Parliament, *Report of the Committee of Inquiry into Inland Waterways*, 25.

36. UK Parliament, *Report of the Committee of Inquiry into Inland Waterways*, 24.

37. UK Parliament, *Report of the Committee of Inquiry into Inland Waterways*, 24.

38. Joseph Boughey, *Charles Hadfield: Canal Man and More* (Stroud: Sutton Publishing, 1998), 58.

39. Bolton, *Race Against Time*, 170.

40. Boughey, *Charles Hadfield: Canal Man and More*, 58.

41. Philip S. Bagwell, *The Transport Revolution* (London: Routledge, 1988), 325.

42. For a chronicling of these efforts, see Roger Squires, *Canals Revived: The Story of the Waterway Restoration Movement* (Bradford-on-Avon: Moonraker Press, 1979). Squires argues that local chapters affiliated with the IWA were instrumental in the success of the national organization.

43. "Operation Flyboat," IWA, *Bulletin* 55 (January 1958), 11.

44. "Voyage Through the Sixteen," IWA, *Bulletin* 61 (April 1960): 14–17.

45. "Voyage Through the Sixteen," 14–17.

46. "Closure in the Midlands," IWA, *Bulletin* 59 (July 1959), 49. The preservation and potential restoration of disused lock cottages in which canal employees had

once lived became a prime point of contention between the IWA and other waterways supporters and government officials. In the canal journey he undertook in the early 1960s John Seymour, a leader in the self-sufficiency movement, wrote,

> Yet, these houses and cottages were well-built, in an age of good taste. We, of this age of plastic shabbiness and reach-me-down, have no right to pull them down, or allow them to be destroyed by hooligans. They belong to *us*—the people of England, as the towpath belongs to *us*. (John Seymour, Voyage Into England [New York: August M. Kelley, 1968], 29)

47. A 1967 exhibition on the Black Country held in Dudley emphasized threats to artifacts of the region's heritage. The Black Country Living Museum opened in 1978. See Black Country Living Museum website accessed November 17, 2023, https://bclm.com/about/the-museums-story/. For a further discussion of the history and monuments of the industrial era, see Keith Falconer, "The Industrial Heritage in Britain—the First Fifty Years," *La revue pour l'histoire du CNRS* 14 (2006), accessed May 28, 2024, https://journals.openedition.org/histoire-cnrs/1778.

48. L.T.C. Rolt, *Isambard Kingdom Brunel: A Biography* (London: Longman, Green & Co., 1957), *Thomas Telford* (London: Longmans, Green, 1958), *George and Robert Stephenson: The Railway Revolution* (London: Longmans, 1960), *Thomas Newcomen: The Prehistory of the Steam Engine* (London: MacDonald, 1963).

49. L.T.C. Rolt, *Landscape with Figures: The Final Part of his Autobiography* (Stroud: Alan Sutton, 1992), 144, 148.

50. L.T.C. Rolt, *The Railway Revolution: George and Robert Stephenson* (New York: St. Martin's Press, 1960), viii.

51. L.T.C. Rolt, *Victorian Engineering* (Harmondsworth: Penguin Books, 1970), 14.

52. Rolt, *Victorian Engineering*, 52.

53. Rolt, *Victorian Engineering*, 118.

54. Rolt, *Victorian Engineering*, 162.

55. R. A. Buchanan, "Lionel Thomas Caswall Rolt, (1910 1974)," *Technology and Culture* 16, no. 3 (July 1975): 455. Rolt would later serve as the first president of the Association of Industrial Archaeology, as it was established in 1973. See also, Rolt, *Landscape With Figures*, 216. For further details on the survey, see Keith Falconer, "The industrial heritage in Britain—the first fifty years." Falconer has argued that the developments in industrial archaeology in Britain in the 1960s and 1970s went on to have effects on the future of the field overall.

56. Rex Wailes, "Industrial Archeology: The Industrial Monuments Survey," *Technology and Culture* 8, no. 2 (April 1967): 199.

57. Wailes, "Industrial Archeology: The Industrial Monuments Survey," 199.

58. Rolt, *Landscape With Figures*, 216.

59. Rolt, *Landscape With Figures*, 218.

60. Rolt, *Landscape With Figures*, 221. Rolt also served on the council of the Centre for the Study of the History of Technology at the University of Bath. See Keith Falconer, "A Celebration of L.T.C. Rolt and "Industrial' World Heritage Sites," in

Angus Buchanan, Neil Cossons, et al., "Remembering Rolt: A Symposium in Honour of the 100th Anniversary of L.T.C. Rolt's Birth—A Pioneer of Industrial Archaeology," *Industrial Archaeology Review* 33, no. 1 (2011): 11.

61. Rolt, *Victorian Engineering*, 115.

62. I. G. Simmons, *An Environmental History of Great Britain; From 10,000 Years Ago to the Present* (Edinburgh: Edinburgh Univesity Press, 2001), 193.

63. Simmons, *An Environmental History of Great Britain*, 193.

64. "Motor Moloch," IWA, *Bulletin* 61 (April 1960), 3.

65. "Motor Moloch," 3.

66. Lydia Jackson and Richard Cracknell, *House of Commons Briefing Paper: Road Accident Casualties in Britain and the World*, Number 7615 (April 2018), 5. (https://researchbriefings.files.parliament.uk/documents/CBP-7615/CBP-7615.pdf). Subsequently, according to Jackson and Cracknell, they began a steady decline, and reached 2,000 or less per year by the second decade of the twenty-first century.

67. "Motor Moloch," 4.

68. L. T. C. Rolt, *Railway Adventure* (Stroud: The History Press, 2010 (1953)), 82.

69. National Executive Committee, *Leisure for Living* (London: Labour Party, 1959), 5.

70. National Executive Committee, *Leisure for Living*, 49.

71. National Executive Committee, *Leisure for Living*, 49.

72. National Parks and Access to the Countryside Act 1949, *c.* 97.

73. National Executive Committee, *Leisure for Living*, 49.

74. National Executive Committee, *Leisure for Living*, 50.

75. This was in accord with Rolt's overall outlook. He was willing to consider a process of filling in canals where physical decline was significant and maintenance costs were high, and the potential gains of reconstruction were difficult to assess. Rolt later emphasized that many projects proposed by the BTC for infilling were rejected as too expensive or impracticable—the committee instead recommended either a partial or full restoration as either cheaper or more desirable for multiple uses. (Rolt, *Landscape With Figures*, 214.)

76. IWA, *Bulletin* 63 (January 1961), 9–10.

77. Report No. 1. Report of the Inland Waterways Redevelopment Committee on a Scheme for the Redevelopment of the Stratford-Upon-Avon Canal (Southern Section), 1. Committees, Conferences, and Working Parties. National Trust. Restoration of Stratford-Upon-Avon Canal (Southern Section), 1959–1964, MT 88/31, National Archives, United Kingdom.

78. Report No. 1. Report of the Inland Waterways Redevelopment Committee on a Scheme for the Redevelopment of the Stratford-Upon-Avon Canal (Southern Section), 4.

79. Bolton, *Race Against Time*, 181.

80. *Daily Telegraph*, October 15, 1962, clipping, in Committees Conferences and Working Parties. National Trust. Restoration of Stratford-Upon-Avon Canal (Southern Section), 1959–1964, MT 88/31, National Archives, United Kingdom.

81. *Daily Telegraph*, October 15, 1962.

82. IWA, *Bulletin* 50 (April 1956), 32.

83. UK Parliament, *Report of the Committee of Inquiry into Inland Waterways*, 52.

84. Report of the Inland Waterways Redevelopment Committee on the redevelopment of the Kennet and Avon Canal, Report No. 19 (prepared by the IWRC), in NTB'23'1'04 PT2 (MT/88/200), National Archives, United Kingdom.

85. Report of the Inland Waterways Redevelopment Committee on the redevelopment of the Kennet and Avon Canal, Report No. 19.

86. IWA, *Bulletin* 63 (January 1961), 9–10.

87. IWA, *Bulletin* 63 (January 1961), 13.

88. IWA, *Bulletin* 63 (January 1961), 13.

89. British Waterways, *Six Years of Progress* (London: The Curwen Press, 1961), 14, pamphlet in On Waterways in Great Britain . . . (Part of British Transport Historical Records Office library: Canals and Inland Waterways, ZLIB 3), ZLIB 3/71, National Archives, United Kingdom.

90. British Waterways, *Six Years of Progress*, 14.

91. British Waterways, *Six Years of Progress*, 15.

92. British Waterways, *Six Years of Progress*, 15.

93. British Waterways, *Six Years of Progress*, 15.

94. Transport Act 1962, *c.* 46.

95. Transport Act 1962, *c.* 46.

96. Transport Act 1962, *c.* 46.

97. Boughey, *Charles Hadfield: Canal Man and More*, 59.

98. Robert Aickman, *The River Runs Uphill* (Burton on Trent, J.M. Pearson, 1986), 71.

99. Hugh Malet, *Voyage in a Bowler Hat* (London: M&M Baldwin, 1985 (1960)), 97. The second half of Malet's work records his canal journey across Ireland, completed just after the voyage across Britain.

100. Malet, *Voyage in a Bowler Hat*, 103.

101. Leslie Morton, "The Willow Wren Group," IWA, *Bulletin* 74 (July 1965), 5.

102. "Future of Narrow Boats," IWA, *Bulletin* 68 (May 1963), 16.

103. David Blagrove, *Bread Upon the Waters* (Burton-on-Trent: J.M. Pearson & Son, 1984), 23, 39.

104. Blagrove, *Bread Upon the Waters*, 34.

105. Blagrove, *Bread Upon the Waters*, 42, 44, 50.

106. Blagrove, *Bread Upon the Waters*, 61.

107. Blagrove, *Bread Upon the Waters*, 63.

108. Blagrove, *Bread Upon the Waters*, 77.

109. Blagrove, *Bread Upon the Waters*, 85.

110. For an example of the images of Joe and Rose Skinner, see "Rally Awards," IWA, *Bulletin* 75 (November 1965), 43.

111. Blagrove, *Bread Upon the Waters*, 101.

112. Blagrove, *Bread Upon the Waters*, 185.

113. "Future of the Narrow boats," 16; "Stop Press," IWA, *Bulletin* 71 (June 1964), 80; and Morton, "The Willow Wren Group," 6. In the second, 1965, article, Morton indicated the BWB boats were transferred to Willow Wren in July 1963.

114. Morton, "The Willow Wren Group," 6.

115. Seymour, *Voyage into England*, 131.

116. Boughey, *Charles Hadfield: Canal Man and More*, 77.

117. Nicholas Tomalin, "The Pleasures of Floating Nostalgia," IWA, *Bulletin* 74 (July 1965), 32.

118. Tomalin, "The Pleasures of Floating Nostalgia," 33–34.

119. Tomalin, "The Pleasures of Floating Nostalgia," 34. Debates regarding the treatment of post-industrial landscapes would increase as the 1960s continued. John Barr, writing in *Derelict Britain* in the aftermath of the 1966 Aberfan disaster, rejected what he saw as a romanticization of industrial ruins by Ian Nairn and others, but encouraged government actions in reclaiming wasteland, a process in which he believed canals could play a role. (See *Derelict Britain* (Harmondsworth: Penguin Books, 1970 (1969)), 25–26, 58.)

120. Tomalin, "The Pleasures of Floating Nostalgia," 34.

121. Tomalin, "The Pleasures of Floating Nostalgia," 35.

122. Tomalin, "The Pleasures of Floating Nostalgia," 35.

123. Tomalin, "The Pleasures of Floating Nostalgia," 36.

124. British Waterways Board, *The Future of the Waterways: Interim Report of the Board* (London: HMSO, 1964), 16.

125. British Waterways Board, *The Future of the Waterways*, 17.

126. British Waterways Board, *The Future of the Waterways*, 36–37.

127. "Hawton Interim Report," IWA, *Bulletin* 70 (March 1964), 5.

128. "Chairman's Report for the Year 1963–1964," IWA, *Bulletin*, 71 (June 1964), 72. Joseph Boughey has argued that in his time on the BWB, Hadfield was instrumental in this change, having drawn up a detailed map of an amenity network in 1963 and proposed a standardization of some equipment. (See Boughey, *Charles Hadfield: Canal Man and More*, 64–65.)

129. Ian Nairn, "Do-it-Yourself," IWA *Bulletin*, 68 (May 1963), 3.

130. Nairn, "Do-it-Yourself," 4.

131. Nairn, "Do-it-Yourself," 5.

132. Nairn, "Do-it-Yourself," 7.

133. Nairn, "Do-it-Yourself," 7–8.

134. "Brief for Meeting Between Parl. SEC. (C) and Chairman of Inland Waterways Association 26th April 1965" in Inland Waterways Association (IWA) publication, New Waterways: proposals on future of waterway network; parliamentary and other correspondence, MT 88/121, National Archives, United Kingdom.

135. "Brief for Meeting Between Parl. SEC. (C) and Chairman of Inland Waterways Association on 26th April 1965."

136. "Note of a meeting between Parl. SEC. (C) and the Inland Waterways Association on 26th April 1965."

137. "Brief for Meeting Between Parl, SEC. (C) and Chairman of Inland Waterways Association on 26th April 1965."

138. "Brief for Meeting Between Parl. SEC. (C) and Chairman of Inland Waterways Association on 26th April 1965."

139. "Brief for Meeting Between Parl. SEC. (C) and Chairman of Inland Waterways Association 26th April 1965."

140. Parliamentary Inland Waterways Group, Inland Waterways Association, National Trust, *Proposals for a Waterways Trust* (London: July 1965), 5, in Inland Waterways Association (IWA) publication, New Waterways: National Waterways Trust proposals, MT 88/122, National Archives, United Kingdom.

141. Other groups included the Association of Pleasure Craft Operators, British Waterways Board, Council for the Preservation of Rural England, Inland Waterways Association, Kennet and Avon Canal Trust, Local Authorities, National Farmers Union, National Parks Commission, National Trust, Ramblers Association, River Authorities, Royal Institute of British Architects, Royal Yachting Association, and the Ship and Boat Builders National Federation. Parliamentary Inland Waterways Group, Inland Waterways Association, National Trust, *Proposals for a Waterways Trust* (London: July 1965), 5, in MT 88/122, National Archives, United Kingdom.

142. British Waterways Board, *The Facts About the Waterways* (London: December, 1965), 1.

143. British Waterways Board, *The Facts About the Waterways*, 32. More specifically, the major financial conclusions regarding the waterways were the following:

> (iii) the most serious and over-riding fact emerging is that, even if the whole of the 'rest of the system' were to be ruthlessly treated (without any regard to social welfare) and every possible canal were to be either eliminated altogether or reduced to minimum water channeling flow—whichever was the cheaper—the exchequer would still be saddled with a bill equivalent to about £600,000 a year. And this, the survey reveals, is not an optional 'subsidy' but an inescapable minimum exchequer charge. (iv) it follows that in hard reality the field open to discussion on 'amenity' grounds of pleasure boating and leisure use lies between (a) a rock-bottom starting point not of zero but of £600,000 a year; and (b) a figure for using most of the non-commercial system as it is for pleasure boating, which is some £300,000/£350,000 a year higher. The true room for manoeuvre (subsidy, use of volunteer monies, etc.) lies in that latter range. The £600,000 is not optional!, 40

144. Michael Dower, *The Challenge of Leisure: A Civic Trust Survey on Planning for Increased Leisure Facilities, First Published under the Title "Fourth Wave'* ["Reprinted from the *Architects' Journal* of January 20, 1965, by permission of the editor, by Harrison & Sons, Ltd."] London: Civic Trust, 1967[?], 10.

145. John Sheail, "Leisure in the English Countryside: Policy Making in the 1960s," *Planning Perspectives* 16, no. 1 (2001): 71.

146. Sheail, "Leisure in the English Countryside," 71.

147. Sheail, "Leisure in the English Countryside," 72–73.

148. See E. M. Nicholson, "The Countryside in 1970," *Nature* 208 (December 25, 1965): 1265–1266.

149. Sheail, "Leisure in the English Countryside," 75.

150. Ministry of Land and Natural Resources, *Leisure in the Countryside, England and Wales*, Cmnd. 2928 (London: HMSO, 1966), 7.

151. Memo to the Minister of Transport from R. D. Poland, July 4, 1966, in Inland Waterways Association (IWA) publication, New Waterways: National Waterways . . . , MT 88/122, National Archives, United Kingdom.

152. "Note of the Minister's meeting with Officers of the Parliamentary Inland Waterways Group and representatives of the Inland Waterways Association and the National Trust in Room 8/72, St. Christopher House, at 3 p.m. on Tuesday, July 5, 1966, in Inland Waterways Association (IWA) publication, New Waterways: National Waterways . . . , MT 88/122, National Archives, United Kingdom.

153. Ministry of Transport, *Transport Policy*, Cmnd. 3057 (London: HMSO, 1966), 28.

154. Ministry of Transport, *Transport Policy*, 28. An acknowledgment of leisure was still present,

> If the non-commercial waterways at present used by these boats were to be closed to them the annual loss could be cut from about £940,000 a year to the inescapable minimum of about £600,000 a year at present costs. But once they were taken out of use for pleasure cruising and standards of maintenance lowered it is unlikely that it would ever be possible to restore them. It would be wrong to allow this to happen to those waterways which seem likely to have a valuable part to play in meeting the country's leisure-time needs in years to come, 28.

155. Ministry of Transport, *Transport Policy*, 28.

156. Memo to the Minister of Transport from R. D. Poland, July 4, 1966, in Inland Waterways Association (IWA) publication, New Waterways: National Waterways . . . , MT 88/122, National Archives, United Kingdom.

157. "Infamous' White Paper on Transport," August 29, 1966, *Lloyd's Daily List* in Inland Waterways Association (IWA publication, New Waterways: Interim report on new . . . , MT 88/123, National Archives, United Kingdom.

158. "Infamous' White Paper on Transport."

159. Bolton, *Race Against Time*, 231.

160. Geoffrey Dibb, ed., *The Canals Book: Navigator Guide Directory* (Brundall and Norwich: Geoffrey Dibb, Ltd., 1966), 5.

161. Dibb, *The Canals Book*, 6.

162. Dibb, *The Canals Book*, 7.

163. Dibb, *The Canals Book*, 7.

164. Inland Waterways Association [IWA], *"The Way Ahead for the Amenity Waterways": A Memorandum to the Minister of Transport on Behalf of the Inland Waterways Association, Ltd.* London: 1967), 3.

165. IWA, *"The Way Ahead for the Amenity Waterways,"* 3.

166. IWA, *"The Way Ahead for the Amenity Waterways,"* 3.

167. IWA, *"The Way Ahead for the Amenity Waterways,"* 4.

168. IWA, *"The Way Ahead for the Amenity Waterways,"* 4.

169. IWA, *"The Way Ahead for the Amenity Waterways,"* 5.

170. IWA, *"The Way Ahead for the Amenity Waterways,"* 12.

171. IWA, *"The Way Ahead for the Amenity Waterways,"* 13.

172. IWA, *"The Way Ahead for the Amenity Waterways,"* 13.

173. Hugh McKnight, "Townscape and the Waterway," IWA, *Bulletin* 79 (January 1967), 49.

174. McKnight, "Townscape and the Waterway," 50.

175. McKnight, "Townscape and the Waterway," 50.

176. Regent's Canal Group, *Regent's Canal: A Policy for its Future* (Sevenoaks: Caxton & Holmesdale Press, 1967), 17.

177. Regent's Canal Group, *Regent's Canal: A Policy for its Future*, 19.

178. McKnight, "Townscape and the Waterway," 52.

179. McKnight, "Townscape and the Waterway," 52.

180. IWA, *Bulletin* 85 (November 1968), 28.

181. Bolton, *Race Against Time*, 229.

182. Michael Dower, "The Recreational Potential of the Waterways," IWA, *Bulletin* 80 (September 1967), 10.

183. Dower, "The Recreational Potential of the Waterways," 11.

184. Dower, "The Recreational Potential of the Waterways," 16.

185. J. J. Richardson and Richard Kimber, "The British Waterways Board: A Neglected Asset?" *Public Administration* 52, no. 3 (1974): 306.

186. Ministry of Transport, *British Waterways: Recreation and Amenity*. Cmd. 3401 (London: HMSO, 1968), 2.

187. Ministry of Transport, *British Waterways: Recreation and Amenity*, 2.

188. Ministry of Transport, *British Waterways: Recreation and Amenity*, 3.

189. Ministry of Transport, *British Waterways: Recreation and Amenity*, 3.

190. Ministry of Transport, *British Waterways: Recreation and Amenity*, 3.

191. Ministry of Transport, *British Waterways: Recreation and Amenity*, 4.

192. See, Vincent Powell-Smith, *The Transport Act 1968* (London: Butterworths, 1969), 150–151.

193. Powell-Smith, *The Transport Act 1968*, 155.

194. Letter from Lionel Munk to Stephen Swingler, April 23, 1968, in Inland Waterways Association: amendments to Bill; departmental consideration, MT/88/207, National Archives, United Kingdom.

195. See, for instance, IWA, "Press Release: Repeal of the Public Right of Navigation;" "Waterways could close to the public," *Guardian*, July 1968; Letter from J.R. Donie, NTB Division to Wilfred Moore, General Secretary, IWA; and Letter from Wilfred Moore to Stephen Swingler, September 25, 1968, in Inland Waterways Association: amendments to Bill; departmental consideration, MT/88/208, National Archives, United Kingdom.

196. Ministry of Transport: Press Notice: Waterways Amenity Council Appointed (November 21, 1968) in Inauguration of the Inland Waterways Amenity Advisory Council: press notices and speeches, MT 88/287, National Archives, United Kingdom.

197. Ministry of Transport: Press Notice: Inland Waterways Amenity Advisory Council, May 1, 1968, in Inauguration of the Inland Waterways Amenity Advisory Council: press notices and speeches, MT 88/287, National Archives, United Kingdom.

198. Countryside Act 1968, *c*. 41.

199. Countryside Act 1968, *c*. 41.

200. As quoted in Horace Herring, "The Conservation Society: Harbinger of the 1970s Environment Movement in the United Kingdom," *Environment and History* 7, no. 4 (November 2001): 385.

201. Herring, "The Conservation Society: Harbinger of the 1970s Environment Movement in the UK," 382.

202. Herring, "The Conservation Society: Harbinger of the 1970s Environment Movement in the UK," 382–385.

203. B.W. Clapp, *An Environmental History of Britain since the Industrial Revolution* (London: Routledge 2013 (1994)), 8. See also the discussion of the awakening of the environmental movement in the late 1960s in Glen O'Hara, *The Politics of Water in Post-War Britain* (London: Palgrave Macmillan, 2017), 19.

204. See Squires, *Canals Revived.* As noted by Squires, the efforts of these volunteers, often referred to as new "navvies," reached a crescendo between 1968 and 1970 (95–98).

205. "Canal Dumping Obstacle to Weekend Boat Rally," *Walsall Observer*, July 18, 1969, 7. BNA.

206. "Canal Dumping Obstacle to Weekend Boat Rally."

207. "Canal Dumping Obstacle to Weekend Boat Rally."

208. "Big Boat Trek Starts to Midland Water Rally," *The Birmingham Post*, July 16, 1969, 2, and "A Splash of Colour for the Inland Boat Rally," *The Birmingham Post*, July 26, 1969, 2. BNA.

209. "Cruising Down the Years," *Sunday Mercury*, July 27, 1969, 16–17. BNA.

210. "Ditching the Industrial Image of our Canals," *The Birmingham Post: Friday Magazine*, July 25, 1969, 1. BNA.

211. "Ditching the Industrial Image of our Canals," 1.

212. "Ditching the Industrial Image of our Canals," 1.

213. "Ditching the Industrial Image of our Canals," 1.

214. O'Hara, *The Politics of Water in Post-War Britain*, 19.

215. Inland Waterways Association, accessed August 28, 2015, https://www.waterways.org.uk/waterways/history/timeline/1977_1979.

216. BTC, *Report of the Board of Survey*, 65.

217. See the contrasting perspectives offered by Patrick Wright, *On Living in an Old Country: The National Past in Contemporary Britain* (London: Verso, 1985), and Merlin Waterson, *The National Trust: the First Hundred Years* (London: BBC Books, 1994).

218. Tomalin, "The Pleasures of Floating Nostalgia," 35.

Conclusion

Canals remain a point of convergence in understanding the evolving place of technology, culture, and the environment in Britain. Recent historiography has focused upon rivers as organic and evolving entities—whether recognized for their roles in maintaining spirituality, national boundaries, agriculture, transportation, or recreation. Similar to the land itself, many have been subject to human intervention for thousands of years. In these ways, they have become central to the economic, cultural, and ecological interpretations of nations. Canals can share many of these qualities but are marked for their intensified application of engineering and the need for continued human intervention to sustain them. Questions regarding how and whether they shall continue to be used are, to an even greater degree than rivers, subject to deliberative choices of either their private or public guardians. Yet, once constructed as part of a major transit route or system, canals are not easily ignored or removed. Their construction can raise questions about the limits of human engineering, their significance in creating human communities along their banks can raise questions regarding responses to mobile populations, and the choice for their abandonment or continued use can present issues of national economic importance. Moreover, once constructed, they become part of an evolving landscape. The question of human intervention having already been decisively settled, debates about their ensuing aesthetic and ecological significance, as well as their evolving relationship with the communities along their banks, can become essential. This study argues that the story of Britain's canals between the 1760s and 1968 offers not only a lens through which to view changes in British life but also a means of understanding the enormously complex evolution of a technological network.

An understanding of the forces which led to the revitalization of canals in the late twentieth century must be seen in the context of the social and commercial forces which effected their construction. Between the late eighteenth and early nineteenth centuries, canals represented the application of early modern technology to the management of water flows so that smooth pathways might be created for the transport of people, resources, and products between the towns and cities of Britain's expanding commercial and industrial world. In their rapid creation at the dawn of accelerated industrialization, they integrated innovations in water engineering from elsewhere with Britain's new commercial and capitalist drive. This led to the rapid and unprecedented creation of a network supporting the wider acceleration of industrialization.

Though canals entered the nineteenth century first and foremost as industrial resources, their waters wound around hillsides and their banks could be lined with trees. Substantive systems of locks and the presence of aqueducts testified to the fact that their initial focus was industry, but they remained dependent upon mechanical operations and sources of power resistant to innovation. Traffic relied upon horsepower, speeds were limited to just a few miles per hour, and the operation of locks and other equipment were dependent upon human labor. Their use could also be affected by water shortages. The resolve of many owners weakened when steam arrived on the rails. Though some canals remained prosperous, many were soon considered holdovers from another era, a static technology whose day had passed. For most who pressed concerns regarding the preservation of the countryside, the authentic landscape was one without canals. Yet, significant meanings surrounding canals were in the process of being formed that would dramatically affect their fate in the twentieth century. The characteristics that had undermined their ability to compete—their winding paths and regional and local focuses, individualized engineering, as well as the necessity of slow speeds—helped to ensure their emergence as a recreational resource offering stark contrasts to the contemporary world of the late 1960s. For their later supporters, they could be easily linked with the natural environment and an initial, and much simpler, stage of the modern world.

In the nineteenth century, Britons' relationship with canals became intimately linked with their place in the social and cultural worlds of the Victorian era through the populations whose identities were integrated into the canals' waters. Though workers on the banks might be in the direct employ of the canal companies, many others were associated with a diverse array of carrying firms—or, perhaps taken on only through a boat captain. They thus formed organic living and working communities not dissimilar from their counterparts on the rivers. In the early nineteenth century, those working canal boats could be associated with crime, violence, and death, as were the spaces along canal banks through which they traveled. Itinerant male boatmen established an early reputation for vice and drinking that cast an extended shadow over their reputation in the nineteenth century. Boatmen came to the attention of missionaries and drew special attention within parliamentary investigations focusing upon crime and sobriety. In the mid-1870s, George Smith ("of Coalville") pursued a campaign focusing upon the grim and sordid conditions he believed were to be found on the waters. Smith put his case forcefully, and although drawing the critique, if not the ire, of others associated with the boat population, influenced both the Canal Boats Act and the Canal Boats Amendment Act. The outcome was likely not what Smith had anticipated. John Brydone, and later Owen J. Llewellyn, who undertook their responsibilities as chief canal boats inspectors seriously, offered supportive overall assessments of the health of the canal boat population. Nonetheless, despite the waning of Victorian moral condemnation, the canal boat population had become marked as insular, poorly educated, and resistant to innovation. The latter quality was probably linked at least in part to the technical capacities of the canals they worked. At the same time, the endurance of the canal boat population and their maintenance of traditions of work that stood in stark contrast to the technical developments of life on shore made them a valued reminder of an earlier age, particularly for those unnerved by the pace of change in the late nineteenth century. These ambivalent visions of life afloat in the nineteenth century would go on to affect views of canals and responses to calls to preserve them in the twentieth century.

The first half of the twentieth century represented a period of conflicting interpretations of canals. As the century opened, they were no longer associated with the prowess of a commercial nation, but rather with its struggles. Increasing economic competition, diplomatic isolation, and fears that the nation could no longer martial the human or material resources to outbid its continental partners were all topics of political and social debate. On inland waters, the building of the Manchester Ship Canal had signaled the ability to imagine a new canal project, but across the system more widely, company ownership remained fractured, as did that of the transport companies. The Royal Commission on Canals and Inland Waterways suggested there was hope, but only with significant upgrades, a task left hanging at the opening of World War I, which significantly undermined the political and economic feasibility of government investment. Aside from the system's structural health, in the context of post-World War I investigations, the worst imaginings of Victorian reformers regarding the health and morality of the canal boat population were put to rest. Some observers suggested life on the boats was, in fact, a contributor to overall physical robustness. Education remained a concern, but there was little political will to disrupt the family life of the boats as overall numbers were diminishing. Despite Neville Chamberlain's attention to canals in the immediate postwar years and the creation of a more modest plan of upgrades, no significant governmental action was forthcoming.

Moreover, when World War II broke out, governmental attention to canals again lagged, despite the Pick report which in quick fashion outlined how the government might take the initiative. In the aftermath of World War II, there was little room for canals as attention turned to a variety of social and economic concerns brought by the war and the prospect of a "people's peace."

However, twentieth-century understandings of canals were affected by interpretations of their place in the landscape, expressed most directly through ruralist and organicist literature. The changes wrought by increasing commerce and wealth, along with the expansion of both urban populations and industrial production into previously rural areas, helped to enshrine visions of the rural—particularly rural southern England—as the cultural ideal of the nation by the late nineteenth century. Although the degree to which visions of southern England had overwhelmed those of northern England has been subject to debate, commemoration of the implements of industrialization, of which canals were a significant example, was by and large sparse among those who sought conservation of the countryside. And yet, among those who traveled the canals and found in them a world in stark contrast to their own, was laid the groundwork for canals' association with the countryside and nature. In the nineteenth century, John Hassell's journey on the waterways, and his implicit comparison of them to railways and stagecoaches, emphasized their links to the slower pace of a rural world. Later, explorations by William Black and Reginald Blunt highlighted the degree to which a continuing loss of traffic in the south of England had succeeded in integrating them into an experience of nature which, for travelers, was both close at hand and a world apart from the city. The re-framing of the canal boat population, beginning in the very late nineteenth century, also offered a more comfortable linking of their population with ruralist and organicist understandings for the new century. Many canals in southern England had been increasingly divorced from their industrial focus, and the lack of traffic as well as the increasing presence of natural flora and fauna along their banks indicated that they could be interpreted as part of a non-industrial landscape. Those who still worked the waterways were becoming associated with the nearby life of rural England in their dress, manners, and modes of labor.

Nostalgia for a lost world of rural Britain crescendoed as mass consumption and mass marketing appeared to undo the sense of local identity that rural advocates believed had been essential to a larger national identity. Still a part of the industrial world for most, even canals where industrial transport had lapsed were the haunts of only a few hearty travelers. In the writings of C. J. Aubertin, E. Temple Thurston, and later L.T.C. Rolt, interpretations of canals' engineering, place within the landscape, and patterns of labor laid the groundwork for the artificial waterways' integration within ruralist and organicist themes. In particular, canal boat families became a foil for contemporary anxieties regarding twentieth-century life. Maintaining independence in their patterns of work, they traveled the byways of England and Wales at a pace reminiscent of the eighteenth century. Rolt, who had been unnerved by the drastic technological changes brought as a result of World War II, felt experiences on canals could be seen as a riposte to a new era of accelerating change and collectivism. The early endorsement of Rolt's work by H. J. Massingham, the analysis of canal boats' "roses and castles" by Barbara Jones, and the involvement of Viscount Lymington in the activities of the Inland Waterways Association (IWA) early in the postwar era testified to connections of canal boat life to understandings of rural Britain.

In establishing the IWA, Rolt and Robert Aickman owed much to the examples of organizations such as the National Trust and the Council for the Preservation of Rural England. Arguing that a reliance upon planning held the potential to sweep away the individuality of many canals and the unique cultural world of the boatmen, the IWA pressed the British Transport Commission (BTC) to keep all waterways under its control open, even while decrying the expansion of a government they saw as flattening British society and culture. The collective state, however, had its attractions. One of these for the IWA was the accommodation of holidays and recreation, now seen as linked to governmental responsibility in a more democratic era. IWA rhetoric increasingly emphasized recreation's potential as a means of justifying continued use and revivification of canals. For Rolt, such a step was disconcerting, and in reflecting upon his decision to leave the organization, he later emphasized the close connection between utility, work, and aesthetic appreciation that was woven into ruralist and organicist views of Britain at mid-century.

The early success of the IWA indicates how ruralist and organicist observers could absorb significant change within the countryside if that change reflected the persistence of pre-industrial rhythms of work and an apparently peaceful integration of technological change within the landscape. The working life of canals offered compelling images of family units determining their own rhythms of work and guarding their unique working and living environments within the tradition of boat painting. Though canal boat people's identification as Romani ("Gypsies"/"gypsies") was noted as a mark of alienation, it also indicated a lack of absorption into a homogenized working and consuming class. Those who traveled the canals, particularly in southern England, recognized the ways in which the paths of canals had been governed by the physical spaces through which they passed and the ways in which an appreciation of the natural world could emerge as industrial traffic departed. These understandings reflected the heart of ruralist and organicist observations—a position which Rolt was instrumental in securing. For contemporary historians, the transformation of artificial waterways from cuts in the land in service to industry to corridors of natural beauty and recreation can offer an additional window into how natural and man-made worlds can become blurred.

It was within the realm of political and social change after 1951 that the place of canals in modern British society was reshaped and renewed. Although Rolt's *Narrow Boat* had sparked interest among a core group of early canal aficionados, the IWA, as led by Aickman, still faced significant headwinds. The BTC and its successive executive bodies saw the IWA as obstructionist and preferred the assistance and advice of those who seemed ready to compromise regarding what canals could be saved. The early embrace of holiday traffic by the IWA had begun the process of integrating the postwar era of leisure with arguments emphasizing canals' utility and cultural significance. Attitudes of governing officials, through both the Rusholme survey and the Bowes Committee, remained functional, focusing on the financial losses the canal system continued to accrue. As the 1960s opened, some of the last remaining narrowboat workers, whose lives had become synonymous with the cultural heritage of the waterways, were departing. As they did so, there were some ready to preserve a portion of this world by writing their own heartfelt memoirs of travels on the water that bemoaned the loss of narrowboat industrial traffic. Their works coincided with the arrival of a new interest in industrial artifacts. As represented by the popularity of Rolt's biographies of early engineers, a new focus on rescuing the heritage of the industrial world had arrived. Having begun to embrace the world of leisure, canals could now also encompass direct engagement with the history of industrial change.

The fate of canals was also significantly affected by automobiles, whose place in British life and society had become steadily more complex as their numbers increased. Since their first significant appearance in the countryside in the interwar period, frustration had ensued as many worried automobiles' presence would undermine what travelers had come to experience. By the 1960s, canals could offer themselves as an antithesis to the world of the auto through their slow speed, minimal pollution, and ability to engage with the countryside lock by lock. Throughout the decade, reliance on automobiles was called into question as the expenses of road construction intensified with little indication that their expansion would ultimately solve the problem of traffic congestion. Several reinforcing strands of interest ultimately helped to usher in the Ministry of Transport's new direction on waterways by 1968. Most significantly, government officials found the preservation of canals a worthwhile financial endeavor as the potential losses that would follow upon closure to navigation or abandonment now appeared to be at least partially balanced by fees from recreational boaters and the gains to be made from the new interests in nature, heritage, and leisure. Yet, cultural understandings played a significant role, as did emerging understandings of canals as a portion of the industrial environment whose value could be reclaimed through renewed use. In 1969, Jestyn Philipps, Viscount St. Davids, who was active in the House of Lords on behalf of canal issues, may have been the first to raise the question of global warming in Parliament, thus signaling the opening of a new era in which the cause of canals would be linked with environmental issues.[1]

The challenges created by leisure were sobering, but government officials found themselves with a unique and powerful resource. Canals had become embedded within the natural world into which they had first intruded, and their presence was associated with the simplified engineering of the early industrial era as well as the one-time cultural presence of canal boat communities. They offered an extended network of space that could be integrated into a linear engagement with both the rural and urban worlds through which they passed. They challenged trends of suburbanization, which had seemed to bring homogenization and to erase the distinctive experiences of local towns and villages. They emphasized instead a slow discovery of urban and rural space and both the distinctive technical features of the waterways and the histories of the regions through which they passed. Moreover, they were a means of experiencing an amalgam of sites centered on nature as well as early industrial and commercial enterprises. Canals now paid dividends far different than those of 200 years previously. No longer abandoned and a lost artifact of industrialization, they had been re-purposed in a more democratic era, one in which peace appeared to have been made with a system that had once seemed overtaken by more powerful technologies and could never regain its footing. Rather than symbols of the refuse of industrialization, they were now an enduring product—one which was woven into the landscape, respectful of cultural tradition and minimal in its environmental impact.

The opening of the twenty-first century has continued to see Britons' relationship with canals evolve. Many new canal boaters have embarked upon the waters, making the most of ongoing restoration efforts from the 1970s to the early 2000s led by local volunteer groups and the IWA, as well as the British Waterways Board and its successor administrator of the waterways, the Canal & River Trust (CRT). Moreover, for some, inhabiting the waterways has become a mechanism for surviving life on land in the form of reduced housing costs.[2] The intensification of remote work options after the COVID-19 pandemic, which began in 2020, offered a means to stay connected while afloat.[3] These developments have prompted, once again, a re-examination of what life on the waterways means. Major routes have become crowded, with the number of moored boats sometimes slowing waterways traffic. An increasing number of novice boaters has meant additional stress for old hands. And, the CRT has found it difficult both to enforce requirements for changing moorings and to promote the waterways as a means of escape.[4]

Nonetheless—if in transformed ways—canals and the legacies of their construction and evolving place in cultural life are likely to continue affecting environmental attitudes in Britain.

NOTES

1. Viscount St. Davids. "My Lords, Can My Noble Friend Say Whether He and British Railways Have Taken Account of the Fact That What Were Abnormal Temperatures Last Summer May Not Be Abnormal If We Continue to Discharge Carbon Dioxide into the Air by the Burning of Various Fossil Carbons, So Increasing the Greenhouse Effect?" (House of Lords Debates, November 5, 1969 vol. 305 *c.* 337, accessed May 29, 2024, https://api.parliament.uk/historic-hansard/lords/1969/nov/05/railways-use-of-continuous-welded-rail.)

2. Donna Ferguson, "Pushing the Boater Out: London's Property Crisis Spreads to the Water," *The Guardian*, July 29, 2019, accessed December 30, 2023, https://www.theguardian.com/cities/2019/jul/29/london-property-crisis-canal-narrowboat-gentrification.

3. See Megan Specia, "Living on the Canals, and Loving the Floating Life," *New York Times*, July 21, 2021, A4.

4. Larence Shepherd, "The Housing Crisis No One's Talking About: London's Canals Are Getting Dangerously Over-Crowded," *City Monitor*, July 1, 2016, accessed December 30, 2023, https://citymonitor.ai/community/housing-crisis-no-ones-talking-about-londons-canals-are-getting-dangerously-over-crowded#:~:text=Urban%20canals%20are%20overcrowded%20and,as%20running%20water%20or%20heating.

Bibliography

Aberdeen Evening Express. British Newspaper Archive (https://britishnewspaperarchive.co.uk/).

"An Account of the Duke of Bridgewater's New Inland Navigation." In *The Annual Register or a View of the History, Politics, and Literature for the Year 1763*, 99–100. London: Dodsley, 1764.

Ackroyd, Peter. *Thames: The Biography*. New York: Anchor Books, 2007.

Aetherstone News. British Newspaper Archive (https://britishnewspaperarchive.co.uk/).

Aickman, Robert. "The Case for a National Waterways Conservancy." *Journal of the Royal Society of St. George* (Summer 1956): 9–12.

Aickman, Robert. *The River Runs Uphill: A Story of Success and Failure*. Buton on Trent: J. M. Pearson, 1986.

Aikin, John, and Anna Laetitia Barbauld. "The Canal and the Brook: An Apologue." In *Miscellaneous Pieces of Prose*, Edited by John Aikin and Anna Laeitia Barbauld, 79–87. London: J. Johnson, 1792.

Anderson, *Suicide in Victorian and Edwardian England*. Oxford: Clarendon Press, 1987.

Anonymous. *The Waterway to London, as Explored in the "Wanderer" and "Ranger," with Sail, Paddle, and Oar, in a Voyage on the Mersey, Perry, Severn, and Thames, and Several Canals*. London: Simpkin, Marshall, & Co., 1869.

Archbold, Frederick. *Snowden's Magistrates Assistant*. 4th ed. London: Shaw and Sons, 1859.

Aris's Birmingham Gazette. British Newspaper Archive (https://britishnewspaperarchive.co.uk/).

Arnold, A.J., and S. McCartney. "Veritable Gold Mines Before the Arrival of Railway Competition': But Did Dividends Signal Rates of Return in the English Canal Industry?" *Economic History Review* 64, no. 1 (2011): 214–236.

Ash, Eric H. *The Draining of the Fens: Projectors, Popular Politics, and State Building in Early Modern England*. Baltimore, MD: Johns Hopkins University Press, 2017.

Aubertin, C. J. *A Caravan Afloat*. London: Simpkin, Marshall, Hamilton Kent & Co., Ltd., 1916.

Bagwell, Philip S. *The Transport Revolution*. London: Routledge, 1988.

Bailey, Victor, ed. *Nineteenth-Century Crime and Punishment*. Vol. 1. London: Routledge, 2022.

Baldwin, Mark, and Anthony Burton, eds. *Canals: A New Look*. Chichester: Phillimore & Co., 1984.

Baldwin, Stanley. *On England and Other Addresses*. New York: Frederick A. Stokes Company, 1926.

Barr, John. *Derelict Britain*. Harmondsworth: Penguin Books, 1970.

Becket. "A View of the Advantages of Inland Navigations; With a Plan of a Navigable Canal, Intended for a Communication Between Liverpoole and Hull." In *The Gentleman's Magazine and Historical Chronicle*. Vol. 35 (1765): 532–533.i

Behlmer, George K. *Child Abuse and Moral Reform in England, 1870-1908*. Stanford, CA: Stanford University Press, 1982.

Betjeman, John. "City and Suburban: Ditching the Canals." *The Spectator*, July 22, 1955. Accessed September 1, 2024 (https://archive.spectator.co.uk/article/1st-june-1956/13/city-and-suburban).

Bird, Vivian. *By Lock and Pound*. Cleobury Mortimer: M & M Baldwin, 1988.

Birmingham Journal. British Newspaper Archive (https://britishnewspaperarchive.co.uk/).

Birmingham Mail. British Newspaper Archive (https://britishnewspaperarchive.co.uk/).

Birmingham Post. British Newspaper Archive (https://britishnewspaperarchive.co.uk/).

Black, Malcolm. *"Our Canoe Voyage," How We Launched at Hereford and Landed at Reading*. Manchester: Thomas and Jones, 1876.

Black, William. *The Strange Adventures of a House-Boat*. Vol. 1. London: Sampson Low, Marston, Searle, & Rivington, 1888.

Blackbourn, David. *The Conquest of Nature: Water, Landscape, and the Making of Modern Germany*. New York: W. W. Norton, 2006.

Blackburn Standard. British Newspaper Archive (https://britishnewspaperarchive.co.uk/).

Blagrove, David. *Bread Upon the Waters*. Burton-on-Trent: J. M. Pearson & Son, 1984.

Bliss, William. *The Heart of England by Waterway: A Canoeing Chronicle by River and Canal*. London: H. F. & G. Witherby, 1933.

Blunden, John, and Nigel Curry, eds. *A People's Charter*? London: Countryside Commission, 1990.

Blunt, Reginald. "In Tow." *Pall Mall Magazine* 1, no. 4 (May–October 1893): 504–518.

Bogart. Dan. "Did Turnpike Trusts Increase Transportation Investment in Eighteenth-Century England?" *The Journal of Economic History* 65, no. 2 (June 2002): 439–468.

Bogart, Dan, Michael Lefors, and A. E. M. Satchell. "Canal Carriers and Creative Destruction in English Transport." *Explorations in Economic History* 71 (2019): 1–24.

Bogart, Dan, Max Satchell, et al. "Turnpikes, Canals, and Economic Growth in England and Wales, 1800-1850." In *Dan Bogart: Research: Research on Transport and the English Industrial Revolution.* University of California, Irvine, 2017 (https://sites.socsci.uci.edu/~dbogart/turnpike_canal_growth_may202017.pdf).

Bolton, David. *Race Against Time: How Britain's Waterways were Saved.* London: Methuen, 1990.

Boughey, Joseph. *Charles Hadfield: Canal Man and More.* Stroud: Sutton Publishing, 1998.

Boughey, Joseph. "From Transport's Golden Ages to an Age of Tourism: L.T.C. Rolt, Waterway Revival and Railway Preservation in Britain, 1944-54." *The Journal of Transport History* 34, no. 1 (June 2013): 22–38.

Boyd, Charles W., ed. *Mr. Chamberlain's Speeches.* Vol. 1. London: Constable and Company, 1914.

Boyes, John H. "Brindley, James." In *A Biographical Dictionary of Civil Engineers in Great Britain and Ireland,* edited by A. W. Skempton, et. al., 75–80. London: Thomas Telford, 2002.

Boyes, John, and Ronald Russell. *The Canals of Eastern England.* London: David & Charles, 1977.

Brindley, James. *The History of Inland Navigations. Particularly Those of the Duke of Bridgwater[sic.] in Lancashire and Cheshire.* 2nd ed. London: T. Lowndes, 1769.

British Waterways. *Six Years of Progress.* London: The Curwen Press, 1961. Pamphlet in On Waterways in Great Britain . . . Part of British Transport Historical Records Office library: Canals and Inland Waterways, ZLIB 3, ZLIB 3/71. National Archives, UK.

Briton. British Newspaper Archive (https://britishnewspaperarchive.co.uk/).

Broich, John. *London: Water and the Making of the Modern City.* Pittsburgh, PA: University of Pittsburgh Press, 2013.

Buchanan, R. A. "Lionel Thomas Caswall Rolt, (1910-1974)." *Technology and Culture* 16, no. 3 (July 1975): 454–457.

Buckingham, J. S., *Evidence on Drunkenness, Presented to the House of Commons, by the Select Committee Appointed by the House to Inquire Into This Subject, and Report the Minutes of Evidence, with Their Opinions Thereupon* ("J. S. Buckingham Esq., M.P., in the Chair"). London: Samuel Bagster, 1834.

Buckinghamshire Advertiser and Aylesbury News. British Newspaper Archive (https://britishnewspaperarchive.co.uk/).

Bunce, M. F. *The Countryside Ideal: Anglo-American Images of Landscape.* London: Routledge, 1994.

Burton, Anthony. *The Canal Builders.* Stroud: Tempus, 2005 (1972).

Burton, Anthony. "Canals in the Landscape." In *Canals: A New Look,* edited by Mark Baldwin and Anthony Burton, 9–22. Chichester: Phillimore, 1984.

Burton, Anthony, and Derek Pratt. *The Anatomy of Canals: Decline and Renewal.* Stroud: Tempus Publishing, 2004.

Burton, Anthony, and Derek Pratt. *The Anatomy of Canals: The Early Years.* Stroud: Tempus Publishing, 2001.

Burton, Anthony, and Derek Pratt. *The Anatomy of Canals: The Mania Years*. Stroud: Tempus Publishing, 2002.

Caffyn, Douglas. *The Right of Navigation on Non-tidal Rivers and the Common Law*. Self-Published Thesis, Master of Laws, 2004.

The Canal Boatman's Magazine. Volume 1. April–December 1829.

"Canals." In *The Cyclopaedia: or, Universal Dictionary of Arts, Sciences, and Literature*, Vol. VI, edited by Abraham Rees. London: Longman, Hurst, Rees, Orme, & Brown, 1819.

Carlisle Journal. British Newspaper Archive (https://britishnewspaperarchive.co.uk/).

Certificates of Registration, BW 117/2/2. Waterways Archive, Canal & River Trust, Ellesmere Port, UK.

Chapman, David. *Holidays and the State*. London: Fabian Publications, Ltd., 1949.

Chase, Malcolm. "'This Is No Claptrap: This Is Our Heritage.'" In *The Imagined Past: History and Nostalgia*, edited by Christopher Shaw and Malcolm Chase, 128–146. Manchester: Manchester University Press, 1989.

Chase, Malcolm. "Wallop, Gerard Vernon, Ninth Earl of Portsmouth (1898–1984), Politician and Environmentalist." In *Oxford Dictionary of National Biography*, edited by Lawrence Goldman. Oxford: Oxford University Press, 2004.

Cheltenham Journal and Gloucestershire Fashionable Weekly Gazette. British Newspaper Archive (https://britishnewspaperarchive.co.uk/).

Chester Courant. British Newspaper Archive (https://britishnewspaperarchive.co.uk/).

Chester Gazette. British Newspaper Archive (https://britishnewspaperarchive.co.uk/).

Clapp, B. W. *An Environmental History of Britain since the Industrial Revolution*. London: Routledge, 2013 (1994).

Clarke, Mike. "John Rennie and the Rochdale Canal, UK." *Engineering History and Heritage* 164, no. EH3 (August 2011): 143–154.

Cohen, Stanley. *Folk Devils and Moral Panics*. 3rd ed. London: Routledge, 2002.

A Collection of Rules and Standing Orders of the House of Commons, Relative to the Applying for and Passing Bills for Inclosing and Draining of Lands; making turnpike Roads, Navigations, and Aqueducts;. New Edition. London: Luke Hansard, 1799.

Conford, Philip. *The Origins of the Organic Movement*. Edinburgh: Floris Books, 2001.

Coopey, Richard. "The Severn, 1750-1850." In *A History of Water*, edited by Tvedt and Coopey, Series 2, Vol. 2, 395–423. London: I. B. Tauris, 2006–2014.

Cork Constitution. British Newspaper Archive (https://britishnewspaperarchive.co.uk/).

Correspondence and Papers of Robert Aickman, PRO 30/82/1. National Archives, UK.

County of Middlesex Independent. British Newspaper Archive (https://britishnewspaperarchive.co.uk/).

Coventry Herald and Observer. British Newspaper Archive (https://britishnewspaperarchive.co.uk/).

Cregan-Reid, Vybarr. "Water Defences: 'The Arts of Swimming in Nineteenth-Century Culture.'" *Critical Survey* 16, no. 3 (2004): 33–47.

Crichton, Charles, dir. *Painted Boats*. 1945. Ealing Studios. Also released as *The Girl of the Canal* in 1947 in the U.S.

Crompton, Gerald. "The Tortoise and the Economy." *Journal of Transport History* 25, no. 2 (2004): 1–22.

Cubbon, T. W. *Only a Little Cockboat*. London: George Roberts, 1928.

Curtis, Wilfrid Henry. "Canals." *Journal of the Royal Society of Arts* 83 (July 19, 1935): 837–857.

Daily Herald. British Newspaper Archive (https://britishnewspaperarchive.co.uk/).

Daily News (London). British Newspaper Archive (https://britishnewspaperarchive .co.uk/).

De Maré, Eric. *The Canals of England*. London: The Architectural Press, 1950.

De Maré, Eric. "Canals: A Special Number of *The Architectural Review* for July, 1949." London: *The Architectural Review*, 1949.

De Salis, Henry Rodolph. *Bradshaw's Canals and Navigable Rivers of England and Wales*. London: Henry Blacklock & Co., Ltd., 1904.

De Salis, Henry Rodolph. *A Handbook of Inland Navigation (Bradshaw's Canals and Navigable Rivers of England and Wales)*. London: Henry Blacklock & Co., Ltd., 1928.

De Vries, Jacobus. "Water Control in the Netherlands with a Focus on the Period 1750-1900." In *A History of Water,* edited by Terje Tvedt and Richard Coopey, Series 2, Vol. 2, 239–262. London: I. B. Tauris, 2010.

Debrett, J. *The Parliamentary Register* 33 [1792]; 37 [1794]; 2 [1802]. London: J. Debrett, 1792, 1794, 1802.

Defoe, Daniel. *A Tour Through England and Wales*. Vol. 2. London J. M. Dent & Sons, Ltd., 1928.

Derby Mercury. British Newspaper Archive (https://britishnewspaperarchive.co.uk/).

Derbyshire Advertiser. British Newspaper Archive (https://britishnewspaperarchive .co.uk/).

Dibb, Geoffrey, ed. *The Canals Book: Navigator Guide Directory*. Brundall and Norwich: Geoffrey Dibb, Ltd., 1966.

Dictionary of Irish Architects, 1720–1940. s.v. "Castle, Richard." Accessed December 10, 2023 (https://www.dia.ie/architects/view/347/CASTLE%2C+RICHARD #tab_biography).

Doin, Alexandre. *Mémoire sur le Canal Latéral a la Garonne, Établissant la Jonction Définitive des Deux Mers*. Paris: Éverat. 1832.

Dower, John. *The Case for National Parks in Great Britain*. London: Council for the Preservation of Rural England, Standing Committee on National Parks, 1938.

Dower, John. "Holiday Use of Countryside & Coastline (An Address Given at the R.I.B.A. on 31 March, as the Fourth in the Series of Six Lectures on Town and Country Planning Arranged by the R.I.B.A." *Journal of the Royal Institute of British Architects* 50, no. 8 (June 1943): 181–184.

Dower, Michael. *The Challenge of Leisure: A Civic Trust Survey on Planning for Increased Leisure Facilities, First Published Under the Title 'Fourth Wave'*

["Reprinted from the *Architects' Journal* of 20 January 1965, by Permission of the Editor, by Harrison & Sons, Ltd."]. London: Civic Trust, 1967[?].

Dower, Michael. "The Recreational Potential of the Waterways." *Inland Waterways Association (IWA) Bulletin* 80 (September 1967): 10.

Duckham, Baron F. "Canals and River Navigations." In *Transport in the Industrial Revolution*, edited by Derek H. Aldcroft and Michael J. Freeman, 199–141. Manchester: Manchester University Press, 1983.

Dudley Chronicle. British Newspaper Archive (https://britishnewspaperarchive.co.uk/).

Dugdale, William. *The History of Imbanking and Drayning of Divers Fenns and Marshes, Both in Forein Parts, and in this Kingdom; and of the Improvements thereby*. London: Alice Warren, 1652.

Dutton, Julian. *Water Gypsies: A History of Life on Britain's Rivers and Canals*. Cheltenham: The History Press, 2021.

English Chronicle and Whitehall Evening Post. British Newspaper Archive (https://britishnewspaperarchive.co.uk/).

Evening Standard. British Newspaper Archive (https://britishnewspaperarchive.co.uk/).

Examiner. British Newspaper Archive (https://britishnewspaperarchive.co.uk/).

Falconer, Keith. "A Celebration of L.T.C. Rolt and 'Industrial' World Heritage Sites." In "Remembering Rolt: A Symposium in Honour of the 100th Anniversary of L.T.C. Rolt's Birth – A Pioneer of Industrial Archaeology," by Angus Buchanan, Neil Cossons, et. al. *Industrial Archaeology Review* 33, no. 1 (2011): 3–17.

Falconer, Keith. "The Industrial Heritage in Britain – The First Fifty Years." *La revue pour l'histoire du CNRS* 14 (2006). https://doi.org/10.4000/histoire-cnrs.1778.

Ferguson, Donna. "Pushing the Boater Out: London's Property Crisis Spreads to the Water." *The Guardian*, July 29, 2019 (https://www.theguardian.com/cities/2019/jul/29/london-property-crisis-canal-narrowboat-gentrification).

Field, The country Gentleman's Newspaper. British Newspaper Archive (https://britishnewspaperarchive.co.uk/).

Firth, J. B. "Water Transport or Rail." *Fortnightly Review* 87, New Series (January to June, 1910): 755–764.

Francis, John Francis, *A History of the English Railway; Its Social Relations and Revelations, 1820-1845*. Vol. 1. London: Longman, Brown, Green, & Longmans, 1851.

Freeman, Michael J. *Railways and the Victorian Imagination*. New Haven, CT: Yale University Press, 1999.

Freer, Wendy. "Canal Boat People, 1840-1970." PhD Dissertation, Vol. 1, University of Nottingham, 1991 (https://eprints.nottingham.ac.uk/10946/).

Freer, Wendy. *Women & Children of the Cut*. Gwernymynydd: Railway & Canal Historical Society, 1995.

Freer, Wendy, and Gill Foster. *Canal Boatmen's Missions*. Oxford: Railway and Canal Historical Society, 2004.

Gamble, Andrew. "Theories and Explanations of British Decline." In *Rethinking British Decline,* edited by Richard English and Michael Kenny, 1–22. London: Macmillan, 2000.

Gehrke, Jules. "Countryside, Recreation, and the Transformation of Canals in Britain in the Mid-Twentieth Century." *Journal of Tourism History* 11, no. 2 (2019): 167–186.

General Evening Post. British Newspaper Archive (https://britishnewspaperarchive.co.uk/).

Glen, Alexander. *The Rivers Pollution Prevention Act, 1876, 39&40 Vict. c. 75; with Introduction, Notes, and Index*. London: Knight & Co. 1876.

Gloucester Journal. British Newspaper Archive (https://britishnewspaperarchive.co.uk/).

Gordon, Mrs. Gordon. *The Life and Correspondence of William Buckland, D.D., F.R.S., Sometime Dean of Westminster, Twice President of the Geological Society, and First President of the British Association*. London: John Murray, 1894.

Graphic. British Newspaper Archive (https://britishnewspaperarchive.co.uk/).

Gray, Annie. *The Old Lock Farm: A Story of Canal Life*. London: Sunday School Union, 1883.

Gray, Thomas. *Observations on A General Iron Rail-Way: (With Plates and Map Ilustrative [sic.] of the Plan,) Showing Its Great Superiority, By the General Introduction of Mechanic Power, Over All the Present Methods of Conveyance by Turnpike Roads and Canals; And Claiming the Particular Attention of Merchants, Manufacturers, Farmers, and Indeed, Every Class of Society*. 4th ed. London: Baldwin, Cradock, and Joy, 1823.

Great Canal Journeys. Series 1–5. London: Spirit Entertainment Ltd., 2018.

Hadfield, Charles. *The Canals of the East Midlands*. 2nd ed. Newton Abbot: David & Charles, 1970.

Hadfield, Charles. *The Canals of South and South East England*. New York: Augustus M. Kelley, 1967.

Hadfield, Charles. *The Canals of the West Midlands*. Newton Abbott: David & Charles, 1969.

Hadfield, Charles. *World Canals: Inland Navigation Past and Present*. New York: Facts on File, 1986.

Hanson, Harry. *The Canal Boatmen, 1760-1914*. Manchester: Manchester University Press, 1975.

Hadfield Collection (IWA Papers), HADFIELD/10/16a. London School of Economics, Archives, London, UK.

Hadfield Collection (IWA Papers), HADFIELD/10/16b. London School of Economics, Archives, London, UK.

Hampshire Chronicle. British Newspaper Archive (https://britishnewspaperarchive.co.uk/).

Hartlepool Northern Daily Mail. British Newspaper Archive (https://britishnewspaperarchive.co.uk/).

Harvey-Fishenden, Alice, and Neil Macdonald. "The Development of Early Reservoirs to Supply Water to Arterial Canals in England and Wales." *Landscape History* 42, no. 2 (2021): 79–98.

Herbert, A. P., *Water Gipsies*. Garden City, NY: Doubleday, Doran & Co., Inc., 1930.

Herring, Horace. "The Conservation Society: Harbinger of the 1970s Environment Movement in the UK." *Environment and History* 7, no. 4 (November 2001): 381–401.

Hill, Reginald Hill. "The Future of Our Inland Waterways." *Journal of the Institute of Transport* 25, no. 9 (March 1954): 343–347, 349.

Hoare, Edward Newenham (E.N.). *Notable Workers in Humble Life*. London: Thomas Nelson and Sons, 1889.

Hodder, Edwin. *George Smith (of Coalville): The Story of an Enthusiast*. London: James Nisbet and Co., Ltd., 1896.

Hollingshead, John. *Odd Journeys In and Out of London*. London: Groombridge and Sons, 1860.

Hopkinson, G. G. "The Development of Inland Navigation in South Yorkshire and North Derbyshire, 1697-1850." *Transactions of the Hunter Archaeological Society* 7 (1951): 229–251.

Hopkinson, G. G. "The Inland Navigations of the Derbyshire and Nottinghamshire Coalfield." *Derbyshire Archaeological Journal* 79 (1959): 22–41.

Hoskins, W. G. *The Making of the English Landscape*. London: Hodder and Stoughton, 1955.

Howitt, William. *The Rural Life of England*. 3rd ed. London: Longman, 1838; Philadelphia, PA: Carey and Hart, 1841.

Howkins, Alun. *The Death of Rural England*. London: Routledge, 2003.

Howkins, Alun. *Reshaping Rural England*. London: HarperCollins, 1991.

Hull Advertiser. British Newspaper Archive (https://britishnewspaperarchive.co.uk/).

Hull Packet. British Newspaper Archive (https://britishnewspaperarchive.co.uk/).

Inland Waterways Association (IWA). *Bulletin*, 1 (November 1946) – 85 (November 1968). London: Inland Waterways Association, 1946–1968.

Inland Waterways Association. *A National Waterways Conservancy: Proposals by the Inland Waterways Association Limited*. London: W. S. Cowell, 1958.

Inland Waterways Association. *"The Way Ahead for the Amenity Waterways": A Memorandum to the Minister of Transport on Behalf of the Inland Waterways Association, Ltd.* (London, 1967) (https://waterways.org.uk/wp-content/uploads/2023/05/The-Way-Ahead-for-the-Amenity-Waterways-1967-greyscale-compressed-OCR.pdf).

Ipswich Journal. British Newspaper Archive (https://britishnewspaperarchive.co.uk/).

Ives, W. L. *The Problem of our Inland Waterways*. London: British Transport Commission, 1957.

Jackson, Lydia, and Richard Cracknell. *House of Commons Briefing Paper: Road Accident Casualties in Britain and the World*. Number 7615. April 2018 (https://researchbriefings.files.parliament.uk/documents/CBP-7615/CBP-7615.pdf).

Jefferies, Matthew, and Mike Tyldesley, eds. *Rolf Gardiner: Folk, Nature and Culture in Interwar Britain*. Farnham: Ashgate, 2011.

Johnson, Ben. "The Manchester Ship Canal." *History Magazine*. Accessed August 29, 2023 (http://www.historic-uk.com/HistoryMagazine/DestinationsUK/The-Manchester-Ship-Canal/).

Johnson's Sunday Monitor. British Newspaper Archive (https://britishnewspaperarchive.co.uk/).

Jones, Barbara. "The Rose & Castle." *The Architectural Review* 100 (December 1946): 161–168.

Jones, Barbara. *The Unsophisticated Arts*. London: The Architectural Press, 1951.

King-Hele, Desmond. "The 1997 Wilkins Lecture: Erasmus Darwin, the Lunaticks and Evolution." *Notes and Records of the Royal Society of London* 52, no. 1 (January 1998): 153–180.

Knights, Ben, "In Search of England: Travelogue and Nation Between the Wars." In *Landscape and Englishness*, edited by Robert Burden and Stephan Kohl, 165–184. New York: Rodopi, 2006.

Kohl, Stephan. "Rural England: An Invention of the Motor Industries?" In *Landscape and Englishness*, edited by Robert Burden and Stephan Kohl, 185–205. New York: Rodopi, 2006.

Labour Party National Executive Committee. *Leisure for Living*. London: Labour Party, 1959.

Lancashire Evening Post. British Newspaper Archive (https://britishnewspaperarchive.co.uk/).

Lancaster Gazette. British Newspaper Archive (https://britishnewspaperarchive.co.uk/).

Leeds Intelligencer. British Newspaper Archive (https://britishnewspaperarchive.co.uk/).

Leicester Chronicle [*Supplement to the Leicester Chronicle and Leicestershire Mercury*]. British Newspaper Archive (https://britishnewspaperarchive.co.uk/).

Leicester Chronicle and Leicestershire Mercury. British Newspaper Archive (https://britishnewspaperarchive.co.uk/).

Leicester Herald. British Newspaper Archive (https://britishnewspaperarchive.co.uk/).

Leicester Journal. British Newspaper Archive (https://britishnewspaperarchive.co.uk/).

[Letter]. *The Gentleman's Magazine, and Historical Chronicle* Vol. 30 (1760): 623.

Liverpool Echo. British Newspaper Archive (https://britishnewspaperarchive.co.uk/).

Liverpool Standard and General Commercial Advertiser. British Newspaper Archive (https://britishnewspaperarchive.co.uk/).

Local Government Board, MH 32. National Archives, UK.

London Observer. British Newspaper Archive (https://britishnewspaperarchive.co.uk/).

Luckin, Bill. *Questions of Power: Electricity and Environment in Inter-War Britain*. Manchester: Manchester University Press, 1990).

Lymington, Viscount. *Famine in England*. London: The "Right" Book Club, 1938.

Mackersey, Ian. *Tom Rolt and the Cressy Years*. London: M&M Baldwin, 1985.

MacLeod, Roy M. "Social Policy and the 'Floating Population: The Administration of the Canal Boats Acts, 1877-1899." *Past & Present* 35 (December 1966): 101–132.

Malet, Hugh. *Voyage in a Bowler Hat*. London: Hutchinson, 1960; M&M Baldwin, 1985.

Manchester Evening News. British Newspaper Archive (https://britishnewspaperarchive.co.uk/).

Manchester Mercury. British Newspaper Archive (https://britishnewspaperarchive.co.uk/).

Marx, Leo. *The Machine in the Garden: Technology and the Pastoral Ideal in America*. London: Oxford University Press, 1964.

Massingham, H. J. *Remembrance: An Autobiography*. London: Batsford, 1941.

Massingham, H. J., ed. *The Small Farmer*. London: Collins, 1947.

Matless, David. *Landscape and Englishness*. 2nd ed. London: Reaktion Books, 2016.

Matthews, Jodie. *The British Industrial Canal: Reading the Waterways from the Eighteenth Century to the Anthropocene*. Cardiff: University of Wales Press, 2023.

Matthews, Jodie. "Canals and Transatlantic Slavery: A Preliminary Literature Review." Canal & River Trust. Accessed December 10, 2023 (https://canalrivertrust.org.uk/media/original/42453-canals-and-transatlantic-slavery.pdf?v=445116).

Matthews, Jodie. *The Gypsy Woman: Representations in Literature and Visual Culture*. London: I. B. Tauris, 2018.

Mauch, Christof, and Thomas Zeller. *Rivers in History, Perspectives on Waterways in Europe and North America*. Pittsburgh, PA: University of Pittsburgh Press, 2008.

Maw, Peter. *Transport and the Industrial City: Manchester and the Canal Age, 1750-1850*. Manchester: Manchester University Press, 2013.

Mayall, David. *Gypsy-Travellers in Nineteenth-Century Society*. Cambridge: Cambridge University Press, 1988.

McIvor, Liz. *Canals: The Making of a Nation*. London: Penguin Random House, 2015.

McNeill, J. R. "Observations on the Nature and Culture of Environmental History." *History and Theory* 42, no. 4 (December 2003): 5–43.

Meade, L. T. (Elizabeth Thomasina). *Water Gipsies, or the Adventures of Tag, Rag, and Bobtail*. London: John F. Shaw and Co., 1883?.

Mearns, Andrew. *The Bitter Cry of Outcast London*. London: James Clarke, 1883.

Merchant, Carolyn. *The Death of Nature: Women, Ecology, and the Scientific Revolution*. New York: Harper and Row, 1980.

Ministry of Health, HLG 52. National Archives, UK.

Ministry of Transport, MT 52. National Archives, UK.

Ministry of Transport: Nationalised Transport, MT 88. National Archives, UK.

Mokyr, Joel. *The Enlightened Economy*. New Haven, CT: Yale University Press, 2009.

Moore, H. C. "Among the Water Gipsies." *The Churchman* 16, no. 160 (January 1902): 213–219.

Morning Advertiser. British Newspaper Archive (https://britishnewspaperarchive.co.uk/).

Morning Post. British Newspaper Archive (https://britishnewspaperarchive.co.uk/).

Morton, H. V. *In Search of England*, 22nd ed. London Methuen & Co., 1935.
Moulton, H. G. "England's Waterway Revival." *Journal of Political Economy* 19, no. 1 (January 1911): 1–16.
Mukerji, Chandra. *Impossible Engineering: Technology and Territoriality on the Canal du Midi*. Princeton, NJ: Princeton University Press, 2009.
National Register. British Newspaper Archive (https://britishnewspaperarchive.co.uk/).
Neal, Austin E. *Canals, Cruises, and Contentment*. London: Heath Cranton, 1921.
Newcastle Current. British Newspaper Archive (https://britishnewspaperarchive.co.uk/).
Newcastle Daily Chronicle. British Newspaper Archive (https://britishnewspaperarchive.co.uk/).
Nicholson, E. M. "The Countryside in 1970." *Nature* 208 (December 25, 1965): 1265–1266.
Nijman, Jan, ed. *Amsterdam's Canal District: Origins, Evolution, and Future Prospects*. Toronto: University of Toronto Press, 2020.
Northampton Mercury. British Newspaper Archive (https://britishnewspaperarchive.co.uk/).
Notcutt, George Jarvis. *The Law Relating to Factories and Workshops with Introduction and Explanatory Notes*. 2nd ed. London: Stevens and Sons, 1879.
Nottingham Evening Post. British Newspaper Archive (https://britishnewspaperarchive.co.uk/).
Nottingham Post. British Newspaper Archive (https://britishnewspaperarchive.co.uk/).
Oddy, Joshua Jepson. *A Sketch for the Improvement of the Political, Commercial, and Local, Interests of Britain, as Exemplified by the Inland Navigations of Europe in General, and of England in Particular; Including Details Relative to the Intended Stamford Junction Navigation, to Unite the Eastern With the Midland and Western Counties of the Kingdom*. London: J. J. Stockdale, 1810.
Paddington, Kensington, and Bayswater Chronicle. British Newspaper Archive (https://britishnewspaperarchive.co.uk/).
Patriot. British Newspaper Archive (https://britishnewspaperarchive.co.uk/).
Patterson, A. Temple. "The Making of the Leicestershire Canals." *Transactions of the*
Leicestershire Archaeological Society 27 (1951): 66–99.
Peberdy, R. B. "Navigation on the River Thames between London and Oxford in the Late Middle Ages: A Reconsideration." *Oxoniensia* 61, no. LXI (1996): 311–340.
Pennant, Thomas. *The Journey from Chester to London*. London: Printed for Wilkie and Robinson, et. al., 1811.
Pepper, Simon, and Peter Richmond. "Homes Unfit for Heroes: The Slum Problem in London and Neville Chamberlain's Unhealthy Areas Committee, 1919-21." *TPR: Town Planning Review* 80, no. 2 (2009): 143–171.
Phillips, John. *The General History of Inland Navigation; Containing a Complete Account of All the Canals of the United Kingdom, with Their Variations and Extensions, According to the Amendments of Acts of Parliament to June 1803,*

and a Brief History of the Canals of Foreign Countries. 4th ed. London: C. and R. Baldwin, 1803.

Philpott, Hugh B. "Water Gipsies." *Methodist Magazine and Review* 61 (January to June 1905): 509–513.

Powell-Smith, Vincent. *The Transport Act 1968*. London: Butterworths, 1969.

Pratt, Edwin A. *British Railways and the Great War: Organisation, Efforts, Difficulties and Achievements*. Vol. I. London: Selwyn and Blount, 1921.

Pratt, Edwin A. *A History of Inland Transport and Communication in England*. London: Kegan Paul, Trench, Trübner & Co., Ltd., 1912.

Priestley, Joseph. *Historical Account of the Navigable Rivers, Canals, and Railways, of Great Britain*. London: Longman, Rees, Orme, Brown, and Green, 1831.

Pritchard, Sara B., and Thomas Zeller. "The Nature of Industrialization." In *The Illusory Boundary*, edited by Martin Reuss and Stephen H. Cutcliffe, 69–100. Charlottesville, VA: University of Virgina Press, 2010.

Pugh, Martin. *State and Society: A Social and Political History of Britain since 1870*. 3rd ed. London: Hodder Education, 2008.

Readman, Paul. *Storied Ground: Landscape and the Shaping of English National Identity*. Cambridge: Cambridge University Press, 2018.

Regent's Canal Group. *Regent's Canal: A Policy for its Future*. Sevenoaks: Caxton & Holmesdale Press, 1967.

Register of Inspections, Birmingham, BW 162/10/5/12. Waterways Archive, Canal & River Trust, Ellesmere Port, UK.

"Report." *A Supplementary Number of the Paddington Canal Boatman's Magazine* 1 (May 20, 1829): 35–48.

Reuss, Martin, and Stephen H. Cutcliffe, eds. *The Illusory Boundary: Environment and Technology in History*. Charlottesville, VA: University of Virginia Press, 2010.

"Richard Castles." *Anthologia Hibernica* 2 (October 1793): 242.

Richardson, J. J., and Richard Kimber. "The British Waterways Board: A Neglected Asset?" *Public Administration* 52, no. 3 (1974): 306.

Rolt, L. T. C. *George and Robert Stephenson: The Railway Revolution*. London: Longmans, 1960.

Rolt, L. T. C. *High Horse Riderless*. London: George Allen & Unwin, 1947.

Rolt, L. T. C. *The Inland Waterways of England*. London: George Allen & Unwin, 1950.

Rolt, L. T. C. *Isambard Kingdom Brunel: A Biography*. London: Longman, Green & Co., 1957.

Rolt, L. T. C. *Landscape with Canals*. London: Allen Lane, 1977.

Rolt, L. T. C. *Landscape with Figures: The Final Part of his Autobiography*. Stroud: Alan Sutton, 1992.

Rolt, L. T. C. *Landscape with Machines: An Autobiography*. London: Longman, 1971.

Rolt, L. T. C. *Narrow Boat*. London: Eyre Methuen, 1944.

Rolt, L. T. C. *Narrow Boat*. With a Foreword by H.J. Massingham. 2nd rev. ed. London: Methuen, 1972.

Rolt, L. T. C. *Railway Adventure*. London: Constable, 1953; Stroud: The History Press, 1993.

Rolt, L. T. C. *Thomas Newcomen. The Prehistory of the Steam Engine*. London: MacDonald, 1963.

Rolt, L. T. C. *Thomas Telford*. London: Longmans, 1958: Harmondswoth: Penguin Books, 1985.

Rolt, L. T. C. *Victorian Engineering*. Harmondsworth: Penguin Books, 1970.

Rubinstein, W. D. *Capitalism, Culture, and Decline in Britain 1750-1990*. London: Routledge, 1993.

Runcorn Weekly News and District Reporter. British Newspaper Archive (https://britishnewspaperarchive.co.uk/).

Ryan, Yann. "A Short Guide to Historical Newspaper Data, Using R." *Bookdown*, November 27, 2021. https://bookdown.org/yann_ryan/r-for-newspaper-data/.

Saint James's Chronicle. British Newspaper Archive (https://britishnewspaperarchive.co.uk/).

Salisbury Journal. British Newspaper Archive (https://britishnewspaperarchive.co.uk/).

Sargent, Andrew. *The Story of the Thames*. Stroud: Amberley, 2015.

Sargent, G. H. *Ned the Barge Boy*. London: The Religious Tract Society, 1885.

Satchell, Max. "Navigable Waterways and the Economy of England and Wales: 1600-1835." In *The Online Historical Atlas of Transport, Urbanization and Economic Development in England and Wales c.1680-1911*. Cambridge Group for the History of Population, 2017. (https://www.campop.geog.cam.ac.uk/research/projects/transport/onlineatlas/waterways.pdf).

Saunders's News-Letter, and Daily Advertiser. British Newspaper Archive (https://britishnewspaperarchive.co.uk/).

Savage, C. I. *Inland Transport*. London: HMSO, 1957.

Schivelbusch, Wolfgang. *The Railway Journey: The Industrialization of Time and Space in the 19th Century*. Munich: Carl Hanser Verlag, 1977; Berkeley, CA: University of California Press, 1986.

Scotsman. British Newspaper Archive (https://britishnewspaperarchive.co.uk/).

Scott, Anthony, and Georgina Coustalin. "The Evolution of Water Rights." *Natural Resources Journal* 35, no. 4 (Fall 1995): 821–979.

Scott, E. Kilburn. "Three Joseph Priestleys." *Transactions of the Newcomen Society* 20, no. 1 (1939): 147–149.

Searle, G. R. *The Quest for National Efficiency: A Study in Politics and Political Thought*. Oxford: Oxford University Press, 1971.

Seymour, John. *Voyage into England*. New York: Augustus M. Kelley, 1966.

Sharp, Thomas. *The Anatomy of the Village*. Harmondsworth: Penguin Books, 1946.

Shaw, Rev. S. *A Tour to the West of England in 1788*. London: Robson and Clarke, 1789.

Sheail, John. "Dower, John Gordon (1900–1947)." In *Oxford Dictionary of National Biography*, edited by David Cannadine. Oxford: Oxford University Press, May 2006; online ed., ed. David Cannadine, January 2008.

Sheail, John. *An Environmental History of Twentieth-Century Britain*. London: Palgrave Macmillan, 2002.

Sheail, John. "Leisure in the English Countryside: Policy Making in the 1960s." *Planning Perspectives* 16, no. 1 (2001): 67–84.

Sheail, John. "Scott Revisited: Post-War Agriculture, Planning and the British Countryside." *Journal of Rural Studies* 13, no. 4 (1997): 387–398.

Shepherd, Larence. "The Housing Crisis No One's Talking About: London's Canals Are Getting Dangerously Over-Crowded." *City Monitor*, July 1, 2016 (https://www.citymonitor.ai/analysis/housing-crisis-no-ones-talking-about-londons-canals-are-getting-dangerously-over-crowded/).

Simmons, I. G. *An Environmental History of Great Britain: From 10,0000 Years Ago to the Present*. Edinburgh: Edinburgh University Press, 2001.

Sindall, Rob. *Street Violence in the Nineteenth Century: Media Panic or Real Danger?* Leicester: Leicester University Press, 1990.

Smiles, Samuel. *Lives of the Engineers, with an Account of Their Principal Works; Comprising Also a History of Inland Communication in Britain*, Vol. 1. London: John Murray, 1862.

Smith, George ("of Coalville"). *Gipsy Life: Being an Account of our Gipsies and Their Children, with Suggestions for Their Improvement*. London: Haughton, 1880.

Smith, George ("of Coalville"). *Our Canal Population: The Sad Condition of the Women and Children, -- with Remedy*. London: Haughton, 1875.

Smith, [Rodney] "Gipsy." *His Life and Work*. National Council of the Free Evangelical Churches, 1901. New York: Fleming H. Revell Co., 1902.

Snell, K. D. M. "Bell, Adrian Hanbury (1901–1980)." In *Oxford Dictionary of National Biography*, edited by Lawrence Goldman. Oxford: Oxford University Press, 2004.

Specia, Megan. "Living on the Canals, and Loving the Floating Life." *New York Times*, July 21, 2021, A4.

Squires, Roger W. *Canals Revived: The Story of the Waterway Restoration Movement*. Bradford-on-Avon: Moonraker Press, 1979.

Stafford, William. "This Once Happy Country': Nostalgia for Pre-Modern Society." In *The Imagined Past*, edited by Christopher Shaw and Malcolm Chase, 33–45. Manchester: Manchester University Press, 1989.

Staffordshire Advertiser. British Newspaper Archive (https://britishnewspaperarchive.co.uk/).

Staffordshire Daily Sentinel. British Newspaper Archive (https://britishnewspaperarchive.co.uk/).

Staffordshire Gazette and County Standard. British Newspaper Archive (https://britishnewspaperarchive.co.uk/).

Staffordshire Sentinel. British Newspaper Archive (https://britishnewspaperarchive.co.uk/).

Stamp, Sir Dudley. *The Land of Britain: Its Use and Misuse*. London: Longmans, Green and Co., 1948.

Stamp, Sir Dudley. *Nature Conservation in Britain*. London: Collins, 1969.

Stamp, Sir Dudley. “The Scott Report.” *The Geographical Journal* 101, no. 1 (January 1943): 16–17.

Stine, Jeffrey K., and Joel A. Tarr. “At the Intersection of Histories: Technology and the Environment.” *Technology and Culture* 39, no. 4 (October 1998): 601–640.

Sturt, George. *The Wheelwright’s Shop*. Cambridge: Cambridge University Press, 1923.

Suffolk & Essex Free Press. British Newspaper Archive (https://britishnewspaperarchive.co.uk/).

Sun (London). British Newspaper Archive (https://britishnewspaperarchive.co.uk/).

“Sunday, February 26,” *Gentleman’s Magazine: And Historical Chronicle* Vol. 84 (Part I): 192.

Sunday Mirror. British Newspaper Archive (https://britishnewspaperarchive.co.uk/).

Sutcliffe, John. *A Treatise on Canals and Reservoirs, and The Best Mode of Designing and Executing Them. . . .* Rochdale: John Sutcliffe, 1816.

Taunton Courier. British Newspaper Archive (https://britishnewspaperarchive.co.uk/).

Thomas, A. L. “The Duke of Bridgewater and the Canal Era.” *Geography* 21, no. 4 (December 1936): 297–300.

Thomas, Keith. *Man and the Natural World*. Oxford: Oxford University Press, 1983.

Thompson, Flora. *Lark Rise to Candleford*. With an introduction by H. J. Massingham. London: Oxford University Press, 1945.

Thompson, H. Gordon. *The Canal System of England*. London: T. Fisher Unwin, 1902.

Thurston, Temple. *The “Flower of Gloster*.” New York: Dodd, Mead and Co., 1912.

Tilley, Christopher. *A Phenomenology of Landscape*. Oxford: Berg, 1994.

Trentmann, Frank. "Gardiner, (Henry) Rolf (1902–1971), Ecological Campaigner and Youth Leader." In *Oxford Dictionary of National Biography*, edited by Lawrence Goldman. Oxford: Oxford University Press, 2009.

Tvedt, Terje, Terje Oestigaard, et. al., eds. *A History of Water*. Series 1, vols. 1–3; Series 2, vols. 1–3. London: I. B. Tauris, 2006–2014.

“Unfamiliar Ministries. IV.—On Barge and Monkey Boat.” *The London City Mission Magazine* 82 (1917): 87–89.

[United Kingdom.] “An Act for the Bringying of the Vier of Lee, to the North Side of the Citie of London.” 13 Eliz. 1, c. XVIII. *At the Parliament begunne and Holden at Westminster the Second of Apryll, in the xiii yere of the raigne of Our Most Gratious Soveraigne Lady Elizabeth, . . .* (1571).

United Kingdom. An Act Enabling Francis Duke of Bridgewater to Make a Canal from Longford Bridge in Stretford (Lancashire) to the River Mersey at a Place Called The Hemp Stones in Halton (Cheshire). 2 Geo. 3, c. 11.

United Kingdom. An Act for the Establishment of County and District Constables by the Authority of the Justices of the Peace. 2&3 Vict., c. 93.

United Kingdom. An Act for Making and Maintaining a Navigable Cut or Canal from Leeds Bridge to North Lady's Walk Liverpool and from Thence to the River Mersey. 10 Geo. III, c. 114 [1770].

United Kingdom. An Act for Making and Maintaining a Navigable Cut or Canal from the River Severn at Hawford to Chapel Bridge Droitwich. 8 Geo. III, c. 36.

United Kingdom. An Act for Making and Maintaining a Navigable Cut or Canal from the River Severn between Bewdley and Titton Brook to cross the River Trent near Haywood Mill and to Communicate with a Canal Intended to be Made between the River Trent and the River Mersey. 6 Geo. III, c. 97 [1766].

United Kingdom. An Act for Making a Navigable Cut, or Canal, from the River Trent, at or near Wilden Ferry in the County of Derby, to the River Mersey at or Near Runcorn Gap. 6 Geo. 3, c. 96 [1766].

United Kingdom. An Act to Consolidate and Amend the Law Relating to Factories and Workshops. 41 Vict., c. 16.

United Kingdom. An Act to Enable the Company of Proprietors of the Navigation from the Trent to the Mersey, to Make a Navigable Canal from the Said Navigation, on the South Side of Harecastle, in the County of Stafford, to Froghall, and a Railway from Thence to or Near Caldon, in the Said County; and to Make Other Railways. [1776].

United Kingdom. An Act to Provide for Keeping the Peace on Canals and Navigable Rivers. 3&4 Vict., c. 50.

United Kingdom. An Act to Provide for the Registration and Regulation of Canal Boats Used as Dwellings. 40&41 Vict., c. 60.

United Kingdom. Board of Agriculture. *Communication to the Board of Agriculture; on Subjects Relative to the Husbandry, and Internal Improvement of the Country.* Vol. 1, Parts III and IV. London: W. Bulmer and Co., 1797.

United Kingdom. Board of Education. Reports. 1905–1906; 1909–1910; 1910–1911; 1911–1912; 1912–1913; 1913–1914; 1919–1920. London: HMSO.

United Kingdom. British Transport Commission (BTC). *Canals and Inland Waterways: Report of the Board of Survey.* London: British Transport Commission, 1955.

United Kingdom. British Waterways Board. *The Facts About the Waterways.* London: December, 1965.

United Kingdom. British Waterways Board. *The Future of the Waterways: Interim Report of the Board.* London: HMSO, 1964.

United Kingdom. Committee of Council on Education. *Report of the Committee of Council on Education* (England and Wales). London: HMSO, 1893.

United Kingdom. Countryside Act 1968. c. 41.

United Kingdom. Factory and Workshops Acts Commission. *Report of the Commissioners Appointed to Inquire into the Working of the Factory and Workshops Acts, with a View to their Consolidation and Amendment*. Vol. 1. London: Eyre and Spottiswoode, 1876, 1:124.

United Kingdom. *First Report of the Commissioners Appointed to Inquire as to the Best Means of Establishing an Efficient Constabulary Force in the Counties of England and Wales.* London: HMSO, 1839.

United Kingdom. Hansard Parliamentary Debates 234 (1930). Accessed September 1, 2024 (https://hansard.parliament.uk/commons/1930-01-31/debates/95b23363-d36a-4f9b-be5a-f96acbb2c21d/CommonsChamber)

United Kingdom. House of Lords. *Report from the Select Committee of the House of Lords Appointed to Inquire Into the Expediency of Restraining the Practice of Carrying Goods and Merchandize on Canals, Navigable Rivers and Railways on Sundays; . . . with the Minutes of Evidence*. Sessional Papers, 1841. Reports From Select Committees of the House of Lords and Evidence. vol. 2. London: House of Lords, 1841.

United Kingdom. Local Government Board. *Annual Reports*. 1884–1885; 1885–1886; 1887–1888; 1888–1889; 1896–1897; 1900–1901; 1905–1906; 1910–1911. London: HMSO.

United Kingdom. Ministry of Health. Committee on Living-in on Canal Boats. *Minutes of Evidence Taken Before the Committee*. London: HMSO, 1921.

United Kingdom. Ministry of Health. Committee on Living-in on Canal Boats. *Report of the Departmental Committee Appointed to inquire into the practice of Living in on Canal Boats in England and Wales and to report whether any alteration in the practice is desirable*. London: HMSO, 1921.

United Kingdom. Ministry of Land and Natural Resources. *Leisure in the Countryside, England and Wales*. Cmnd. 2928. London: HMSO, 1966.

United Kingdom. Ministry of Transport. *British Waterways: Recreation and Amenity*. Cmd. 3401. London: HMSO, 1968.

United Kingdom. Ministry of Transport. *Second Interim Report of the Committee on Inland Waterways*. Cmnd. 1410. London: HMSO, 1921.

United Kingdom. Ministry of Transport. *Transport Policy*. Cmnd. 3057. London: HMSO, 1966.

United Kingdom. National Parks and Access to the Countryside Act. 1949, 12,13 & 14 Geo. 6., c. 97.

United Kingdom. New Towns Act, 1946. 9&10 Geo. 6, c. 68, 9.

United Kingdom. Parliament. *Report From His Majesty's Commissioners for Inquiring into the Administration and Practical Operation of the Poor Laws*. Vol. 1. London: B. Fellowes, 1834.

United Kingdom. Parliament. *Report of the Committee of Inquiry into Inland Waterways*. London: HMSO, 1958.

United Kingdom. Public Health Act 1875. 38&39 Vict., c.55.

United Kingdom. Railway and Canal Traffic Act 1888.

United Kingdom. Regulation of Railways Act. 36&37 Vict., c. 48.

United Kingdom. Royal Commission into the Condition of the Poorer Classes in Ireland. Appendix G: *Report of the State of the Irish Poor in Great Britain*. London: W. Clowes and Sons, 1835.

United Kingdom. Royal Commission on Canals and Inland Waterways. *First Report of the Royal Commission Appointed to Enquire into and to Report on the Canals and Inland Navigations of the United Kingdom*. Vol. 1. London: Printed for H. M. Stationery Office by Wyman and Sons, Ltd., 1906.

United Kingdom. Royal Commission on Canals and Inland Waterways. *Fourth and Final Report the Royal Commission Appointed to Enquire into and to Report on the Canals and Inland Navigations of the United Kingdom*. Vol. VII. London: Printed for H. M. Stationary Office by Wyman and Sons, Ltd., 1909.

United Kingdom. Royal Commission on Canals and Inland Waterways. *Report of the Royal Commission Appointed to Enquire into and to Report on The Canals and Inland Navigations of the United Kingdom, Fourth & Final Report*. Vol. 7. London: Printed for H. M. Stationery Office by Wyman and Sons, Ltd., 1909.

United Kingdom. Royal Commission on Labour. *Minutes of Evidence, With Appendices, Taken Before Group "B." Transport and Agriculture . . . of the Royal Commission on Labour*. Vol. 3. London: HMSO, 1893.

United Kingdom. Royal Commission on Transport. *Final Report: The Co-Ordination and Development of Transport*. London: HMSO, 1931.

United Kingdom. *Third Report of the Commissioners Appointed to Inquire into the Best Means of Preventing the Pollution of Rivers* (Rivers Aire and Calder.) Report, Appendix, Plans. London: HMSO, 1867.

United Kingdom. Town and Country Planning Act, 1947. 10&11 Geo. 6, c.51.

United Kingdom. Transport Act, 1947. 10&11 Geo. 6, c. 49.

United Kingdom. Transport Act, 1962. 10&11 Eliz. 2, c. 46.

United Kingdom. War Cabinet. The War Cabinet, *Report for the Year 1918*. London: HMSO, 1919.

[United Kingdom.] "William III, 1698: An Act for the Makeing and Keeping Navigable the Rivers of Aire and Calder in the County of Yorke. [Chapter XXV. Rot. Parl. 10 Gul. III. p. 4. n. 10]." In *Statutes of the Realm: Volume 7, 1695-1701*, edited by John Raithby, 534–538. s.l: Great Britain Record Commission, 1820). British History Online. http://www.british-history.ac.uk/statutes-realm/vol7/pp534-538.

Vernon-Harcourt, Leveson Francis. *Rivers and Canals*. 2nd ed. Vol. 2. Oxford: Clarendon Press, 1896.

Wailes, Rex. "Industrial Archeology: The Industrial Monuments Survey." *Technology and Culture* 8, no. 2 (April 1967): 199–203.

Wallace, Andrew, and Katy Wright. "The Post-Industrial English Canalscape: From Enclosure to Enlacement." *Cultural Geographies* 29, no. 2 (2021): 185–202.

Walsall Observer. British Newspaper Archive (https://britishnewspaperarchive.co.uk/).

Ward, J. R. *The Finance of Canal Building in Eighteenth-Century England*. Oxford: Oxford University Press, 1974.

Waterson, Merlin. *The National Trust: The First Hundred Years*. London: BBC Books, 1994.

Webster, Robert. *The Law Relating to Canals: Comprising a Treatise on Navigable Rivers and Canals; and Including All Legislation to the Close of the Last Session of Parliament, Together with the Procedure and Practice in Private Bill Legislation; with a Map of the Existing Canals in England and Wales*. London: Stevens and Sons, 1885.

Wiener, Martin J. *English Culture and the Decline of the Industrial Sprit, 1850-1980*. Rev. ed. Cambridge: Cambridge University Press, 2004.

Willan, T. S. "The River Navigation and Trade of the Severn Valley, 1600-1750." *The Economic History Review* 8, no. 1 (November 1937): 68–79.

Willan, T. S. *River Navigation in England, 1600-1750*. Oxford: Oxford University Press, 1936.

Williams, James C. "The Technology Junction: Exploring Technology and the Environment." *ICON: Journal of the International Committee for the History of Technology* 6 (2000): 7–20.

Williams, Raymond. *The Country and the City*. New York: Oxford University Press, 1973.

Williams-Ellis, Clough. *Britain and the Beast*. London: J. M. Dent and Sons, 1937.

Williams-Ellis, Clough. *England and the Octopus*. London: G. Bles, 1928.

Winchester, Simon. *The Map That Changed the World: William Smith and the Birth of Modern Geology*. New York: HarperCollins, 2001.

Windsor and Eton Express. British Newspaper Archive (https://britishnewspaperarchive.co.uk/).

Winter, James. *Secure from Rash Assault: Securing the Victorian Environment*. Berkeley, CA: University of California Press, 1999.

Worcester Journal. British Newspaper Archive (https://britishnewspaperarchive.co.uk/).

Worcestershire Chronicle. British Newspaper Archive (https://britishnewspaperarchive.co.uk/).

Wright, Patrick. *On Living in an Old Country: The National Past in Contemporary Britain*. London: Verso Books, 1985.

Yorkshire Gazette. British Newspaper Archive (https://britishnewspaperarchive.co.uk/).

Young, Arthur. *A Six Months Tour Through the North of England*. Vol. IV. London: W. Strahan, 1770.

Index

Abercrombie, Sir Leslie Patrick, 158, 176, 179, 186
Aickman, Robert, 167–70, 172, 174, 176–81, 183, 186–87, 204, 215, 217, 222–23, 227, 231, 233, 238, 253–54
Aire and Calder Navigation, 5, 16, 24, 117, 177, 201
amenity waterways. *See* recreation
aqueducts, 13, 17–18, 32, 36, 146, 206, 221, 228, 250
archaeology (and canals). *See* industrial archaeology; Survey of Industrial Monuments
Ashby Canal, 86–87, 123, 151–52
Aubertin, C.J., 112–13, 252
automobiles (impact of), 4, 152–54, 159–60, 163, 175–76, 184, 203, 211–12, 224–25, 227–28, 230, 237, 254

Baldwin, Stanley, 153
Banbury, 55, 160–61, 230
Barbauld, Anna Laetitia [Aiken], 32, 150–51
Beveridge Report. *See* Social Insurance and Allied Services
Birmingham, 20–21, 26, 28, 55, 64–65, 69, 73, 88, 113–15, 117, 119, 126–27, 129, 134, 155, 169, 182, 202, 207, 216–18, 227, 230, 234–35, 238
Birmingham Canal Navigations (BCN) (including Birmingham Canal), 20, 25–26, 30, 53, 117, 119, 155, 177, 213, 217–18, 227, 230, 234–35, 238
Black, William, 147–48, 252
Black Country, 115, 148, 155, 209, 227
Blagrove, David, 218–20, 238
Bliss, William, 156–58, 184
Blunt, Reginald, 148–49, 252
Board of Survey (Rusholme Report, 1954), 200–201, 205–7, 236, 254
boatmen (boat people), 3–4, 7, 23, 25–27, 32, 49–54, 56–57, 59, 61–65, 71–72, 76, 83–85, 88, 92–94, 109, 116–17, 119–22, 127, 134–35, 145, 150, 156, 162, 168, 171, 181, 184, 186–87, 211, 218–19, 234, 236, 238–39, 250–51, 253. *See also* "Number Ones"; women
boat people. *See* boatmen
Bowes Report. *See* Report of the Committee of Inquiry Into Inland Waterways
Braunston, 69, 161
Bridgewater Canal, 18–22, 26, 35, 52, 61, 182
Brindley, James, 18–22, 26–28, 30, 114, 161, 182

British Transport Commission (BTC), 4, 177–78, 186, 199–201, 204–5, 207–8, 213–16, 218, 236, 253–54
British Transport Waterways (BTW), 201, 215–16
British Waterways Board (BWB), 216–17, 220–26, 228–29, 231–34, 238, 255
Brydone, John, 79–82, 84, 86–87, 94, 112, 251
Bulkeley-Johnson, Captain Vivian, 218, 220

Canal & River Trust (CRT), 255
Canal Association, 117, 124
canal boats acts: (1877), 3, 7, 73–79, 82, 85–86, 89, 94, 251
  (1884, amendment act), 79–82, 85–86, 89, 94, 251
Canal Boats bill [1929], 123–26. *See also* Gosling, Harry; Wilson, Lilian
Canals (Offences) Act (1840), 62. *See also* crime: constabulary force(s)
canoe(s), 70–72, 156–57, 225, 228, 232, 237
cars. *See* automobiles
Castle, Barbara, 225
castles (as decoration on boats). *See* "roses and castles"
Chamberlain, Joseph, 73
Chamberlain, Neville, 7, 117–21, 134, 251
children (on or alongside canals), 4, 50, 54, 60, 68, 72–84, 87–90, 94, 119–28, 131, 135, 156, 183, 201–3, 208, 215, 234. *See also* boatmen; women
Collins, Christina, 60–65, 92. *See also* crime: rape
Committee on Inland Waterways (1920), 118–20, 134
conservancy (waterways), 204–5, 236
conservation (waterways and nature), 145, 159, 173, 178, 213, 225, 233–34, 252
Council for the Preservation of Rural England (CPRE), 153, 158, 167, 175, 253
countryside, 1, 7–8, 17, 20, 35, 66–68, 71–72, 88–89, 93–94, 132–33, 158, 160–61, 163–65, 167, 169–70, 175–76, 178–80, 182, 184, 187, 199, 212–13, 216, 219, 222, 224, 226–28, 230, 232–33, 237–38, 250, 252–54
Coventry Canal, 20, 26, 29–30, 128, 159, 219
crime (on or alongside canals), 51–60, 73, 84, 92, 128, 135, 211, 250–51
  assault, 51, 53, 60
  constabulary force(s), 58–60, 62. *See also* Canals (Offences) Act; infanticide, 51, 54–55
  property crime, 51–53, 58–60, 62–65
  rape, 53, 60–61, 65. *See also* Collins, Christina
  suicide, 50, 54–55
The "Cross," 19–20, 30, 114
Cubbon, T.W., 155, 184

death (accidental). *See* drowning
declines in traffic (canals). *See* traffic, declines in
Defence of the Realm Act (DORA, 1914). *See* World War I
de Maré, Eric, 174, 178, 183, 186
Departmental Committee on Living-In on Canal Boats, 120–24, 134
diesel (canal boats), 91, 93, 110
disease (on or alongside canals). *See* health
Docks and Inland Waterways Executive, 177–78, 180, 199, 201, 236
Dower, John, 175–76, 186
Dower, Michael, 224, 230, 238
drowning (canals), 4, 53–55, 60–61, 126–28, 155, 187, 201–3, 215
drunkenness (on or alongside canals), 3, 25, 53, 58–62, 67, 69, 76–77, 92, 161

education (on or alongside canals), 3, 7, 56, 75–79, 82–85, 87–88, 90, 92, 119–25, 131, 162–63, 184, 211, 251
Ellesmere Canal, 70–71, 151
environment (and environmental concerns), 1–9, 13–14, 20, 28, 35–37, 49, 66, 71–73, 83, 93, 145, 147, 149, 153, 172, 174, 199, 213, 218–19, 222, 233–35, 237, 249–50, 253, 255–56
Exe River, 14–15

Factory Acts, 57, 76–77
Fellows Morton & Clayton, 91, 112, 124
Fens, 15, 35–36, 228
flyboats, 26, 34, 50, 60, 64, 67, 72, 91. *See also* narrowboats
Forth and Clyde Canal, 29, 91
France (canals), 14, 18, 110–11

Gosling, Harry, 123–26. *See also* Canal boats bill; Wilson, Lilian
Grand Junction Canal, 25–26, 28, 30–31, 34, 52–53, 55, 65, 67–69, 83, 119, 121, 126, 128, 146–47, 156
Grand Union [Old] Canal, 30, 34
Grand Union Canal, 119, 126–29, 134, 156, 159, 161, 177, 182, 201, 214, 218–19
"Gypsies"/"gypsies." *See* Romani

Hadfield, Charles, 5, 72, 166–69, 183, 217, 232
Hassell, John, 146–47, 252
health: on or alongside industrial canals, 3, 6–7, 23, 51, 72–76, 78–80, 82–86, 88, 94, 120–27, 130–32, 134–35, 147, 202, 251
  and twentieth-century touring on canals, 147, 152–53, 159, 173, 178, 228
Herbert, A.P., 155–56, 172, 184
Hill, Sir Reginald, 180, 199–200, 236
Hinckley. *See* Ashby Canal
Hollingshead, John, 67–71, 93

industrial archaeology, 208–10, 231, 236. *See also* Survey of Industrial Monuments
Inland Waterways Amenity Advisory Council (IWAAC), 232, 238
Inland Waterways Association (IWA), 4, 7–8, 145, 156, 168–80, 182–87, 200–205, 207–8, 212–18, 222–23, 225–38, 253–55
  protest cruises, 174–75, 208, 213, 215
Inland Waterways Redevelopment Committee (IWRC), 207–8, 212–15, 236
inspection (canal boats), 7, 57, 76, 78–83, 85–88, 92, 94, 120, 123, 131, 151, 251

Jones, Barbara, 170–71, 177, 186, 253

Kennet and Avon Canal, 30, 54, 91, 116, 147–48, 169, 187, 215, 223

Labour (Party, Government), 8, 123, 125–26, 133, 168, 172, 176–77, 185, 212, 222, 237
Lake District, 66–67, 158, 224, 228
Lancaster Canal, 146, 228
landscape, 1–2, 4–8, 13, 18, 21–22, 28–29, 33, 37, 65–66, 70, 93–94, 109, 133, 146–48, 150–51, 153, 160–62, 164–65, 172–75, 179, 181–82, 184, 186–87, 199, 202, 206, 209, 211, 218, 220, 227, 230, 233, 237–39, 249–50, 252–53, 255
Leeds and Liverpool Canal, 24, 29, 54, 91, 128
leisure (on or alongside canals). *See* recreation
Llewellyn, Owen J., 86–88, 94, 123, 151, 251
Local Government Board (LGB), 76, 78–79, 81, 84, 86–88, 94

locks (canals), 2, 8, 13, 15, 18, 36
Lymington, Viscount (Gerard Vernon Wallop), 159, 253

Malet, Hugh, 217, 238
Manchester Ship Canal, 94, 110–11, 133, 211, 251
"mania" (canal building), 24–25
Market Harborough, 182–83
Marx, Enid, 178, 186
Massingham, H.J., 153, 157–59, 164–65, 171, 233, 253
McKnight, Hugh, 223, 226–27, 229–30
Mersey and Irwell Navigation, 16, 61, 70–71
Ministry of Transport, 7–8, 117–19, 123–24, 128, 132, 145, 172, 177, 216, 223, 225, 227, 231–32, 238, 254. *See also* Transport Acts
missionaries (on or alongside canals), 3, 5, 7, 50, 56–57, 64–65, 76–77, 84–85, 88–89, 91–92, 94, 119–20, 135, 251
Morton, H.V., 154–55, 163, 184
Morton, Leslie, 218, 220
Munk, Lionel, 207, 222–23, 226, 232–33

Nairn, Ian, 203–4, 212, 222
Narrow Boat. *See* Rolt, L.T.C
narrowboats, 26, 81, 119–20, 126, 132, 148, 159, 165, 170–71, 182, 218, 220, 234, 237, 254. *See also* flyboats
National Parks, 133, 175–76, 186–87, 213, 217, 224–25, 228, 233
National Society for the Prevention of Cruelty to Children (NSPCC), 124–25
National Trust, 149, 158, 204, 207, 213–14, 223, 225, 235, 237, 253
navvies (navigators), 3, 25, 34, 49, 51
Neal, Austin E., 151–52, 184
Netherlands, 14
"Number Ones," 162, 171, 182, 185–86, 237. *See also* boatmen

organicism, 7, 153, 156–57, 159, 164–65, 171, 173, 180–82, 184, 249–50, 252
Oxford Canal, 20, 26, 28, 30, 53–54, 69, 91, 130, 147, 160, 218, 230

Paddington Basin, 30, 53, 55–56, 76, 83, 88, 124, 128, 220
Painted Boats (movie), 166–67, 185
Parliamentary Inland Waterways Group, 223, 225
Pick, Frank, 129–32, 135, 252. *See also* World War II
planning (twentieth century), 133, 158, 164, 171, 174–77, 179–81, 185–86, 203, 212, 222, 224–25, 230–32, 253
pollution (on or alongside canals), 6, 69, 72–76, 83–84, 164, 169, 211, 234, 237–38, 254
Poor Law(s), 57–58, 92

railways, locomotive: construction and development, 2, 28, 33–34, 65–67, 69–71, 74, 93, 160, 181, 183, 208–9
  impact on canals, 35, 91, 109–10, 112–16, 119, 128, 130, 172–74, 177, 205, 212, 214, 232
recreation (on or alongside canals), 4–5, 7, 91, 109, 146, 149, 172, 177–78, 182–84, 199–200, 207, 213–16, 222, 224–25, 228, 231–33, 235, 237–38, 249–50, 253–54
  amenity waterways, 131, 173, 203, 213–15, 221–22, 225, 227–33
  holidays, 71, 148–49, 173, 175–76, 178–79, 183, 187, 216, 224, 227–28, 231, 235–36, 253–54
  leisure, 1, 4, 8, 35, 158, 169, 173, 175–76, 178, 180, 182–83, 186, 199, 206, 212–13, 221, 224–25, 230–31, 235, 237–38, 254–55
Regent's Canal, 30, 53, 83, 119, 126, 220–21, 229
Rennie, John, 26, 28

Report of the Committee of Inquiry Into Inland Waterways (Bowes Report, 1958), 205–7, 212–13, 215, 218, 236, 254
Rolt, L.T.C. (Lionel Thomas Caswall), 4–5, 7, 145, 162, 164–69, 171–72, 174, 176–77, 180, 182–87, 207–12, 217, 221, 227, 235–36, 252–54
  and *Narrow Boat*, 159–63, 165–68, 180, 185, 208, 254
Romani ("Gypsies"/"gypsies") (perceptions of association with canals), 4, 51, 66, 76–78, 82, 84–86, 92–94, 150, 156, 162, 168, 170, 173–74, 182, 186, 253
"roses and castles," 3, 156, 162, 170, 174, 182, 253
Royal Commission on Canals and Inland Waterways (1906), 3, 7, 109, 111–14, 118, 133–35, 177, 251
ruralism, 4, 7, 94, 152–53, 155–56, 163–64, 180, 182–85, 235, 237, 252–53
Rusholme Report. *See* Board of Survey

sanitation (on or alongside canals). *See* health
Sankey Brook, 15
Scott, Peter, 163–64, 172
Severn River, 14–15, 19–20, 33, 50, 59, 61, 64, 70–71, 114, 147, 155, 201, 204, 228
Skinner, Joe and Rose, 219–20
Smeaton, John, 26–27, 29
Smiles, Samuel, 5, 208
Smith, George ("of Coalville"), 7, 73–79, 82–86, 89, 93–94, 251
Social Insurance and Allied Services (Beveridge Report, 1942), 132–33
Soil Association, 164, 233
steam (canal boats), 91, 93, 110–11
Stratford-upon-Avon Canal, 147, 155, 161, 169, 174, 187, 213–15, 222, 237
suburbs and suburbanization, 4, 158, 160, 186–87, 203, 212, 224, 227, 255
Sunday trading (on or alongside canals), 61–65, 72, 83
Survey of Industrial Monuments, 210, 236–37. *See also* industrial archaeology

Talyllyn Railway, 183, 212
Telford, Thomas, 20, 27, 30, 208
Thames and Severn Canal, 30, 52, 70–72, 157, 160
Thames River, 14–15, 19, 30, 52, 56, 64, 75, 114, 124, 147–48, 155, 169, 204, 221, 223, 228
Thurston, Temple, 5, 150–51, 160, 184, 252
tolls (canals), 2, 13, 16, 21, 25, 31, 33, 72, 91, 115, 118, 127–28, 180, 200, 205, 207, 215, 218, 229, 235
Tomalin, Nicholas, 220–21, 238
traffic, declines in (canals), 2, 34, 36, 86, 91, 116–17, 119, 123, 134, 183–84, 200, 207, 235, 237
Transport Acts: (1919), 117
  (1947), 177
  (1962), 216, 238
  (1968), 2–4, 232, 238. *See also* Ministry of Transport
Trent and Mersey Canal, 20, 23, 26, 29–30, 52, 54, 60–64, 151, 159, 177
Trent River, 14, 54, 118–19, 126–27, 177, 201, 204
trust (waterways), 223, 226
tunnels, 13, 18, 21, 27, 29–30, 32, 34, 36, 65–66, 91, 146–47, 151, 180, 208, 213, 215, 221, 228, 230
turnpikes, 16–17, 70, 146

Williams-Ellis, Clough, 153, 158–59, 179–80
Willow Wren, 218–20, 222
Wilson, Lilian, 125–26, 135. *See also* Gosling, Harry; Canal boats bill
Wiltshire and Berkshire Canal, 91, 148

women (on or alongside canals), 50, 53–55, 58, 68, 72–74, 76–78, 81, 83–84, 87, 119–21, 124–27, 129, 156, 168, 170, 174, 183. *See also* boatmen; children
Wordsworth, William, 66–67
World War I, 3, 7, 94, 109, 115–16, 123, 128, 134–35, 151, 154, 184, 251
World War II, 4, 7, 109, 128–33, 135, 155–56, 161–63, 166, 185, 211–12, 252–53. *See also* Pick, Frank

About the Author

Jules P. Gehrke received a PhD in History at the University of Minnesota in 2006 and joined the Saginaw Valley State University Department of History in 2007, where he was promoted to Professor of History in 2023. He has researched, published, and presented on Britain's canals since 2015.

www.ingramcontent.com/pod-product-compliance
Lightning Source LLC
LaVergne TN
LVHW011809200926
848484LV00005B/1119
* 9 7 8 1 7 9 3 6 5 2 3 3 1 *